CW00687503

Fast Fourier Transforms

A catalogue record for this book is available from the Hong Kong Public Libraries.

Published in Hong Kong by Samurai Media Limited.

Email: info@samuraimedia.org

ISBN 978-988-8407-52-1

Table of Contents

Chapter 1

Preface: Fast Fourier Transforms[1]

This book focuses on the discrete Fourier transform (DFT), discrete convolution, and, particularly, the fast algorithms to calculate them. These topics have been at the center of digital signal processing since its beginning, and new results in hardware, theory and applications continue to keep them important and exciting.

As far as we can tell, Gauss was the first to propose the techniques that we now call the fast Fourier transform (FFT) for calculating the coefficients in a trigonometric expansion of an asteroid's orbit in 1805 [174]. However, it was the seminal paper by Cooley and Tukey [88] in 1965 that caught the attention of the science and engineering community and, in a way, founded the discipline of digital signal processing (DSP).

The impact of the Cooley-Tukey FFT was enormous. Problems could be solved quickly that were not even considered a few years earlier. A flurry of research expanded the theory and developed excellent practical programs as well as opening new applications [94]. In 1976, Winograd published a short paper [403] that set a second flurry of research in motion [86]. This was another type of algorithm that expanded the data lengths that could be transformed efficiently and reduced the number of multiplications required. The ground work for this algorithm had be set earlier by Good [148] and by Rader [308]. In 1997 Frigo and Johnson developed a program they called the FFTW (fastest Fourier transform in the west) [130], [135] which is a composite of many of ideas in other algorithms as well as new results to give a robust, very fast system for general data lengths on a variety of computer and DSP architectures. This work won the 1999 Wilkinson Prize for Numerical Software.

It is hard to overemphasis the importance of the DFT, convolution, and fast algorithms. With a history that goes back to Gauss [174] and a compilation of references on these topics that in 1995 resulted in over 2400 entries [362], the FFT may be the most important numerical algorithm in science, engineering, and applied mathematics. New theoretical results still are appearing, advances in computers and hardware continually restate the basic questions, and new applications open new areas for research. It is hoped that this book will provide the

[1] This content is available online at <http://cnx.org/content/m16324/1.10/>.

background, references, programs and incentive to encourage further research and results in this area as well as provide tools for practical applications.

Studying the FFT is not only valuable in understanding a powerful tool, it is also a prototype or example of how algorithms can be made efficient and how a theory can be developed to define optimality. The history of this development also gives insight into the process of research where timing and serendipity play interesting roles.

Much of the material contained in this book has been collected over 40 years of teaching and research in DSP, therefore, it is difficult to attribute just where it all came from. Some comes from my earlier FFT book [59] which was sponsored by Texas Instruments and some from the FFT chapter in [217]. Certainly the interaction with people like Jim Cooley and Charlie Rader was central but the work with graduate students and undergraduates was probably the most formative. I would particularly like to acknowledge Ramesh Agarwal, Howard Johnson, Mike Heideman, Henrik Sorensen, Doug Jones, Ivan Selesnick, Haitao Guo, and Gary Sitton. Interaction with my colleagues, Tom Parks, Hans Schuessler, Al Oppenheim, and Sanjit Mitra has been essential over many years. Support has come from the NSF, Texas Instruments, and the wonderful teaching and research environment at Rice University and in the IEEE Signal Processing Society.

Several chapters or sections are written by authors who have extensive experience and depth working on the particular topics. Ivan Selesnick had written several papers on the design of short FFTs to be used in the prime factor algorithm (PFA) FFT and on automatic design of these short FFTs. Markus P*ü*schel has developed a theoretical framework for "Algebraic Signal Processing" which allows a structured generation of FFT programs and a system called "Spiral" for automatically generating algorithms specifically for an architicture. Steven Johnson along with his colleague Matteo Frigo created, developed, and now maintains the powerful FFTW system: the Fastest Fourier Transform in the West. I sincerely thank these authors for their significant contributions.

I would also like to thank Prentice Hall, Inc. who returned the copyright on The DFT as Convolution or Filtering (Chapter 5) of **Advanced Topics in Signal Processing** [49] around which some of this book is built. The content of this book is in the Connexions (http://cnx.org/content/col10550/) repository and, therefore, is available for on-line use, **pdf** down loading, or purchase as a printed, bound physical book. I certainly want to thank Daniel Williamson, Amy Kavalewitz, and the staff of Connexions for their invaluable help. Additional FFT material can be found in Connexions, particularly content by Doug Jones [205], Ivan Selesnick [205], and Howard Johnson, [205]. Note that this book and all the content in Connexions are copyrighted under the Creative Commons Attribution license (http://creativecommons.org/).

If readers find errors in any of the modules of this collection or have suggestions for improvements or additions, please email the author of the collection or module.

C. Sidney Burrus

Houston, Texas

October 20, 2008

Chapter 2

Introduction: Fast Fourier Transforms[1]

The development of fast algorithms usually consists of using special properties of the algorithm of interest to remove redundant or unnecessary operations of a direct implementation. Because of the periodicity, symmetries, and orthogonality of the basis functions and the special relationship with convolution, the discrete Fourier transform (DFT) has enormous capacity for improvement of its arithmetic efficiency.

There are four main approaches to formulating efficient DFT [50] algorithms. The first two break a DFT into multiple shorter ones. This is done in Multidimensional Index Mapping (Chapter 3) by using an index map and in Polynomial Description of Signals (Chapter 4) by polynomial reduction. The third is Factoring the Signal Processing Operators (Chapter 6) which factors the DFT operator (matrix) into sparse factors. The DFT as Convolution or Filtering (Chapter 5) develops a method which converts a prime-length DFT into cyclic convolution. Still another approach is interesting where, for certain cases, the evaluation of the DFT can be posed recursively as evaluating a DFT in terms of two half-length DFTs which are each in turn evaluated by a quarter-length DFT and so on.

The very important computational complexity theorems of Winograd are stated and briefly discussed in Winograd's Short DFT Algorithms (Chapter 7). The specific details and evaluations of the Cooley-Tukey FFT and Split-Radix FFT are given in The Cooley-Tukey Fast Fourier Transform Algorithm (Chapter 9), and PFA and WFTA are covered in The Prime Factor and Winograd Fourier Transform Algorithms (Chapter 10). A short discussion of high speed convolution is given in Convolution Algorithms (Chapter 13), both for its own importance, and its theoretical connection to the DFT. We also present the chirp, Goertzel, QFT, NTT, SR-FFT, Approx FFT, Autogen, and programs to implement some of these.

Ivan Selesnick gives a short introduction in Winograd's Short DFT Algorithms (Chapter 7) to using Winograd's techniques to give a highly structured development of short prime length FFTs and describes a program that will automaticlly write these programs. Markus Pueschel presents his "Algebraic Signal Processing" in DFT and FFT: An Algebraic View (Chapter 8)

[1] This content is available online at <http://cnx.org/content/m16325/1.10/>.

on describing the various FFT algorithms. And Steven Johnson describes the FFTW (Fastest Fourier Transform in the West) in Implementing FFTs in Practice (Chapter 11)

The organization of the book represents the various approaches to understanding the FFT and to obtaining efficient computer programs. It also shows the intimate relationship between theory and implementation that can be used to real advantage. The disparity in material devoted to the various approaches represent the tastes of this author, not any intrinsic differences in value.

A fairly long list of references is given but it is impossible to be truly complete. I have referenced the work that I have used and that I am aware of. The collection of computer programs is also somewhat idiosyncratic. They are in Matlab and Fortran because that is what I have used over the years. They also are written primarily for their educational value although some are quite efficient. There is excellent content in the Connexions book by Doug Jones [206].

Chapter 3

Multidimensional Index Mapping[1]

A powerful approach to the development of efficient algorithms is to break a large problem into multiple small ones. One method for doing this with both the DFT and convolution uses a linear change of index variables to map the original one-dimensional problem into a multi-dimensional problem. This approach provides a unified derivation of the Cooley-Tukey FFT, the prime factor algorithm (PFA) FFT, and the Winograd Fourier transform algorithm (WFTA) FFT. It can also be applied directly to convolution to break it down into multiple short convolutions that can be executed faster than a direct implementation. It is often easy to translate an algorithm using index mapping into an efficient program.

The basic definition of the discrete Fourier transform (DFT) is

$$C(k) = \sum_{n=0}^{N-1} x(n) \;\; W_N^{nk} \tag{3.1}$$

where n, k, and N are integers, $j = \sqrt{-1}$, the basis functions are the N roots of unity,

$$W_N = e^{-j2\pi/N} \tag{3.2}$$

and $k = 0, 1, 2, \cdots, N-1$.

If the N values of the transform are calculated from the N values of the data, $x(n)$, it is easily seen that N^2 complex multiplications and approximately that same number of complex additions are required. One method for reducing this required arithmetic is to use an index mapping (a change of variables) to change the one-dimensional DFT into a two- or higher dimensional DFT. This is one of the ideas behind the very efficient Cooley-Tukey [89] and Winograd [404] algorithms. The purpose of index mapping is to change a large problem into several easier ones [46], [120]. This is sometimes called the "divide and conquer" approach [26] but a more accurate description would be "organize and share" which explains the process of redundancy removal or reduction.

[1]This content is available online at <http://cnx.org/content/m16326/1.12/>.

3.1 The Index Map

For a length-N sequence, the time index takes on the values

$$n = 0, 1, 2, ..., N-1 \tag{3.3}$$

When the length of the DFT is not prime, N can be factored as $N = N_1 N_2$ and two new independent variables can be defined over the ranges

$$n_1 = 0, 1, 2, ..., N_1 - 1 \tag{3.4}$$

$$n_2 = 0, 1, 2, ..., N_2 - 1 \tag{3.5}$$

A linear change of variables is defined which maps n_1 and n_2 to n and is expressed by

$$n = ((K_1 n_1 + K_2 n_2))_N \tag{3.6}$$

where K_i are integers and the notation $((x))_N$ denotes the integer residue of x modulo N[232]. This map defines a relation between all possible combinations of n_1 and n_2 in (3.4) and (3.5) and the values for n in (3.3). The question as to whether all of the n in (3.3) are represented, i.e., whether the map is one-to-one (unique), has been answered in [46] showing that certain integer K_i always exist such that the map in (3.6) is one-to-one. Two cases must be considered.

3.1.1 Case 1.

N_1 and N_2 are relatively prime, i.e., the greatest common divisor $(N_1, N_2) = 1$.

The integer map of (3.6) is one-to-one if and only if:

$$(K_1 = aN_2) \quad \text{and/or} \quad (K_2 = bN_1) \quad \text{and} \quad (K_1, N_1) = (K_2, N_2) = 1 \tag{3.7}$$

where a and b are integers.

3.1.2 Case 2.

N_1 and N_2 are not relatively prime, i.e., $(N_1, N_2) > 1$.

The integer map of (3.6) is one-to-one if and only if:

$$(K_1 = aN_2) \quad \text{and} \quad (K_2 \neq bN_1) \quad \text{and} \quad (a, N_1) = (K_2, N_2) = 1 \tag{3.8}$$

or

$$(K_1 \neq aN_2) \quad \text{and} \quad (K_2 = bN_1) \quad \text{and} \quad (K_1, N_1) = (b, N_2) = 1 \tag{3.9}$$

Reference [46] should be consulted for the details of these conditions and examples. Two classes of index maps are defined from these conditions.

3.1.3 Type-One Index Map:

The map of (3.6) is called a type-one map when integers a and b exist such that

$$K_1 = aN_2 \quad \text{and} \quad K_2 = bN_1 \tag{3.10}$$

3.1.4 Type-Two Index Map:

The map of (3.6) is called a type-two map when when integers a and b exist such that

$$K_1 = aN_2 \quad \text{or} \quad K_2 = bN_1, \quad \text{but not both.} \tag{3.11}$$

The type-one can be used **only** if the factors of N are relatively prime, but the type-two can be used whether they are relatively prime or not. Good [149], Thomas, and Winograd [404] all used the type-one map in their DFT algorithms. Cooley and Tukey [89] used the type-two in their algorithms, both for a fixed radix $\left(N - R^M\right)$ and a mixed radix [301].

The frequency index is defined by a map similar to (3.6) as

$$k = ((K_3k_1 + K_4k_2))_N \tag{3.12}$$

where the same conditions, (3.7) and (3.8), are used for determining the uniqueness of this map in terms of the integers K_3 and K_4.

Two-dimensional arrays for the input data and its DFT are defined using these index maps to give

$$\hat{x}\,(n_1, n_2) = x((K_1n_1 + K_2n_2))_N \tag{3.13}$$

$$\hat{X}\,(k_1, k_2) = X((K_3k_1 + K_4k_2))_N \tag{3.14}$$

In some of the following equations, the residue reduction notation will be omitted for clarity. These changes of variables applied to the definition of the DFT given in (3.1) give

$$C\,(k) = \sum_{n_2=0}^{N_2-1} \sum_{n_1=0}^{N_1-1} x\,(n)\; W_N^{K_1K_3n_1k_1}\; W_N^{K_1K_4n_1k_2}\; W_N^{K_2K_3n_2k_1}\; W_N^{K_2K_4n_2k_2} \tag{3.15}$$

where all of the exponents are evaluated modulo N.

The amount of arithmetic required to calculate (3.15) is the same as in the direct calculation of (3.1). However, because of the special nature of the DFT, the integer constants K_i can be chosen in such a way that the calculations are "uncoupled" and the arithmetic is reduced. The requirements for this are

$$((K_1K_4))_N = 0 \quad \text{and/or} \quad ((K_2K_3))_N = 0 \tag{3.16}$$

When this condition and those for uniqueness in (3.6) are applied, it is found that the K_i may **always** be chosen such that one of the terms in (3.16) is zero. If the N_i are relatively prime, it is always possible to make **both** terms zero. If the N_i are not relatively prime, only one of the terms can be set to zero. When they are relatively prime, there is a choice, it is possible to either set one or both to zero. This in turn causes one or both of the center two W terms in (3.15) to become unity.

An example of the Cooley-Tukey radix-4 FFT for a length-16 DFT uses the type-two map with $K_1 = 4$, $K_2 = 1$, $K_3 = 1$, $K_4 = 4$ giving

$$n = 4n_1 + n_2 \tag{3.17}$$

$$k = k_1 + 4k_2 \tag{3.18}$$

The residue reduction in (3.6) is not needed here since n does not exceed N as n_1 and n_2 take on their values. Since, in this example, the factors of N have a common factor, only one of the conditions in (3.16) can hold and, therefore, (3.15) becomes

$$\hat{C}\,(k_1, k_2) = C\,(k) = \sum_{n_2=0}^{3}\sum_{n_1=0}^{3} x\,(n)\; W_4^{n_1k_1}\; W_{16}^{n_2k_1}\; W_4^{n_2k_2} \tag{3.19}$$

Note the definition of W_N in (3.3) allows the simple form of $W_{16}^{K_1K_3} = W_4$

This has the form of a two-dimensional DFT with an extra term W_{16}, called a "twiddle factor". The inner sum over n_1 represents four length-4 DFTs, the W_{16} term represents 16 complex multiplications, and the outer sum over n_2 represents another four length-4 DFTs. This choice of the K_i "uncouples" the calculations since the first sum over n_1 for $n_2 = 0$ calculates the DFT of the first row of the data array $\hat{x}\,(n_1, n_2)$, and those data values are never needed in the succeeding row calculations. The row calculations are independent, and examination of the outer sum shows that the column calculations are likewise independent. This is illustrated in Figure 3.1.

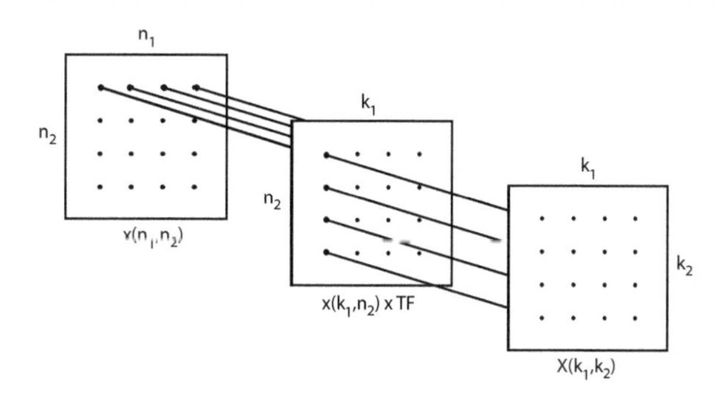

Figure 3.1: Uncoupling of the Row and Column Calculations (Rectangles are Data Arrays)

The left 4-by-4 array is the mapped input data, the center array has the rows transformed, and the right array is the DFT array. The row DFTs and the column DFTs are independent of each other. The twiddle factors (TF) which are the center W in (3.19), are the multiplications which take place on the center array of Figure 3.1.

This uncoupling feature reduces the amount of arithmetic required and allows the results of each row DFT to be written back over the input data locations, since that input row will not be needed again. This is called "in-place" calculation and it results in a large memory requirement savings.

An example of the type-two map used when the factors of N are relatively prime is given for $N = 15$ as

$$n = 5n_1 + n_2 \tag{3.20}$$

$$k = k_1 + 3k_2 \tag{3.21}$$

The residue reduction is again not explicitly needed. Although the factors 3 and 5 are relatively prime, use of the type-two map sets only one of the terms in (3.16) to zero. The DFT in (3.15) becomes

$$X = \sum_{n_2=0}^{4} \sum_{n_1=0}^{2} x \, W_3^{n_1 k_1} \, W_{15}^{n_2 k_1} \, W_5^{n_2 k_2} \tag{3.22}$$

which has the same form as (3.19), including the existence of the twiddle factors (TF). Here the inner sum is five length-3 DFTs, one for each value of k_1. This is illustrated in (3.2) where the rectangles are the 5 by 3 data arrays and the system is called a "mixed radix" FFT.

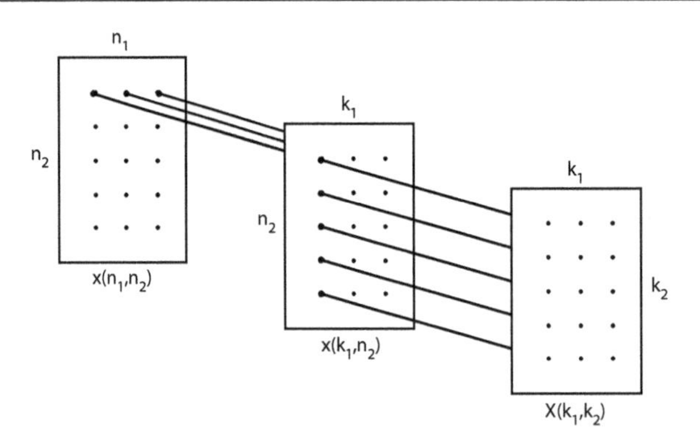

Figure 3.2: Uncoupling of the Row and Column Calculations (Rectangles are Data Arrays)

An alternate illustration is shown in Figure 3.3 where the rectangles are the short length 3 and 5 DFTs.

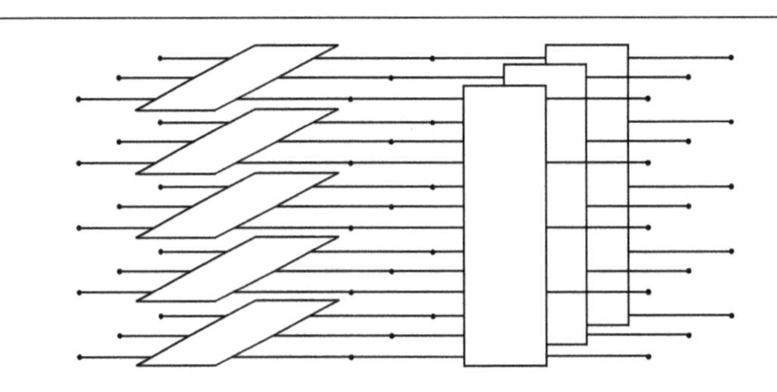

Figure 3.3: Uncoupling of the Row and Column Calculations (Rectangles are Short DFTs)

The type-one map is illustrated next on the same length-15 example. This time the situation of (3.7) with the "and" condition is used in (3.10) using an index map of

$$n = 5n_1 + 3n_2 \tag{3.23}$$

and

$$k = 10k_1 + 6k_2 \tag{3.24}$$

The residue reduction is now necessary. Since the factors of N are relatively prime and the type-one map is being used, both terms in (3.16) are zero, and (3.15) becomes

$$\hat{X} = \sum_{n_2=0}^{4} \sum_{n_1=0}^{2} \hat{x} \; W_3^{n_1 k_1} W_5^{n_2 k_2} \qquad (3.25)$$

which is similar to (3.22), except that now the type-one map gives a pure two-dimensional DFT calculation with no TFs, and the sums can be done in either order. Figures Figure 3.2 and Figure 3.3 also describe this case but now there are no Twiddle Factor multiplications in the center and the resulting system is called a "prime factor algorithm" (PFA).

The purpose of index mapping is to improve the arithmetic efficiency. For example a direct calculation of a length-16 DFT requires 16^2 or 256 real multiplications (recall, one complex multiplication requires 4 real multiplications and 2 real additions) and an uncoupled version requires 144. A direct calculation of a length-15 DFT requires 225 multiplications but with a type-two map only 135 and with a type-one map, 120. Recall one complex multiplication requires four real multiplications and two real additions.

Algorithms of practical interest use short DFT's that require fewer than N^2 multiplications. For example, length-4 DFTs require no multiplications and, therefore, for the length-16 DFT, only the TFs must be calculated. That calculation uses 16 multiplications, many fewer than the 256 or 144 required for the direct or uncoupled calculation.

The concept of using an index map can also be applied to convolution to convert a length $N = N_1 N_2$ one-dimensional cyclic convolution into a N_1 by N_2 two-dimensional cyclic convolution [46], [6]. There is no savings of arithmetic from the mapping alone as there is with the DFT, but savings can be obtained by using special short algorithms along each dimension. This is discussed in Algorithms for Data with Restrictions (Chapter 12) .

3.2 In-Place Calculation of the DFT and Scrambling

Because use of both the type-one and two index maps uncouples the calculations of the rows and columns of the data array, the results of each short length N_i DFT can be written back over the data as it will not be needed again after that particular row or column is transformed. This is easily seen from Figures Figure 3.1, Figure 3.2, and Figure 3.3 where the DFT of the first row of $x(n_1, n_2)$ can be put back over the data rather written into a new array. After all the calculations are finished, the total DFT is in the array of the original data. This gives a significant memory savings over using a separate array for the output.

Unfortunately, the use of in-place calculations results in the order of the DFT values being permuted or scrambled. This is because the data is indexed according to the input map (3.6) and the results are put into the same locations rather than the locations dictated by the output map (3.12). For example with a length-8 radix-2 FFT, the input index map is

$$n = 4n_1 + 2n_2 + n_3 \tag{3.26}$$

which to satisfy (3.16) requires an output map of

$$k = k_1 + 2k_2 + 4k_3 \tag{3.27}$$

The in-place calculations will place the DFT results in the locations of the input map and these should be reordered or unscrambled into the locations given by the output map. Examination of these two maps shows the scrambled output to be in a "bit reversed" order.

For certain applications, this scrambled output order is not important, but for many applications, the order must be unscrambled before the DFT can be considered complete. Because the radix of the radix-2 FFT is the same as the base of the binary number representation, the correct address for any term is found by reversing the binary bits of the address. The part of most FFT programs that does this reordering is called a bit-reversed counter. Examples of various unscramblers are found in [146], [60] and in the appendices.

The development here uses the input map and the resulting algorithm is called "decimation-in-frequency". If the output rather than the input map is used to derive the FFT algorithm so the correct output order is obtained, the input order must be scrambled so that its values are in locations specified by the output map rather than the input map. This algorithm is called "decimation-in-time". The scrambling is the same bit-reverse counting as before, but it precedes the FFT algorithm in this case. The same process of a post-unscrambler or pre-scrambler occurs for the in-place calculations with the type-one maps. Details can be found in [60], [56]. It is possible to do the unscrambling while calculating the FFT and to avoid a separate unscrambler. This is done for the Cooley-Tukey FFT in [192] and for the PFA in [60], [56], [319].

If a radix-2 FFT is used, the unscrambler is a bit-reversed counter. If a radix-4 FFT is used, the unscrambler is a base-4 reversed counter, and similarly for radix-8 and others. However, if for the radix-4 FFT, the short length-4 DFTs (butterflies) have their outputs in bit-revered order, the output of the total radix-4 FFT will be in bit-reversed order, not base-4 reversed order. This means any radix-2^n FFT can use the same radix-2 bit-reversed counter as an unscrambler if the proper butterflies are used.

3.3 Efficiencies Resulting from Index Mapping with the DFT

In this section the reductions in arithmetic in the DFT that result from the index mapping alone will be examined. In practical algorithms several methods are always combined, but it is helpful in understanding the effects of a particular method to study it alone.

The most general form of an uncoupled two-dimensional DFT is given by

$$X(k1,k2) = \sum_{n_2=0}^{N_2-1} \{ \sum_{n_1=0}^{N_1-1} x(n_1,n_2)\ f_1(n_1,n_2,k_1) \}\ f_2(n_2,k_1,k_2) \tag{3.28}$$

where the inner sum calculates N_2 length-N_1 DFT's and, if for a type-two map, the effects of the TFs. If the number of arithmetic operations for a length-N DFT is denoted by $F(N)$, the number of operations for this inner sum is $F = N_2 F(N_1)$. The outer sum which gives N_1 length-N_2 DFT's requires $N_1 F(N_2)$ operations. The total number of arithmetic operations is then

$$F = N_2 F(N_1) + N_1 F(N_2) \tag{3.29}$$

The first question to be considered is for a fixed length N, what is the optimal relation of N_1 and N_2 in the sense of minimizing the required amount of arithmetic. To answer this question, $N1$ and N_2 are temporarily assumed to be real variables rather than integers. If the short length-N_i DFT's in (3.28) and any TF multiplications are assumed to require N_i^2 operations, i.e. $F(N_i) = N_i^2$, "Efficiencies Resulting from Index Mapping with the DFT" (Section 3.3: Efficiencies Resulting from Index Mapping with the DFT) becomes

$$F = N_2 N_1^2 + N_1 N_2^2 = N(N_1 + N_2) = N\left(N_1 + N N_1^{-1}\right) \tag{3.30}$$

To find the minimum of F over N_1, the derivative of F with respect to N_1 is set to zero (temporarily assuming the variables to be continuous) and the result requires $N_1 = N_2$.

$$dF/dN_1 = 0 \quad => \quad N_1 = N_2 \tag{3.31}$$

This result is also easily seen from the symmetry of N_1 and N_2 in $N = N_1 N_2$. If a more general model of the arithmetic complexity of the short DFT's is used, the same result is obtained, but a closer examination must be made to assure that $N_1 = N_2$ is a global minimum.

If only the effects of the index mapping are to be considered, then the $F(N) = N^2$ model is used and (3.31) states that the two factors should be equal. If there are M factors, a similar reasoning shows that all M factors should be equal. For the sequence of length

$$N = R^M \tag{3.32}$$

there are now M length-R DFT's and, since the factors are all equal, the index map must be type two. This means there must be twiddle factors.

In order to simplify the analysis, only the number of multiplications will be considered. If the number of multiplications for a length-R DFT is $F(R)$, then the formula for operation counts in (3.30) generalizes to

$$F = N\sum_{i=1}^{M} F(N_i)/N_i = NMF(R)/R \tag{3.33}$$

for $N_i = R$

$$F = NlnR(N)F(R)/R = (NlnN)(F(R)/(RlnR)) \tag{3.34}$$

This is a very important formula which was derived by Cooley and Tukey in their famous paper [89] on the FFT. It states that for a given R which is called the radix, the number of multiplications (and additions) is proportional to $NlnN$. It also shows the relation to the value of the radix, R.

In order to get some idea of the "best" radix, the number of multiplications to compute a length-R DFT is assumed to be $F(R) = R^x$. If this is used with (3.34), the optimal R can be found.

$$dF/dR = 0 \quad => \quad R = e^{1/(x-1)} \tag{3.35}$$

For $x = 2$ this gives $R = e$, with the closest integer being three.

The result of this analysis states that if no other arithmetic saving methods other than index mapping are used, and if the length-R DFT's plus TFs require $F = R^2$ multiplications, the optimal algorithm requires

$$F = 3Nlog_3N \tag{3.36}$$

multiplications for a length $N = 3^M$ DFT. Compare this with N^2 for a direct calculation and the improvement is obvious.

While this is an interesting result from the analysis of the effects of index mapping alone, in practice, index mapping is almost always used in conjunction with special algorithms for the short length-N_i DFT's in (3.15). For example, if $R = 2$ or 4, there are no multiplications required for the short DFT's. Only the TFs require multiplications. Winograd (see Winorad's Short DFT Algorithms (Chapter 7)) has derived some algorithms for short DFT's that require $O(N)$ multiplications. This means that $F(N_i) = KN_i$ and the operation count F in "Efficiencies Resulting from Index Mapping with the DFT" (Section 3.3: Efficiencies Resulting from Index Mapping with the DFT) is independent of N_i. Therefore, the derivative of F is zero for all N_i. Obviously, these particular cases must be examined.

3.4 The FFT as a Recursive Evaluation of the DFT

It is possible to formulate the DFT so a length-N DFT can be calculated in terms of two length-(N/2) DFTs. And, if $N = 2^M$, each of those length-(N/2) DFTs can be found in terms of length-(N/4) DFTs. This allows the DFT to be calculated by a recursive algorithm with M recursions, giving the familiar order $Nlog\,(N)$ arithmetic complexity.

Calculate the even indexed DFT values from (3.1) by:

$$C\,(2k) = \sum_{n=0}^{N-1} x\,(n)\;W_N^{2nk} \;=\; \sum_{n=0}^{N-1} x\,(n)\;W_{N/2}^{nk} \tag{3.37}$$

$$C\,(2k) = \sum_{n=0}^{N/2-1} x\,(n)\;W_N^{2nk} \;+\; \sum_{n=N/2}^{N-1} x\,(n)\;W_{N/2}^{nk} \tag{3.38}$$

$$C\,(2k) = \sum_{n=0}^{N/2-1} \{x\,(n)\;\;+\;x\,(n+N/2)\}\;W_{N/2}^{nk} \tag{3.39}$$

and a similar argument gives the odd indexed values as:

$$C\,(2k+1) = \sum_{n=0}^{N/2-1} \{x\,(n)\;\;-\;x\,(n+N/2)\}\;W_N^{n}\;W_{N/2}^{nk} \tag{3.40}$$

Together, these are recursive DFT formulas expressing the length-N DFT of $x\,(n)$ in terms of length-N/2 DFTs:

$$C\,(2k) = \mathrm{DFT}_{N/2}\{x\,(n)\;\;+\;x\,(n+N/2)\} \tag{3.41}$$

$$C\,(2k+1) = \mathrm{DFT}_{N/2}\{[x\,(n)\;\;-\;x\,(n+N/2)]\,W_N^{n}\} \tag{3.42}$$

This is a "decimation-in-frequency" (DIF) version since it gives samples of the frequency domain representation in terms of blocks of the time domain signal.

A recursive Matlab program which implements this is given by:

```
        function c = dftr2(x)
% Recursive Decimation-in-Frequency FFT algorithm, csb 8/21/07
L = length(x);
if L > 1
     L2 = L/2;
     TF = exp(-j*2*pi/L).^[0:L2-1];
     c1 = dftr2( x(1:L2) + x(L2+1:L));
     c2 = dftr2((x(1:L2) - x(L2+1:L)).*TF);
     cc = [c1';c2'];
     c = cc(:);
else
     c  = x;
end
```

Listing 3.1: DIF Recursive FFT for $N = 2^M$

A DIT version can be derived in the form:

$$C(k) = \text{DFT}_{N/2}\{x(2n)\} \; + \; W_N^k \text{DFT}_{N/2}\{x(2n+1)\} \tag{3.43}$$

$$C(k+N/2) = \text{DFT}_{N/2}\{x(2n)\} \; - \; W_N^k \text{DFT}_{N/2}\{x(2n+1)\} \tag{3.44}$$

which gives blocks of the frequency domain from samples of the signal.

A recursive Matlab program which implements this is given by:

```
        function c = dftr(x)
% Recursive Decimation-in-Time FFT algorithm, csb
L = length(x);
if L > 1
     L2 = L/2;
     ce = dftr(x(1:2:L-1));
     co = dftr(x(2:2:L));
     TF = exp(-j*2*pi/L).^[0:L2-1];
     c1 = TF.*co;
     c  = [(ce+c1), (ce-c1)];
else
     c  = x;
end
```

Listing 3.2: DIT Recursive FFT for $N = 2^M$

Similar recursive expressions can be developed for other radices and and algorithms. Most recursive programs do not execute as efficiently as looped or straight code, but some can be very efficient, e.g. parts of the FFTW.

Note a length-2^M sequence will require M recursions, each of which will require $N/2$ multiplications. This give the $Nlog\left(N\right)$ formula that the other approaches also derive.

Chapter 4

Polynomial Description of Signals[1]

Polynomials are important in digital signal processing because calculating the DFT can be viewed as a polynomial evaluation problem and convolution can be viewed as polynomial multiplication [27], [261]. Indeed, this is the basis for the important results of Winograd discussed in Winograd's Short DFT Algorithms (Chapter 7). A length-N signal $x(n)$ will be represented by an $N-1$ degree polynomial $X(s)$ defined by

$$X(s) = \sum_{n=0}^{N-1} x(n)\ s^n \tag{4.1}$$

This polynomial $X(s)$ is a single entity with the coefficients being the values of $x(n)$. It is somewhat similar to the use of matrix or vector notation to efficiently represent signals which allows use of new mathematical tools.

The convolution of two finite length sequences, $x(n)$ and $h(n)$, gives an output sequence defined by

$$y(n) = \sum_{k=0}^{N-1} x(k)\ h(n-k) \tag{4.2}$$

$n = 0, 1, 2, \cdots, 2N-1$ where $h(k) = 0$ for $k < 0$. This is exactly the same operation as calculating the coefficients when multiplying two polynomials. Equation (4.2) is the same as

$$Y(s) = X(s)\ H(s) \tag{4.3}$$

In fact, convolution of number sequences, multiplication of polynomials, and the multiplication of integers (except for the carry operation) are all the same operations. To obtain cyclic convolution, where the indices in (4.2) are all evaluated modulo N, the polynomial multiplication in (4.3) is done modulo the polynomial $P(s) = s^N - 1$. This is seen by noting that $N = 0 \bmod N$, therefore, $s^N = 1$ and the polynomial modulus is $s^N - 1$.

[1] This content is available online at <http://cnx.org/content/m16327/1.8/>.

4.1 Polynomial Reduction and the Chinese Remainder Theorem

Residue reduction of one polynomial modulo another is defined similarly to residue reduction for integers. A polynomial $F(s)$ has a residue polynomial $R(s)$ modulo $P(s)$ if, for a given $F(s)$ and $P(s)$, a $Q(S)$ and $R(s)$ exist such that

$$F(s) = Q(s)\,P(s) + R(s) \tag{4.4}$$

with $degree\{R(s)\} < degree\{P(s)\}$. The notation that will be used is

$$R(s) = ((F(s)))_{P(s)} \tag{4.5}$$

For example,

$$(s+1) = \left(\left(s^4 + s^3 - s - 1\right)\right)_{(s^2-1)} \tag{4.6}$$

The concepts of factoring a polynomial and of primeness are an extension of these ideas for integers. For a given allowed set of coefficients (values of $x(n)$), any polynomial has a unique factored representation

$$F(s) = \prod_{i=1}^{M} F_i(s)^{k_i} \tag{4.7}$$

where the $F_i(s)$ are relatively prime. This is analogous to the fundamental theorem of arithmetic.

There is a very useful operation that is an extension of the integer Chinese Remainder Theorem (CRT) which says that if the modulus polynomial can be factored into relatively prime factors

$$P(s) = P_1(s)\ P_2(s) \tag{4.8}$$

then there exist two polynomials, $K_1(s)$ and $K_2(s)$, such that any polynomial $F(s)$ can be recovered from its residues by

$$F(s) = K_1(s)\,F_1(s) + K_2(s)\,F_2(s) \mod P(s) \tag{4.9}$$

where F_1 and F_2 are the residues given by

$$F_1(s) = ((F(s)))_{P_1(s)} \tag{4.10}$$

and

$$F_2(s) = ((F(s)))_{P_2(s)} \tag{4.11}$$

if the order of $F(s)$ is less than $P(s)$. This generalizes to any number of relatively prime factors of $P(s)$ and can be viewed as a means of representing $F(s)$ by several lower degree polynomials, $F_i(s)$.

This decomposition of $F(s)$ into lower degree polynomials is the process used to break a DFT or convolution into several simple problems which are solved and then recombined using the CRT of (4.9). This is another form of the "divide and conquer" or "organize and share" approach similar to the index mappings in Multidimensional Index Mapping (Chapter 3).

One useful property of the CRT is for convolution. If cyclic convolution of $x(n)$ and $h(n)$ is expressed in terms of polynomials by

$$Y(s) = H(s)X(s) \mod P(s) \tag{4.12}$$

where $P(s) = s^N - 1$, and if $P(s)$ is factored into two relatively prime factors $P = P_1P_2$, using residue reduction of $H(s)$ and $X(s)$ modulo P_1 and P_2, the lower degree residue polynomials can be multiplied and the results recombined with the CRT. This is done by

$$Y(s) = ((K_1H_1X_1 + K_2H_2X_2))_P \tag{4.13}$$

where

$$H_1 = ((H))_{P_1}, \quad X_1 = ((X))_{P_1}, \quad H_2 = ((H))_{P_2}, \quad X_2 = ((X))_{P_2} \tag{4.14}$$

and $K1$ and $K2$ are the CRT coefficient polynomials from (4.9). This allows two shorter convolutions to replace one longer one.

Another property of residue reduction that is useful in DFT calculation is polynomial evaluation. To evaluate $F(s)$ at $s = x$, $F(s)$ is reduced modulo $s - x$.

$$F(x) = ((F(s)))_{s-x} \tag{4.15}$$

This is easily seen from the definition in (4.4)

$$F(s) = Q(s)(s - x) + R(s) \tag{4.16}$$

Evaluating $s = x$ gives $R(s) = F(x)$ which is a constant. For the DFT this becomes

$$C(k) = ((X(s)))_{s-W^k} \tag{4.17}$$

Details of the polynomial algebra useful in digital signal processing can be found in [27], [233], [261].

4.2 The DFT as a Polynomial Evaluation

The Z-transform of a number sequence $x(n)$ is defined as

$$X(z) = \sum_{n=0}^{\infty} x(n)\ z^{-n} \tag{4.18}$$

which is the same as the polynomial description in (4.1) but with a negative exponent. For a finite length-N sequence (4.18) becomes

$$X(z) = \sum_{n=0}^{N-1} x(n)\ z^{-n} \tag{4.19}$$

$$X(z) = x(0) + x(1)\,z^{-1} + x(2)\,z^{-2} + \cdot + x(N-1)\,z^{-N+1} \tag{4.20}$$

This $N-1$ order polynomial takes on the values of the DFT of $x(n)$ when evaluated at

$$z = e^{j2\pi k/N} \tag{4.21}$$

which gives

$$C(k) = X(z)\,|_{z=e^{j2\pi k/N}} = \sum_{n=0}^{N-1} x(n)\ e^{-j2\pi nk/N} \tag{4.22}$$

In terms of the positive exponent polynomial from (4.1), the DFT is

$$C(k) = X(s)\,|_{s=W^k} \tag{4.23}$$

where

$$W = e^{-j2\pi/N} \tag{4.24}$$

is an Nth root of unity (raising W to the Nth power gives one). The N values of the DFT are found from $X(s)$ evaluated at the N N^{th} roots of unity which are equally spaced around the unit circle in the complex s plane.

One method of evaluating $X(z)$ is the so-called Horner's rule or nested evaluation. When expressed as a recursive calculation, Horner's rule becomes the Goertzel algorithm which has some computational advantages especially when only a few values of the DFT are needed. The details and programs can be found in [272], [61] and The DFT as Convolution or Filtering: Goertzel's Algorithm (or A Better DFT Algorithm) (Section 5.3: Goertzel's Algorithm (or A Better DFT Algorithm))

Another method for evaluating $X(s)$ is the residue reduction modulo $\left(s - W^k\right)$ as shown in (4.17). Each evaluation requires N multiplications and therefore, N^2 multiplications for the N values of $C(k)$.

$$C(k) = ((X(s)))_{\left(s - W^k\right)} \tag{4.25}$$

A considerable reduction in required arithmetic can be achieved if some operations can be shared between the reductions for different values of k. This is done by carrying out the residue reduction in stages that can be shared rather than done in one step for each k in (4.25).

The N values of the DFT are values of $X(s)$ evaluated at s equal to the N roots of the polynomial $P(s) = s^N - 1$ which are W^k. First, assuming N is even, factor $P(s)$ as

$$P(s) = \left(s^N - 1\right) = P_1(s)\ P_2(s) = \left(s^{N/2} - 1\right)\left(s^{N/2} + 1\right) \tag{4.26}$$

$X(s)$ is reduced modulo these two factors to give two residue polynomials, $X_1(s)$ and $X_2(s)$. This process is repeated by factoring P_1 and further reducing X_1 then factoring P_2 and reducing X_2. This is continued until the factors are of first degree which gives the desired DFT values as in (4.25). This is illustrated for a length-8 DFT. The polynomial whose roots are W^k, factors as

$$P(s) = s^8 - 1 \tag{4.27}$$

$$= \left[s^4 - 1\right]\left[s^4 + 1\right] \tag{4.28}$$

$$= \left[\left(s^2 - 1\right)\left(s^2 + 1\right)\right]\left[\left(s^2 - j\right)\left(s^2 + j\right)\right] \tag{4.29}$$

$$= \left[(s - 1)(s + 1)(s - j)(s + j)\right]\left[(s - a)(s + a)(s - ja)(s + ja)\right] \tag{4.30}$$

where $a^2 = j$. Reducing $X(s)$ by the first factoring gives two third degree polynomials

$$X(s) = x_0 + x_1 s + x_2 s^2 + ... + x_7 s^7 \tag{4.31}$$

gives the residue polynomials

$$X_1(s) = ((X(s)))_{(s^4 - 1)} = (x_0 + x_4) + (x_1 + x_5)\, s + (x_2 + x_6)\, s^2 + (x_3 + x_7)\, s^3 \tag{4.32}$$

$$X_2(s) = ((X(s)))_{(s^4 + 1)} = (x_0 - x_4) + (x_1 - x_5)\, s + (x_2 - x_6)\, s^2 + (x_3 - x_7)\, s^3 \tag{4.33}$$

Two more levels of reduction are carried out to finally give the DFT. Close examination shows the resulting algorithm to be the decimation-in-frequency radix-2 Cooley-Tukey FFT [272], [61]. Martens [227] has used this approach to derive an efficient DFT algorithm.

Other algorithms and types of FFT can be developed using polynomial representations and some are presented in the generalization in DFT and FFT: An Algebraic View (Chapter 8).

Chapter 5

The DFT as Convolution or Filtering[1]

A major application of the FFT is fast convolution or fast filtering where the DFT of the signal is multiplied term-by-term by the DFT of the impulse (helps to be doing finite impulse response (FIR) filtering) and the time-domain output is obtained by taking the inverse DFT of that product. What is less well-known is the DFT can be calculated by convolution. There are several different approaches to this, each with different application.

5.1 Rader's Conversion of the DFT into Convolution

In this section a method quite different from the index mapping or polynomial evaluation is developed. Rather than dealing with the DFT directly, it is converted into a cyclic convolution which must then be carried out by some efficient means. Those means will be covered later, but here the conversion will be explained. This method requires use of some number theory, which can be found in an accessible form in [234] or [262] and is easy enough to verify on one's own. A good general reference on number theory is [259].

The DFT and cyclic convolution are defined by

$$C(k) = \sum_{n=0}^{N-1} x(n)\ W^{nk} \tag{5.1}$$

$$y(k) = \sum_{n=0}^{N-1} x(n)\ h(k-n) \tag{5.2}$$

For both, the indices are evaluated modulo N. In order to convert the DFT in (5.1) into the cyclic convolution of (5.2), the nk product must be changed to the $k-n$ difference. With real numbers, this can be done with logarithms, but it is more complicated when working in a finite set of integers modulo N. From number theory [28], [234], [262], [259], it can be shown that if the modulus is a prime number, a base (called a primitive root) exists such

[1] This content is available online at <http://cnx.org/content/m16328/1.9/>.

that a form of integer logarithm can be defined. This is stated in the following way. If N is a prime number, a number r called a primitive roots exists such that the integer equation

$$n = ((r^m))_N \tag{5.3}$$

creates a unique, one-to-one map of the $N-1$ member set $m = \{0, ..., N-2\}$ and the $N-1$ member set $n = \{1, ..., N-1\}$. This is because the multiplicative group of integers modulo a prime, p, is isomorphic to the additive group of integers modulo $(p-1)$ and is illustrated for $N = 5$ below.

r	m=	0	1	2	3	4	5	6	7
1		1	1	1	1	1	1	1	1
2		1	2	4	3	1	2	4	3
3		1	3	4	2	1	3	4	2
4		1	4	1	4	1	4	1	4
5		*	0	0	0	*	0	0	0
6		1	1	1	1	1	1	1	1

Table 5.1: Table of Integers $n = ((r^m))$ modulo 5, [* not defined]

Table 5.1 is an array of values of r^m modulo N and it is easy to see that there are two primitive roots, 2 and 3, and (5.3) defines a permutation of the integers n from the integers m (except for zero). (5.3) and a primitive root (usually chosen to be the smallest of those that exist) can be used to convert the DFT in (5.1) to the convolution in (5.2). Since (5.3) cannot give a zero, a new length-(N-1) data sequence is defined from $x(n)$ by removing the term with index zero. Let

$$n = r^{-m} \tag{5.4}$$

and

$$k = r^s \tag{5.5}$$

where the term with the negative exponent (the inverse) is defined as the integer that satisfies

$$\left((r^{-m} r^m)\right)_N = 1 \tag{5.6}$$

If N is a prime number, r^{-m} always exists. For example, $((2^{-1}))_5 = 3$. (5.1) now becomes

$$C(r^s) = \sum_{m=0}^{N-2} x\left(r^{-m}\right) \; W^{r^{-m} r^s} + \; x(0), \tag{5.7}$$

for $s = 0, 1, .., N-2$, and

$$C(0) = \sum_{n=0}^{N-1} x(n) \tag{5.8}$$

New functions are defined, which are simply a permutation in the order of the original functions, as

$$x'(m) = x\left(r^{-m}\right), \quad C'(s) = C\left(r^{s}\right), \quad W'(n) = W^{r^n} \tag{5.9}$$

(5.7) then becomes

$$C'(s) = \sum_{m=0}^{N-2} x'(m) \; W'(s-m) \; + \; x(0) \tag{5.10}$$

which is cyclic convolution of length N-1 (plus $x(0)$) and is denoted as

$$C'(k) = x'(k) * W'(k) + x(0) \tag{5.11}$$

Applying this change of variables (use of logarithms) to the DFT can best be illustrated from the matrix formulation of the DFT. (5.1) is written for a length-5 DFT as

$$\begin{bmatrix} C(0) \\ C(1) \\ C(2) \\ C(3) \\ C(4) \end{bmatrix} = \begin{bmatrix} 0 & 0 & 0 & 0 & 0 \\ 0 & 1 & 2 & 3 & 4 \\ 0 & 2 & 4 & 1 & 3 \\ 0 & 3 & 1 & 4 & 2 \\ 0 & 4 & 3 & 2 & 1 \end{bmatrix} \begin{bmatrix} x(0) \\ x(1) \\ x(2) \\ x(3) \\ x(4) \end{bmatrix} \tag{5.12}$$

where the square matrix should contain the terms of W^{nk} but for clarity, only the exponents nk are shown. Separating the $x(0)$ term, applying the mapping of (5.9), and using the primitive roots $r = 2$ (and $r^{-1} = 3$) gives

$$\begin{bmatrix} C(1) \\ C(2) \\ C(4) \\ C(3) \end{bmatrix} = \begin{bmatrix} 1 & 3 & 4 & 2 \\ 2 & 1 & 3 & 4 \\ 4 & 2 & 1 & 3 \\ 3 & 4 & 2 & 1 \end{bmatrix} \begin{bmatrix} x(1) \\ x(3) \\ x(4) \\ x(2) \end{bmatrix} + \begin{bmatrix} x(0) \\ x(0) \\ x(0) \\ x(0) \end{bmatrix} \tag{5.13}$$

and

$$C(0) = x(0) + x(1) + x(2) + x(3) + x(4) \tag{5.14}$$

which can be seen to be a reordering of the structure in (5.12). This is in the form of cyclic convolution as indicated in (5.10). Rader first showed this in 1968 [234], stating that a prime length-N DFT could be converted into a length-(N-1) cyclic convolution of a permutation of the data with a permutation of the W's. He also stated that a slightly more complicated version of the same idea would work for a DFT with a length equal to an odd prime to a power. The details of that theory can be found in [234], [169].

Until 1976, this conversion approach received little attention since it seemed to offer few advantages. It has specialized applications in calculating the DFT if the cyclic convolution is done by distributed arithmetic table look-up [77] or by use of number theoretic transforms [28], [234], [262]. It and the Goertzel algorithm [273], [62] are efficient when only a few DFT values need to be calculated. It may also have advantages when used with pipelined or vector hardware designed for fast inner products. One example is the TMS320 signal processing microprocessor which is pipelined for inner products. The general use of this scheme emerged when new fast cyclic convolution algorithms were developed by Winograd [405].

5.2 The Chirp Z-Transform (or Bluestein's Algorithm)

The DFT of $x(n)$ evaluates the Z-transform of $x(n)$ on N equally spaced points on the unit circle in the z plane. Using a nonlinear change of variables, one can create a structure which is equivalent to modulation and filtering $x(n)$ by a "chirp" signal. [34], [306], [298], [273], [304], [62].

The mathematical identity $(k-n)^2 = k^2 - 2kn + n^2$ gives

$$nk = \left(n^2 - (k-n)^2 + k^2\right)/2 \tag{5.15}$$

which substituted into the definition of the DFT in Multidimensional Index Mapping: Equation 1 (3.1) gives

$$C(k) = \{\sum_{n=0}^{N-1} \left[x(n)\; W^{n^2/2}\right] \; W^{-(k-n)^2/2}\} \; W^{k^2/2} \tag{5.16}$$

This equation can be interpreted as first multiplying (modulating) the data $x(n)$ by a chirp sequence $(W^{n^2/2}$, then convolving (filtering) it, then finally multiplying the filter output by the chirp sequence to give the DFT.

Define the chirp sequence or signal as $h(n) = W^{n^2/2}$ which is called a chirp because the squared exponent gives a sinusoid with changing frequency. Using this definition, (5.16) becomes

$$C(n) = \{[x(n)\; h(n)] \; * \; h^{-1}\}\; h(n) \tag{5.17}$$

We know that convolution can be carried out by multiplying the DFTs of the signals, here we see that evaluation of the DFT can be carried out by convolution. Indeed, the convolution represented by $*$ in (5.17) can be carried out by DFTs (actually FFTs) of a larger length.

This allows a prime length DFT to be calculated by a very efficient length-2^M FFT. This becomes practical for large N when a particular non-composite (or N with few factors) length is required.

As developed here, the chirp z-transform evaluates the z-transform at equally spaced points on the unit circle. A slight modification allows evaluation on a spiral and in segments [298], [273] and allows savings with only some input values are nonzero or when only some output values are needed. The story of the development of this transform is given in [304].

Two Matlab programs to calculate an arbitrary length DFT using the chirp z-transform is shown in p. **??**.

```
        function y = chirpc(x);
% function y = chirpc(x)
% computes an arbitrary-length DFT with the
% chirp z-transform algorithm.  csb.  6/12/91
%
N  = length(x);  n = 0:N-1;      %Sequence length
W  = exp(-j*pi*n.*n/N);          %Chirp signal
xw = x.*W;                       %Modulate with chirp
WW = [conj(W(N:-1:2)),conj(W)]; %Construct filter
y  = conv(WW,xw);                %Convolve w filter
y  = y(N:2*N-1).*W;              %Demodulate w chirp
```

```
function y = chirp(x);
% function y = chirp(x)
% computes an arbitrary-length Discrete Fourier Transform (DFT)
% with the chirp z transform algorithm. The linear convolution
% then required is done with FFTs.
% 1988: L. Arevalo; 11.06.91 K. Schwarz, LNT Erlangen; 6/12/91 csb.
%
N   = length(x);                 %Sequence length
L   = 2^ceil(log((2*N-1))/log(2)); %FFT length
n   = 0:N-1;
W   = exp(-j*pi*n.*n/N);         %Chirp signal
FW  = fft([conj(W), zeros(1,L-2*N+1), conj(W(N:-1:2))],L);
y   = ifft(FW.*fft(x.'.*W,L));   %Convolve using FFT
y   = y(1:N).*W;                 %Demodulate
```

Figure 5.1

5.3 Goertzel's Algorithm (or A Better DFT Algorithm)

Goertzel's algorithm [144], [62], [269] is another methods that calculates the DFT by converting it into a digital filtering problem. The method looks at the calculation of the DFT as the evaluation of a polynomial on the unit circle in the complex plane. This evaluation is done by Horner's method which is implemented recursively by an IIR filter.

5.3.1 The First-Order Goertzel Algorithm

The polynomial whose values on the unit circle are the DFT is a slightly modified z-transform of x(n) given by

$$X(z) = \sum_{n=0}^{N-1} x(n)\, z^{-n} \tag{5.18}$$

which for clarity in this development uses a positive exponent . This is illustrated for a length-4 sequence as a third-order polynomial by

$$X(z) = x(3)\, z^3 + x(2)\, z^2 + x(1)\, z + x(0) \tag{5.19}$$

The DFT is found by evaluating (5.18) at $z = W^k$, which can be written as

$$C(k) = X(z)\,|_{z=W^k} = DFT\{x(n)\} \tag{5.20}$$

where

$$W = e^{-j2\pi/N} \tag{5.21}$$

The most efficient way of evaluating a general polynomial without any pre-processing is by “Horner's rule" [208] which is a nested evaluation. This is illustrated for the polynomial in (5.19) by

$$X(z) = \{[x(3)\, z + x(2)]\, z + x(1)\}z + x(0) \tag{5.22}$$

This nested sequence of operations can be written as a linear difference equation in the form of

$$y(m) = z\, y(m-1) + x(N-m) \tag{5.23}$$

with initial condition $y(0) = 0$, and the desired result being the solution at $m = N$. The value of the polynomial is given by

$$X(z) = y(N)\,. \tag{5.24}$$

(5.23) can be viewed as a first-order IIR filter with the input being the data sequence in reverse order and the value of the polynomial at z being the filter output sampled at $m = N$. Applying this to the DFT gives the Goertzel algorithm [283], [269] which is

$$y(m) = W^k y(m-1) + x(N-m) \tag{5.25}$$

with $y(0) = 0$ and

$$C(k) = y(N) \tag{5.26}$$

where

$$C(k) = \sum_{n=0}^{N-1} x(n)\, W^{nk}. \tag{5.27}$$

The flowgraph of the algorithm can be found in [62], [269] and a simple FORTRAN program is given in the appendix.

When comparing this program with the direct calculation of (5.27), it is seen that the number of floating-point multiplications and additions are the same. In fact, the structures of the two algorithms look similar, but close examination shows that the way the sines and cosines enter the calculations is different. In (5.27), new sine and cosine values are calculated for each frequency and for each data value, while for the Goertzel algorithm in (5.25), they are calculated only for each frequency in the outer loop. Because of the recursive or feedback nature of the algorithm, the sine and cosine values are "updated" each loop rather than recalculated. This results in $2N$ trigonometric evaluations rather than $2N^2$. It also results in an increase in accumulated quantization error.

It is possible to modify this algorithm to allow entering the data in forward order rather than reverse order. The difference (5.23) becomes

$$y(m) = z^{-1} y(m-1) + x(m-1) \tag{5.28}$$

if (5.24) becomes

$$C(k) = z^{N-1}\, y(N) \tag{5.29}$$

for $y(0) = 0$. This is the algorithm programmed later.

5.3.2 The Second-Order Goertzel Algorithm

One of the reasons the first-order Goertzel algorithm does not improve efficiency is that the constant in the feedback or recursive path is complex and, therefore, requires four real multiplications and two real additions. A modification of the scheme to make it second-order removes the complex multiplications and reduces the number of required multiplications by two.

Define the variable $q(m)$ so that

$$y(m) = q(m) - z^{-1}\, q(m-1). \tag{5.30}$$

This substituted into the right-hand side of (5.23) gives

$$y(m) = z\, q(m-1) - q(m-2) + x(N-m). \tag{5.31}$$

Combining (5.30) and (5.31) gives the second order difference equation

$$q(m) = \left(z + z^{-1}\right) q(m-1) - q(m-2) + x(N-m) \tag{5.32}$$

which together with the output (5.30), comprise the second-order Goertzel algorithm where

$$X(z) = y(N) \tag{5.33}$$

for initial conditions $q(0) = q(-1) = 0$.

A similar development starting with (5.28) gives a second-order algorithm with forward ordered input as

$$q(m) = \left(z + z^{-1}\right) q(m-1) - q(m-2) + x(m-1) \tag{5.34}$$

$$y(m) = q(m) - z\, q(-1) \tag{5.35}$$

with

$$X(z) = z^{N-1} y(N) \tag{5.36}$$

and for $q(0) = q(-1) = 0$.

Note that both difference (5.32) and (5.34) are not changed if z is replaced with z^{-1}, only the output (5.30) and (5.35) are different. This means that the polynomial $X(z)$ may be evaluated at a particular z and its inverse z^{-1} from one solution of the difference (5.32) or (5.34) using the output equations

$$X(z) = q(N) - z^{-1} q(N-1) \tag{5.37}$$

and

$$X(1/z) = z^{N-1} \left(q(N) - z\, q(N-1)\right). \tag{5.38}$$

Clearly, this allows the DFT of a sequence to be calculated with half the arithmetic since the outputs are calculated two at a time. The second-order DE actually produces a solution $q(m)$ that contains two first-order components. The output equations are, in effect, zeros that cancel one or the other pole of the second-order solution to give the desired first-order solution. In addition to allowing the calculating of two outputs at a time, the second-order DE requires half the number of real multiplications as the first-order form. This is because the coefficient of the $q(m-2)$ is unity and the coefficient of the $q(m-1)$ is real if z and z^{-1} are complex conjugates of each other which is true for the DFT.

5.3.3 Analysis of Arithmetic Complexity and Timings

Analysis of the various forms of the Goertzel algorithm from their programs gives the following operation count for real multiplications and real additions assuming real data.

Algorithm	Real Mults.	Real Adds	Trig Eval.
Direct DFT	$4\,N^2$	$4\,N^2$	$2\,N^2$
First-Order	$4\,N^2$	$4\,N^2 - 2N$	$2\,N$
Second-Order	$2\,N^2 + 2N$	$4\,N^2$	$2\,N$
Second-Order 2	$N^2 + N$	$2\,N^2 + N$	N

Table 5.2

Timings of the algorithms on a PC in milliseconds are given in the following table.

Algorithm	$N = 125$	$N = 257$
Direct DFT	4.90	19.83
First-Order	4.01	16.70
Second-Order	2.64	11.04
Second-Order 2	1.32	5.55

Table 5.3

These timings track the floating point operation counts fairly well.

5.3.4 Conclusions

Goertzel's algorithm in its first-order form is not particularly interesting, but the two-at-a-time second-order form is significantly faster than a direct DFT. It can also be used for any polynomial evaluation or for the DTFT at unequally spaced values or for evaluating a few DFT terms. A very interesting observation is that the inner-most loop of the Glassman-Ferguson FFT [124] is a first-order Goertzel algorithm even though that FFT is developed in a very different framework.

In addition to floating-point arithmetic counts, the number of trigonometric function evaluations that must be made or the size of a table to store precomputed values should be considered. Since the value of the W^{nk} terms in (5.23) are iteratively calculate in the IIR filter structure, there is round-off error accumulation that should be analyzed in any application.

It may be possible to further improve the efficiency of the second-order Goertzel algorithm for calculating all of the DFT of a number sequence. Perhaps a fourth order DE could calculate four output values at a time and they could be separated by a numerator that would cancel three of the zeros. Perhaps the algorithm could be arranged in stages to give an $N\, log\,(N)$ operation count. The current algorithm does not take into account any of the symmetries of the input index. Perhaps some of the ideas used in developing the QFT [53], [155], [158] could be used here.

5.4 The Quick Fourier Transform (QFT)

One stage of the QFT can use the symmetries of the sines and cosines to calculate a DFT more efficiently than directly implementing the definition Multidimensional Index Mapping: Equation 1 (3.1). Similar to the Goertzel algorithm, the one-stage QFT is a better N^2 DFT algorithm for arbitrary lengths. See The Cooley-Tukey Fast Fourier Transform Algorithm: The Quick Fourier Transform, An FFT based on Symmetries (Section 9.4: The Quick Fourier Transform, An FFT based on Symmetries).

Chapter 6

Factoring the Signal Processing Operators[1]

A third approach to removing redundancy in an algorithm is to express the algorithm as an operator and then factor that operator into sparse factors. This approach is used by Tolimieri [382], [384], Egner [118], Selesnick, Elliott [121] and others. It is presented in a more general form in DFT and FFT: An Algebraic View (Chapter 8) The operators may be in the form of a matrix or a tensor operator.

6.1 The FFT from Factoring the DFT Operator

The definition of the DFT in Multidimensional Index Mapping: Equation 1 (3.1) can written as a matrix-vector operation by $C = WX$ which, for $N = 8$ is

$$\begin{bmatrix} C(0) \\ C(1) \\ C(2) \\ C(3) \\ C(4) \\ C(5) \\ C(6) \\ C(7) \end{bmatrix} = \begin{bmatrix} W^0 & W^0 & W^0 & W^0 & W^0 & W^0 & W^0 & W^0 \\ W^0 & W^1 & W^2 & W^3 & W^4 & W^5 & W^6 & W^7 \\ W^0 & W^2 & W^4 & W^6 & W^8 & W^{10} & W^{12} & W^{14} \\ W^0 & W^3 & W^6 & W^9 & W^{12} & W^{15} & W^{18} & W^{21} \\ W^0 & W^4 & W^8 & W^{12} & W^{16} & W^{20} & W^{24} & W^{28} \\ W^0 & W^5 & W^{10} & W^{15} & W^{20} & W^{25} & W^{30} & W^{35} \\ W^0 & W^6 & W^{12} & W^{18} & W^{24} & W^{30} & W^{36} & W^{42} \\ W^0 & W^7 & W^{14} & W^{21} & W^{28} & W^{35} & W^{42} & W^{49} \end{bmatrix} \begin{bmatrix} x(0) \\ x(1) \\ x(2) \\ x(3) \\ x(4) \\ x(5) \\ x(6) \\ x(7) \end{bmatrix} \tag{6.1}$$

which clearly requires $N^2 = 64$ complex multiplications and $N(N-1)$ additions. A factorization of the DFT operator, W, gives $W = F_1\ F_2\ F_3$ and $C = F_1\ F_2\ F_3\ X$ or, expanded,

[1] This content is available online at <http://cnx.org/content/m16330/1.8/>.

$$\begin{bmatrix} C(0) \\ C(4) \\ C(2) \\ C(6) \\ C(1) \\ C(5) \\ C(3) \\ C(7) \end{bmatrix} = \begin{bmatrix} 1 & 1 & 0 & 0 & 0 & 0 & 0 & 0 \\ 1 & -1 & 0 & 0 & 0 & 0 & 0 & 0 \\ 0 & 0 & 1 & 1 & 0 & 0 & 0 & 0 \\ 0 & 0 & 1 & -1 & 0 & 0 & 0 & 0 \\ 0 & 0 & 0 & 0 & 1 & 1 & 0 & 0 \\ 0 & 0 & 0 & 0 & 1 & -1 & 0 & 0 \\ 0 & 0 & 0 & 0 & 0 & 0 & 1 & 1 \\ 0 & 0 & 0 & 0 & 0 & 0 & 1 & -1 \end{bmatrix} \begin{bmatrix} 1 & 0 & 1 & 0 & 0 & 0 \\ 0 & 1 & 0 & 1 & 0 & 0 \\ W^0 & 0 & -W^2 & 0 & 0 & 0 \\ 0 & W^0 & 0 & -W^2 & 0 & 0 \\ 0 & 0 & 0 & 0 & 1 & 0 \\ 0 & 0 & 0 & 0 & 0 & 1 \\ 0 & 0 & 0 & 0 & W^0 & 0 \\ 0 & 0 & 0 & 0 & 0 & W^2 \end{bmatrix} \quad (6.2)$$

$$\begin{bmatrix} 1 & 0 & 0 & 0 & 1 & 0 & 0 & 0 \\ 0 & 1 & 0 & 0 & 0 & 1 & 0 & 0 \\ 0 & 0 & 1 & 0 & 0 & 0 & 1 & 0 \\ 0 & 0 & 0 & 1 & 0 & 0 & 0 & 1 \\ W^0 & 0 & 0 & 0 & -W^0 & 0 & 0 & 0 \\ 0 & W^1 & 0 & 0 & 0 & -W^1 & 0 & 0 \\ 0 & 0 & W^2 & 0 & 0 & 0 & -W^2 & 0 \\ 0 & 0 & 0 & W^3 & 0 & 0 & 0 & -W^3 \end{bmatrix} \begin{bmatrix} x(0) \\ x(1) \\ x(2) \\ x(3) \\ x(4) \\ x(5) \\ x(6) \\ x(7) \end{bmatrix} \quad (6.3)$$

where the F_i matrices are sparse. Note that each has 16 (or $2N$) non-zero terms and F_2 and F_3 have 8 (or N) non-unity terms. If $N = 2^M$, then the number of factors is $log(N) = M$. In another form with the twiddle factors separated so as to count the complex multiplications we have

$$\begin{bmatrix} C(0) \\ C(4) \\ C(2) \\ C(6) \\ C(1) \\ C(5) \\ C(3) \\ C(7) \end{bmatrix} = \begin{bmatrix} 1 & 1 & 0 & 0 & 0 & 0 & 0 & 0 \\ 1 & -1 & 0 & 0 & 0 & 0 & 0 & 0 \\ 0 & 0 & 1 & 1 & 0 & 0 & 0 & 0 \\ 0 & 0 & 1 & -1 & 0 & 0 & 0 & 0 \\ 0 & 0 & 0 & 0 & 1 & 1 & 0 & 0 \\ 0 & 0 & 0 & 0 & 1 & -1 & 0 & 0 \\ 0 & 0 & 0 & 0 & 0 & 0 & 1 & 1 \\ 0 & 0 & 0 & 0 & 0 & 0 & 1 & -1 \end{bmatrix} \quad (6.4)$$

$$\begin{bmatrix} 1 & 0 & 0 & 0 & 0 & 0 & 0 & 0 \\ 0 & 1 & 0 & 0 & 0 & 0 & 0 & 0 \\ 0 & 0 & W^0 & 0 & 0 & 0 & 0 & 0 \\ 0 & 0 & 0 & W^2 & 0 & 0 & 0 & 0 \\ 0 & 0 & 0 & 0 & 1 & 0 & 0 & 0 \\ 0 & 0 & 0 & 0 & 0 & 1 & 0 & 0 \\ 0 & 0 & 0 & 0 & 0 & 0 & W^0 & 0 \\ 0 & 0 & 0 & 0 & 0 & 0 & 0 & W^2 \end{bmatrix} \begin{bmatrix} 1 & 0 & 1 & 0 & 0 & 0 & 0 & 0 \\ 0 & 1 & 0 & 1 & 0 & 0 & 0 & 0 \\ 1 & 0 & -1 & 0 & 0 & 0 & 0 & 0 \\ 0 & 1 & 0 & -1 & 0 & 0 & 0 & 0 \\ 0 & 0 & 0 & 0 & 1 & 0 & 1 & 0 \\ 0 & 0 & 0 & 0 & 0 & 1 & 0 & 1 \\ 0 & 0 & 0 & 0 & 1 & 0 & -1 & 0 \\ 0 & 0 & 0 & 0 & 0 & 1 & 0 & -1 \end{bmatrix} \tag{6.5}$$

$$\begin{bmatrix} 1 & 0 & 0 & 0 & 0 & 0 & 0 & 0 \\ 0 & 1 & 0 & 0 & 0 & 0 & 0 & 0 \\ 0 & 0 & 1 & 0 & 0 & 0 & 0 & 0 \\ 0 & 0 & 0 & 1 & 0 & 0 & 0 & 0 \\ 0 & 0 & 0 & 0 & W^0 & 0 & 0 & 0 \\ 0 & 0 & 0 & 0 & 0 & W^1 & 0 & 0 \\ 0 & 0 & 0 & 0 & 0 & 0 & W^2 & 0 \\ 0 & 0 & 0 & 0 & 0 & 0 & 0 & W^3 \end{bmatrix} \begin{bmatrix} 1 & 0 & 0 & 0 & 1 & 0 & 0 & 0 \\ 0 & 1 & 0 & 0 & 0 & 1 & 0 & 0 \\ 0 & 0 & 1 & 0 & 0 & 0 & 1 & 0 \\ 0 & 0 & 0 & 1 & 0 & 0 & 0 & 1 \\ 1 & 0 & 0 & 0 & -1 & 0 & 0 & 0 \\ 0 & 1 & 0 & 0 & 0 & -1 & 0 & 0 \\ 0 & 0 & 1 & 0 & 0 & 0 & -1 & 0 \\ 0 & 0 & 0 & 1 & 0 & 0 & 0 & -1 \end{bmatrix} \begin{bmatrix} x(0) \\ x(1) \\ x(2) \\ x(3) \\ x(4) \\ x(5) \\ x(6) \\ x(7) \end{bmatrix} \tag{6.6}$$

which is in the form $C = A_1 \, M_1 \, A_2 \, M_2 \, A_3 \, X$ described by the index map. A_1, A_2, and A_3 each represents 8 additions, or, in general, N additions. M_1 and M_2 each represent 4 (or $N/2$) multiplications.

This is a very interesting result showing that implementing the DFT using the factored form requires considerably less arithmetic than the single factor definition. Indeed, the form of the formula that Cooley and Tukey derived showing that the amount of arithmetic required by the FFT is on the order of $Nlog\,(N)$ can be seen from the factored operator formulation.

Much of the theory of the FFT can be developed using operator factoring and it has some advantages for implementation of parallel and vector computer architectures. The eigenspace approach is somewhat of the same type [18].

6.2 Algebraic Theory of Signal Processing Algorithms

A very general structure for all kinds of algorithms can be generalized from the approach of operators and operator decomposition. This is developed as "Algebraic Theory of Signal Processing" discussed in the module DFT and FFT: An Algebraic View (Chapter 8) by Püschel and others [118].

Chapter 7

Winograd's Short DFT Algorithms[1]

In 1976, S. Winograd [406] presented a new DFT algorithm which had significantly fewer multiplications than the Cooley-Tukey FFT which had been published eleven years earlier. This new Winograd Fourier Transform Algorithm (WFTA) is based on the type- one index map from Multidimensional Index Mapping (Chapter 3) with each of the relatively prime length short DFT's calculated by very efficient special algorithms. It is these short algorithms that this section will develop. They use the index permutation of Rader described in the another module to convert the prime length short DFT's into cyclic convolutions. Winograd developed a method for calculating digital convolution with the minimum number of multiplications. These optimal algorithms are based on the polynomial residue reduction techniques of Polynomial Description of Signals: Equation 1 (4.1) to break the convolution into multiple small ones [29], [235], [263], [416], [408], [197].

The operation of discrete convolution defined by

$$y(n) = \sum_{k} h(n-k)\ x(k) \tag{7.1}$$

is called a **bilinear** operation because, for a fixed $h(n)$, $y(n)$ is a linear function of $x(n)$ and for a fixed $x(n)$ it is a linear function of $h(n)$. The operation of cyclic convolution is the same but with all indices evaluated modulo N.

Recall from Polynomial Description of Signals: Equation 3 (4.3) that length-N cyclic convolution of $x(n)$ and $h(n)$ can be represented by polynomial multiplication

$$Y(s) = X(s)\ H(s) \mod \left(s^N - 1\right) \tag{7.2}$$

This bilinear operation of (7.1) and (7.2) can also be expressed in terms of linear matrix operators and a simpler bilinear operator denoted by o which may be only a simple element-by-element multiplication of the two vectors [235], [197], [212]. This matrix formulation

[1]This content is available online at <http://cnx.org/content/m16333/1.14/>.

is

$$Y = C\,[AXoBH] \tag{7.3}$$

where X, H and Y are length-N vectors with elements of $x(n)$, $h(n)$ and $y(n)$ respectively. The matrices A and B have dimension M x N , and C is N x M with $M \geq N$. The elements of A, B, and C are constrained to be simple; typically small integers or rational numbers. It will be these matrix operators that do the equivalent of the residue reduction on the polynomials in (7.2).

In order to derive a useful algorithm of the form (7.3) to calculate (7.1), consider the polynomial formulation (7.2) again. To use the residue reduction scheme, the modulus is factored into relatively prime factors. Fortunately the factoring of this particular polynomial, $s^N - 1$, has been extensively studied and it has considerable structure. When factored over the rationals, which means that the only coefficients allowed are rational numbers, the factors are called cyclotomic polynomials [29], [235], [263]. The most interesting property for our purposes is that most of the coefficients of cyclotomic polynomials are zero and the others are plus or minus unity for degrees up to over one hundred. This means the residue reduction will generally require no multiplications.

The operations of reducing $X(s)$ and $H(s)$ in (7.2) are carried out by the matrices A and B in (7.3). The convolution of the residue polynomials is carried out by the o operator and the recombination by the CRT is done by the C matrix. More details are in [29], [235], [263], [197], [212] but the important fact is the A and B matrices usually contain only zero and plus or minus unity entries and the C matrix only contains rational numbers. The only general multiplications are those represented by o. Indeed, in the theoretical results from computational complexity theory, these real or complex multiplications are usually the only ones counted. In practical algorithms, the rational multiplications represented by C could be a limiting factor.

The $h(n)$ terms are fixed for a digital filter, or they represent the W terms from Multidimensional Index Mapping: Equation 1 (3.1) if the convolution is being used to calculate a DFT. Because of this, $d = BH$ in (7.3) can be precalculated and only the A and C operators represent the mathematics done at execution of the algorithm. In order to exploit this feature, it was shown [416], [197] that the properties of (7.3) allow the exchange of the more complicated operator C with the simpler operator B. Specifically this is given by

$$Y = C\,[AXoBH] \tag{7.4}$$

$$Y' = B^T\left[AXoC^T H'\right] \tag{7.5}$$

where H' has the same elements as H, but in a permuted order, and likewise Y' and Y. This very important property allows precomputing the more complicated $C^T H'$ in (7.5) rather than BH as in (7.3).

Because BH or $C^T H$' can be precomputed, the bilinear form of (7.3) and (7.5) can be written as a linear form. If an M x M diagonal matrix D is formed from $d = C^T H$, or in the case of (7.3), $d = BH$, assuming a commutative property for o, (7.5) becomes

$$Y' = B^T DAX \tag{7.6}$$

and (7.3) becomes

$$Y = CDAX \tag{7.7}$$

In most cases there is no reason not to use the same reduction operations on X and H, therefore, B can be the same as A and (7.6) then becomes

$$Y' = A^T DAX \tag{7.8}$$

In order to illustrate how the residue reduction is carried out and how the A matrix is obtained, the length-5 DFT algorithm started in The DFT as Convolution or Filtering: Matrix 1 (5.12) will be continued. The DFT is first converted to a length-4 cyclic convolution by the index permutation from The DFT as Convolution or Filtering: Equation 3 (5.3) to give the cyclic convolution in The DFT as Convolution or Filtering (Chapter 5). To avoid confusion from the permuted order of the data $x(n)$ in The DFT as Convolution or Filtering (Chapter 5), the cyclic convolution will first be developed without the permutation, using the polynomial $U(s)$

$$U(s) = x(1) + x(3)s + x(4)s^2 + x(2)s^3 \tag{7.9}$$

$$U(s) = u(0) + u(1)s + u(2)s^2 + u(3)s^3 \tag{7.10}$$

and then the results will be converted back to the permuted $x(n)$. The length-4 cyclic convolution in terms of polynomials is

$$Y(s) = U(s)\ H(s) \mod \left(s^4 - 1\right) \tag{7.11}$$

and the modulus factors into three cyclotomic polynomials

$$s4 - 1 = \left(s^2 - 1\right)\left(s^2 + 1\right) \tag{7.12}$$

$$= (s-1)(s+1)\left(s^2 + 1\right) \tag{7.13}$$

$$= P_1\ P_2\ P_3 \tag{7.14}$$

Both $U(s)$ and $H(s)$ are reduced modulo these three polynomials. The reduction modulo P_1 and P_2 is done in two stages. First it is done modulo $(s^2 - 1)$, then that residue is further reduced modulo $(s-1)$ and $(s+1)$.

$$U(s) = u0 + u1s + u_2 s^2 + u_3 s^3 \tag{7.15}$$

$$U'(s) = ((U(s)))_{(s^2-1)} = (u_0 + u_2) + (u_1 + u_3)\,s \tag{7.16}$$

$$U1(s) = \left((U'(s))\right)_{P_1} = (u_0 + u_1 + u_2 + u_3) \tag{7.17}$$

$$U2(s) = \left((U'(s))\right)_{P_2} = (u_0 - u_1 + u_2 - u_3) \tag{7.18}$$

$$U3(s) = ((U(s)))_{P_3} = (u_0 - u_2) + (u_1 - u_3)\,s \tag{7.19}$$

The reduction in (7.16) of the data polynomial (7.15) can be denoted by a matrix operation on a vector which has the data as entries.

$$\begin{bmatrix} 1 & 0 & 1 & 0 \\ 0 & 1 & 0 & 1 \end{bmatrix} \begin{bmatrix} u_0 \\ u_1 \\ u_2 \\ u_3 \end{bmatrix} = \begin{bmatrix} u_0 + u_2 \\ u_1 + u_3 \end{bmatrix} \tag{7.20}$$

and the reduction in (7.19) is

$$\begin{bmatrix} 1 & 0 & -1 & 0 \\ 0 & 1 & 0 & -1 \end{bmatrix} \begin{bmatrix} u_0 \\ u_1 \\ u_2 \\ u_3 \end{bmatrix} = \begin{bmatrix} u_0 - u_2 \\ u_1 - u_3 \end{bmatrix} \tag{7.21}$$

Combining (7.20) and (7.21) gives one operator

$$\begin{bmatrix} 1 & 0 & 1 & 0 \\ 0 & 1 & 0 & 1 \\ 1 & 0 & -1 & 0 \\ 0 & 1 & 0 & -1 \end{bmatrix} \begin{bmatrix} u_0 + u_2 \\ u_1 + u_3 \\ u_0 - u_2 \\ u_1 - u_3 \end{bmatrix} = \begin{bmatrix} u_0 + u_2 \\ u_1 + u_3 \\ u_0 - u_2 \\ u_1 - u_3 \end{bmatrix} = \begin{bmatrix} w_0 \\ w_1 \\ v_0 \\ v_1 \end{bmatrix} \tag{7.22}$$

Further reduction of $v_0 + v_1 s$ is not possible because $P_3 = s^2 + 1$ cannot be factored over the rationals. However $s^2 - 1$ can be factored into $P_1 P_2 = (s-1)(s+1)$ and, therefore, $w_0 + w_1 s$ can be further reduced as was done in (7.17) and (7.18) by

$$\begin{bmatrix} 1 & 1 \end{bmatrix} \begin{bmatrix} w_0 \\ w_1 \end{bmatrix} = w_0 + w_1 = u_0 + u_2 + u_1 + u_3 \tag{7.23}$$

$$\begin{bmatrix} 1 & -1 \end{bmatrix} \begin{bmatrix} w_0 \\ w_1 \end{bmatrix} = w_0 - w_1 = u_0 + u_2 - u_1 - u_3 \tag{7.24}$$

Combining (7.22), (7.23) and (7.24) gives

$$\begin{bmatrix} 1 & 1 & 0 & 0 \\ 1 & -1 & 0 & 0 \\ 0 & 0 & 1 & 0 \\ 0 & 0 & 0 & 1 \end{bmatrix} \begin{bmatrix} 1 & 0 & 1 & 0 \\ 0 & 1 & 0 & 1 \\ 1 & 0 & -1 & 0 \\ 0 & 1 & 0 & -1 \end{bmatrix} \begin{bmatrix} u_0 \\ u_1 \\ u_2 \\ u_3 \end{bmatrix} = \begin{bmatrix} r_0 \\ r_1 \\ v_0 \\ v_1 \end{bmatrix} \tag{7.25}$$

The same reduction is done to $H(s)$ and then the convolution of (7.11) is done by multiplying each residue polynomial of $X(s)$ and $H(s)$ modulo each corresponding cyclotomic factor of $P(s)$ and finally a recombination using the polynomial Chinese Remainder Theorem (CRT) as in Polynomial Description of Signals: Equation 9 (4.9) and Polynomial Description of Signals: Equation 13 (4.13).

$$Y(s) = K_1(s)\,U_1(s)\,H_1(s) + K_2(s)\,U_2(s)\,H_2(s) + K_3(s)\,U_3(s)\,H_3(s) \tag{7.26}$$

mod $(s^4 - 1)$

where $U_1(s) = r_1$ and $U_2(s) = r_2$ are constants and $U_3(s) = v_0 + v_1 s$ is a first degree polynomial. U_1 times H_1 and U_2 times H_2 are easy, but multiplying U_3 time H_3 modulo $(s^2 + 1)$ is more difficult.

The multiplication of $U_3(s)$ times $H_3(s)$ can be done by the Toom-Cook algorithm [29], [235], [263] which can be viewed as Lagrange interpolation or polynomial multiplication modulo a special polynomial with three arbitrary coefficients. To simplify the arithmetic, the constants are chosen to be plus and minus one and zero. The details of this can be found in [29], [235], [263]. For this example it can be verified that

$$\left((v0 + v1s)\,(h0 + h1s)\right)_{s^2+1} = (v_0 h_0 - v_1 h_1) + (v_0 h_1 + v_1 h_0)\,s \tag{7.27}$$

which by the Toom-Cook algorithm or inspection is

$$\begin{bmatrix} 1 & -1 & 0 \\ -1 & -1 & 1 \end{bmatrix} \left[\begin{bmatrix} 1 & 0 \\ 0 & 1 \\ 1 & 1 \end{bmatrix} \begin{bmatrix} v_0 \\ v_1 \end{bmatrix} o \begin{bmatrix} 1 & 0 \\ 0 & 1 \\ 1 & 1 \end{bmatrix} \begin{bmatrix} h_0 \\ h_1 \end{bmatrix} \right] = \begin{bmatrix} y_0 \\ y_1 \end{bmatrix} \tag{7.28}$$

where o signifies point-by-point multiplication. The total A matrix in (7.3) is a combination of (7.25) and (7.28) giving

$$AX = A_1 A_2 A_3 X \tag{7.29}$$

$$= \begin{bmatrix} 1 & 0 & 0 & 0 \\ 0 & 1 & 0 & 0 \\ 0 & 0 & 1 & 0 \\ 0 & 0 & 0 & 1 \\ 0 & 0 & 1 & 1 \end{bmatrix} \begin{bmatrix} 1 & 1 & 0 & 0 \\ 1 & -1 & 0 & 0 \\ 0 & 0 & 1 & 0 \\ 0 & 0 & 0 & 1 \end{bmatrix} \begin{bmatrix} 1 & 0 & 1 & 0 \\ 0 & 1 & 0 & 1 \\ 1 & 0 & -1 & 0 \\ 0 & 1 & 0 & -1 \end{bmatrix} \begin{bmatrix} u_0 \\ u_1 \\ u_2 \\ u_3 \end{bmatrix} = \begin{bmatrix} r_0 \\ r_1 \\ v_0 \\ v_1 \end{bmatrix} \quad (7.30)$$

where the matrix A_3 gives the residue reduction s^2-1 and s^2+1, the upper left-hand part of A_2 gives the reduction modulo $s-1$ and $s+1$, and the lower right-hand part of A1 carries out the Toom-Cook algorithm modulo s^2+1 with the multiplication in (7.5). Notice that by calculating (7.30) in the three stages, seven additions are required. Also notice that A_1 is not square. It is this "expansion" that causes more than N multiplications to be required in o in (7.5) or D in (7.6). This staged reduction will derive the A operator for (7.5)

The method described above is very straight-forward for the shorter DFT lengths. For $N=3$, both of the residue polynomials are constants and the multiplication given by o in (7.3) is trivial. For $N=5$, which is the example used here, there is one first degree polynomial multiplication required but the Toom-Cook algorithm uses simple constants and, therefore, works well as indicated in (7.28). For $N=7$, there are two first degree residue polynomials which can each be multiplied by the same techniques used in the $N=5$ example. Unfortunately, for any longer lengths, the residue polynomials have an order of three or greater which causes the Toom-Cook algorithm to require constants of plus and minus two and worse. For that reason, the Toom-Cook method is not used, and other techniques such as index mapping are used that require more than the minimum number of multiplications, but do not require an excessive number of additions. The resulting algorithms still have the structure of (7.8). Blahut [29] and Nussbaumer [263] have a good collection of algorithms for polynomial multiplication that can be used with the techniques discussed here to construct a wide variety of DFT algorithms.

The constants in the diagonal matrix D can be found from the CRT matrix C in (7.5) using $d = C^T H'$ for the diagonal terms in D. As mentioned above, for the smaller prime lengths of 3, 5, and 7 this works well but for longer lengths the CRT becomes very complicated. An alternate method for finding D uses the fact that since the linear form (7.6) or (7.8) calculates the DFT, it is possible to calculate a known DFT of a given $x(n)$ from the definition of the DFT in Multidimensional Index Mapping: Equation 1 (3.1) and, given the A matrix in (7.8), solve for D by solving a set of simultaneous equations. The details of this procedure are described in [197].

A modification of this approach also works for a length which is an odd prime raised to some power: $N=P^M$. This is a bit more complicated [235], [416] but has been done for lengths of 9 and 25. For longer lengths, the conventional Cooley-Tukey type- two index map algorithm seems to be more efficient. For powers of two, there is no primitive root, and therefore, no simple conversion of the DFT into convolution. It is possible to use two generators [235], [263], [408] to make the conversion and there exists a set of length 4, 8, and 16 DFT algorithms of the form in (7.8) in [235].

In Table 7.1 an operation count of several short DFT algorithms is presented. These are practical algorithms that can be used alone or in conjunction with the index mapping to give longer DFT's as shown in The Prime Factor and Winograd Fourier Transform Algorithms (Chapter 10). Most are optimized in having either the theoretical minimum number of multiplications or the minimum number of multiplications without requiring a very large number of additions. Some allow other reasonable trade-offs between numbers of multiplications and additions. There are two lists of the number of multiplications. The first is the number of actual floating point multiplications that must be done for that length DFT. Some of these (one or two in most cases) will be by rational constants and the others will be by irrational constants. The second list is the total number of multiplications given in the diagonal matrix D in (7.8). At least one of these will be unity (the one associated with $X(0)$) and in some cases several will be unity (for $N = 2^M$). The second list is important in programming the WFTA in The Prime Factor and Winograd Fourier Transform Algorithm: The Winograd Fourier Transform Algorithm (Section 10.2: The Winograd Fourier Transform Algorithm).

Length N	Mult Non-one	Mult Total	Adds
2	0	4	4
3	4	6	12
4	0	8	16
5	10	12	34
7	16	18	72
8	4	16	52
9	20	22	84
11	40	42	168
13	40	42	188
16	20	36	148
17	70	72	314
19	76	78	372
25	132	134	420
32	68	-	388

Table 7.1: Number of Real Multiplications and Additions for a Length-N DFT of Complex Data

Because of the structure of the short DFTs, the number of real multiplications required for the DFT of real data is exactly half that required for complex data. The number of real additions required is slightly less than half that required for complex data because $(N-1)$ of the additions needed when N is prime add a real to an imaginary, and that is not actually performed. When $N = 2m$, there are $(N-2)$ of these pseudo additions. The special case for real data is discussed in [101], [177], [356].

The structure of these algorithms are in the form of $X' = CDAX$ or $B^T DAX$ or $A^T DAX$ from (7.5) and (7.8). The A and B matrices are generally M by N with $M \geq N$ and have elements that are integers, generally 0 or ± 1. A pictorial description is given in Figure 7.1.

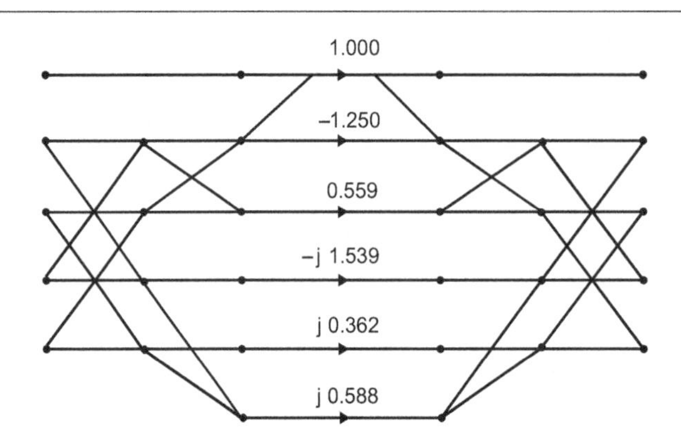

Figure 7.1: Flow Graph for the Length-5 DFT

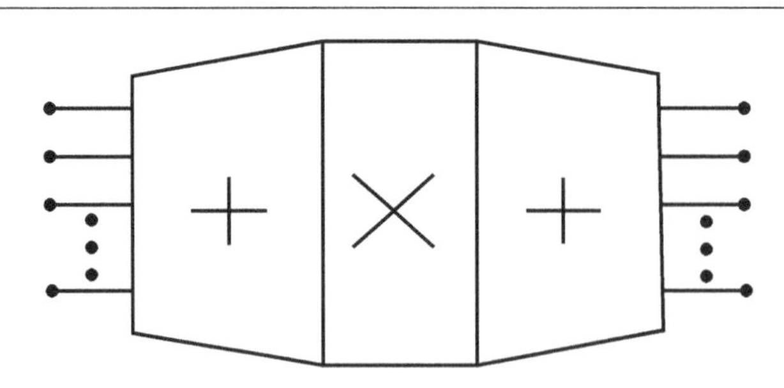

Figure 7.2: Block Diagram of a Winograd Short DFT

The flow graph in Figure 7.1 should be compared with the matrix description of (7.8) and (7.30), and with the programs in [29], [235], [63], [263] and the appendices. The shape in Figure 7.2 illustrates the expansion of the data by A. That is to say, AX has more entries than X because $M > N$. The A operator consists of additions, the D operator gives the M multiplications (some by one) and A^T contracts the data back to N values with additions only. M is one half the second list of multiplies in Table 7.1.

An important characteristic of the D operator in the calculation of the DFT is its entries are either purely real or imaginary. The reduction of the W vector by $\left(s^{(N-1)/2}-1\right)$ and $\left(s^{(N-1)/2}+1\right)$ separates the real and the imaginary constants. This is discussed in [416], [197]. The number of multiplications for complex data is only twice those necessary for real data, not four times.

Although this discussion has been on the calculation of the DFT, very similar results are true for the calculation of convolution and correlation, and these will be further developed in Algorithms for Data with Restrictions (Chapter 12). The A^TDA structure and the picture in Figure 7.2 are the same for convolution. Algorithms and operation counts can be found in [29], [263], [7].

7.1 The Bilinear Structure

The bilinear form introduced in (7.3) and the related linear form in (7.6) are very powerful descriptions of both the DFT and convolution.

$$\text{Bilinear: } Y = C\left[AX \ o \ BH\right] \tag{7.31}$$

$$\text{Linear: } Y = CDA\ X \tag{7.32}$$

Since (7.31) is a bilinear operation defined in terms of a second bilinear operator o , this formulation can be nested. For example if o is itself defined in terms of a second bilinear operator @, by

$$X \ o \ H = C'\left[A'X \ @ \ B'H\right] \tag{7.33}$$

then (7.31) becomes

$$Y = CC'\left[A'AX \ @ \ B'BH\right] \tag{7.34}$$

For convolution, if A represents the polynomial residue reduction modulo the cyclotomic polynomials, then A is square (e.g. (7.25) and o represents multiplication of the residue polynomials modulo the cyclotomic polynomials. If A represents the reduction modulo the cyclotomic polynomials plus the Toom-Cook reduction as was the case in the example of (7.30), then A is NxM and o is term-by- term simple scalar multiplication. In this case AX can be thought of as a transform of X and C is the inverse transform. This is called a rectangular transform [7] because A is rectangular. The transform requires only additions and convolution is done with M multiplications. The other extreme is when A represents reduction over the N complex roots of $s^N - 1$. In this case A is the DFT itself, as in the example of (43), and o is point by point complex multiplication and C is the inverse DFT. A trivial case is where A, B and C are identity operators and o is the cyclic convolution.

This very general and flexible bilinear formulation coupled with the idea of nesting in (7.34) gives a description of most forms of convolution.

7.2 Winograd's Complexity Theorems

Because Winograd's work [29], [235], [416], [408], [413], [419] has been the foundation of the modern results in efficient convolution and DFT algorithms, it is worthwhile to look at his theoretical conclusions on optimal algorithms. Most of his results are stated in terms of polynomial multiplication as Polynomial Description of Signals: Equation 3 (4.3) or (7.11). The measure of computational complexity is usually the number of multiplications, and only certain multiplications are counted. This must be understood in order not to misinterpret the results.

This section will simply give a statement of the pertinent results and will not attempt to derive or prove anything. A short interpretation of each theorem will be given to relate the result to the algorithms developed in this chapter. The indicated references should be consulted for background and detail.

Theorem 1 [416] Given two polynomials, $x(s)$ and $h(s)$, of degree N and M respectively, each with indeterminate coefficients that are elements of a field H, $N+M+1$ multiplications are necessary to compute the coefficients of the product polynomial $x(s)\,h(s)$. Multiplication by elements of the field G (the field of constants), which is contained in H, are not counted and G contains at least $N+M$ distinct elements.

The upper bound in this theorem can be realized by choosing an arbitrary modulus polynomial $P(s)$ of degree $N+M+1$ composed of $N+M+1$ distinct linear polynomial factors with coefficients in G which, since its degree is greater than the product $x(s)\,h(s)$, has no effect on the product, and by reducing $x(s)$ and $h(s)$ to $N+M+1$ residues modulo the $N+M+1$ factors of $P(s)$. These residues are multiplied by each other, requiring $N+M+1$ multiplications, and the results recombined using the Chinese remainder theorem (CRT). The operations required in the reduction and recombination are not counted, while the residue multiplications are. Since the modulus $P(s)$ is arbitrary, its factors are chosen to be simple so as to make the reduction and CRT simple. Factors of zero, plus and minus unity, and infinity are the simplest. Plus and minus two and other factors complicate the actual calculations considerably, but the theorem does not take that into account. This algorithm is a form of the Toom-Cook algorithm and of Lagrange interpolation [29], [235], [263], [416]. For our applications, H is the field of reals and G the field of rationals.

Theorem 2 [416] If an algorithm exists which computes $x(s)h(s)$ in $N+M+1$ multiplications, all but one of its multiplication steps must necessarily be of the form

$$mk = \left(gk^{'} + x(gk)\right)\left(gk" + h(gk)\right) \quad \text{for } k = 0, 1, ..., N+M \tag{7.35}$$

where q_k are distinct elements of G; and $g_k^{'}$ and $g_k"$ are arbitrary elements of G

This theorem states that the structure of an optimal algorithm is essentially unique although the factors of $P(s)$ may be chosen arbitrarily.

Theorem 3 [416] Let $P(s)$ be a polynomial of degree N and be of the form $P(s) = Q(s)k$, where $Q(s)$ is an irreducible polynomial with coefficients in G and k is a positive integer. Let $x(s)$ and $h(s)$ be two polynomials of degree at least $N-1$ with coefficients from H, then $2N-1$ multiplications are required to compute the product $x(s)h(s)$ modulo $P(s)$.

This theorem is similar to Theorem 1 (p. 52) with the operations of the reduction of the product modulo $P(s)$ not being counted.

Theorem 4 [416] Any algorithm that computes the product $x(s)h(s)$ modulo $P(s)$ according to the conditions stated in Theorem 3 and requires $2N-1$ multiplications will necessarily be of one of three structures, each of which has the form of Theorem 2 internally.

As in Theorem 2 (p. 52), this theorem states that only a limited number of possible structures exist for optimal algorithms.

Theorem 5 [416] If the modulus polynomial $P(s)$ has degree N and is not irreducible, it can be written in a unique factored form $P(s) = P_1^{m_1}(s)\, P_2^{m_2}(s) \ldots P_k^{m_k}(s)$ where each of the $P_i(s)$ are irreducible over the allowed coefficient field G. $2N-k$ multiplications are necessary to compute the product $x(s)h(s)$ modulo $P(s)$ where $x(s)$ and $h(s)$ have coefficients in H and are of degree at least $N-1$. All algorithms that calculate this product in $2N-k$ multiplications must be of a form where each of the k residue polynomials of $x(s)$ and $h(s)$ are separately multiplied modulo the factors of $P(s)$ via the CRT.

Corollary: If the modulus polynomial is $P(s) = s^N - 1$, then $2N - t(N)$ multiplications are necessary to compute $x(s)h(s)$ modulo $P(s)$, where $t(N)$ is the number of positive divisors of N.

Theorem 5 (p. 53) is very general since it allows a general modulus polynomial. The proof of the upper bound involves reducing $x(s)$ and $h(s)$ modulo the k factors of $P(s)$. Each of the k irreducible residue polynomials is then multiplied using the method of Theorem 4 (p. 53) requiring $2Ni-1$ multiplies and the products are combined using the CRT. The total number of multiplies from the k parts is $2N-k$. The theorem also states the structure of these optimal algorithms is essentially unique. The special case of $P(s) = s^N - 1$ is interesting since it corresponds to cyclic convolution and, as stated in the corollary, k is easily determined. The factors of $s^N - 1$ are called cyclotomic polynomials and have interesting properties [29], [235], [263].

Theorem 6 [416], [408] Consider calculating the DFT of a prime length real-valued number sequence. If G is chosen as the field of rational numbers, the number of real multiplications necessary to calculate a length-P DFT is $u(DFT(N)) = 2P - 3 - t(P-1)$ where $t(P-1)$ is the number of divisors of $P-1$.

This theorem not only gives a lower limit on any practical prime length DFT algorithm, it also gives practical algorithms for $N = 3, 5$, and 7. Consider the operation counts in Table 7.1 to understand this theorem. In addition to the real multiplications counted by complexity theory, each optimal prime-length algorithm will have one multiplication by a rational constant. That constant corresponds to the residue modulo (s-1) which always exists for the modulus $P(s) = s^{N-1} - 1$. In a practical algorithm, this multiplication must be carried out, and that accounts for the difference in the prediction of Theorem 6 (p. 54) and count in Table 7.1. In addition, there is another operation that for certain applications must be counted as a multiplication. That is the calculation of the zero frequency term $X(0)$ in the first row of the example in The DFT as Convolution or Filtering: Matrix 1 (5.12). For applications to the WFTA discussed in The Prime Factor and Winograd Fourier Transform Algorithms: The Winograd Fourier Transform Algorithm (Section 10.2: The Winograd Fourier Transform Algorithm), that operation must be counted as a multiply. For lengths longer than 7, optimal algorithms require too many additions, so compromise structures are used.

Theorem 7 [419], [171] If G is chosen as the field of rational numbers, the number of real multiplications necessary to calculate a length-N DFT where N is a prime number raised to an integer power: $N = Pm$, is given by

$$u(DFT(N)) = 2N - ((m2 + m)/2)\,t(P-1) - m - 1 \tag{7.36}$$

where $t(P-1)$ is the number of divisors of $(P-1)$.

This result seems to be practically achievable only for $N = 9$, or perhaps 25. In the case of $N = 9$, there are two rational multiplies that must be carried out and are counted in Table 7.1 but are not predicted by Theorem 7 (p. 54). Experience [187] indicates that even for $N = 25$, an algorithm based on a Cooley-Tukey FFT using a type 2 index map gives an over-all more balanced result.

Theorem 8 [171] If G is chosen as the field of rational numbers, the number of real multiplications necessary to calculate a length-N DFT where $N = 2m$ is given by

$$u(DFT(N)) = 2N - m2 - m - 2 \tag{7.37}$$

This result is not practically useful because the number of additions necessary to realize this minimum of multiplications becomes very large for lengths greater than 16. Nevertheless, it proves the minimum number of multiplications required of an optimal algorithm is a linear function of N rather than of $NlogN$ which is that required of practical algorithms. The best practical power-of-two algorithm seems to the Split-Radix [105] FFT discussed in The Cooley-Tukey Fast Fourier Transform Algorithm: The Split-Radix FFT Algorithm (Section 9.2: The Split-Radix FFT Algorithm).

All of these theorems use ideas based on residue reduction, multiplication of the residues, and then combination by the CRT. It is remarkable that this approach finds the minimum number of required multiplications by a constructive proof which generates an algorithm that achieves this minimum; and the structure of the optimal algorithm is, within certain variations, unique. For shorter lengths, the optimal algorithms give practical programs. For longer lengths the uncounted operations involved with the multiplication of the higher degree residue polynomials become very large and impractical. In those cases, efficient suboptimal algorithms can be generated by using the same residue reduction as for the optimal case, but by using methods other than the Toom-Cook algorithm of Theorem 1 (p. 52) to multiply the residue polynomials.

Practical long DFT algorithms are produced by combining short prime length optimal DFT's with the Type 1 index map from Multidimensional Index Mapping (Chapter 3) to give the Prime Factor Algorithm (PFA) and the Winograd Fourier Transform Algorithm (WFTA) discussed in The Prime Factor and Winograd Fourier Transform Algorithms (Chapter 10). It is interesting to note that the index mapping technique is useful inside the short DFT algorithms to replace the Toom-Cook algorithm and outside to combine the short DFT's to calculate long DFT's.

7.3 The Automatic Generation of Winograd's Short DFTs

by Ivan Selesnick, Polytechnic Institute of New York University

7.3.1 Introduction

Efficient prime length DFTs are important for two reasons. A particular application may require a prime length DFT and secondly, the maximum length and the variety of lengths of a PFA or WFTA algorithm depend upon the availability of prime length modules.

This [329], [335], [331], [333] discusses automation of the process Winograd used for constructing prime length FFTs [29], [187] for $N < 7$ and that Johnson and Burrus [197] extended to $N < 19$. It also describes a program that will design any prime length FFT in principle, and will also automatically generate the algorithm as a C program and draw the corresponding flow graph.

Winograd's approach uses Rader's method to convert a prime length DFT into a $P-1$ length cyclic convolution, polynomial residue reduction to decompose the problem into smaller convolutions [29], [263], and the Toom-Cook algorithm [29], [252]. The Chinese Remainder Theorem (CRT) for polynomials is then used to recombine the shorter convolutions. Unfortunately, the design procedure derived directly from Winograd's theory becomes cumbersome for longer length DFTs, and this has often prevented the design of DFT programs for lengths greater than 19.

Here we use three methods to facilitate the construction of prime length FFT modules. First, the matrix exchange property [29], [197], [218] is used so that the transpose of the reduction operator can be used rather than the more complicated CRT reconstruction operator. This is then combined with the numerical method [197] for obtaining the multiplication coefficients rather than the direct use of the CRT. We also deviate from the Toom-Cook algorithm, because it requires too many additions for the lengths in which we are interested. Instead we use an iterated polynomial multiplication algorithm [29]. We have incorporated these three ideas into a single structural procedure that automates the design of prime length FFTs.

7.3.2 Matrix Description

It is important that each step in the Winograd FFT can be described using matrices. By expressing cyclic convolution as a bilinear form, a compact form of prime length DFTs can be obtained.

If y is the cyclic convolution of h and x, then y can be expressed as

$$y = C\left[Ax. * Bh\right] \tag{7.38}$$

where, using the Matlab convention, $.*$ represents point by point multiplication. When A,B, and C are allowed to be complex, A and B are seen to be the DFT operator and C, the inverse DFT. When only real numbers are allowed, A, B, and C will be rectangular. This form of convolution is presented with many examples in [29]. Using the matrix exchange property explained in [29] and [197] this form can be written as

$$y = RB^T\left[C^TRh. * Ax\right] \tag{7.39}$$

where R is the permutation matrix that reverses order.

When h is fixed, as it is when considering prime length DFTs, the term C^TRh can be precomputed and a diagonal matrix D formed by $D = diag\{C^TRh\}$. This is advantageous because in general, C is more complicated than B, so the ability to "hide" C saves computation. Now $y = RB^TDAx$ or $y = RA^TDAx$ since A and B can be the same; they implement a polynomial reduction. The form $y = R^TDAxT$ can also be used for the prime length DFTs, it is only necessary to permute the entries of x and to ensure that the DC term is computed correctly. The computation of the DC term is simple, for the residue of a polynomial modulo $a-1$ is always the sum of the coefficients. After adding the x_0 term of the original input sequence, to the $s-l$ residue, the DC term is obtained. Now $DFT\{x\} = RA^TDAx$. In [197] Johnson observes that by permuting the elements on the diagonal of D, the output can be permuted, so that the R matrix can be hidden in D, and $DFT\{x\} = A^TDAx$. From the knowledge of this form, once A is found, D can be found numerically [197].

7.3.3 Programming the Design Procedure

Because each of the above steps can be described by matrices, the development of a prime length FFTs is made convenient with the use of a matrix oriented programming language such as Matlab. After specifying the appropriate matrices that describe the desired FFT algorithm, generating code involves compiling the matrices into the desired code for execution.

Each matrix is a section of one stage of the flow graph that corresponds to the DFT program. The four stages are:

1. Permutation Stage: Permutes input and output sequence.
2. Reduction Stage: Reduces the cyclic convolution to smaller polynomial products.
3. Polynomial Product Stage: Performs the polynomial multiplications.
4. Multiplication Stage: Implements the point-by-point multiplication in the bilinear form.

Each of the stages can be clearly seen in the flow graphs for the DFTs. Figure 7.3 shows the flow graph for a length 17 DFT algorithm that was automatically drawn by the program.

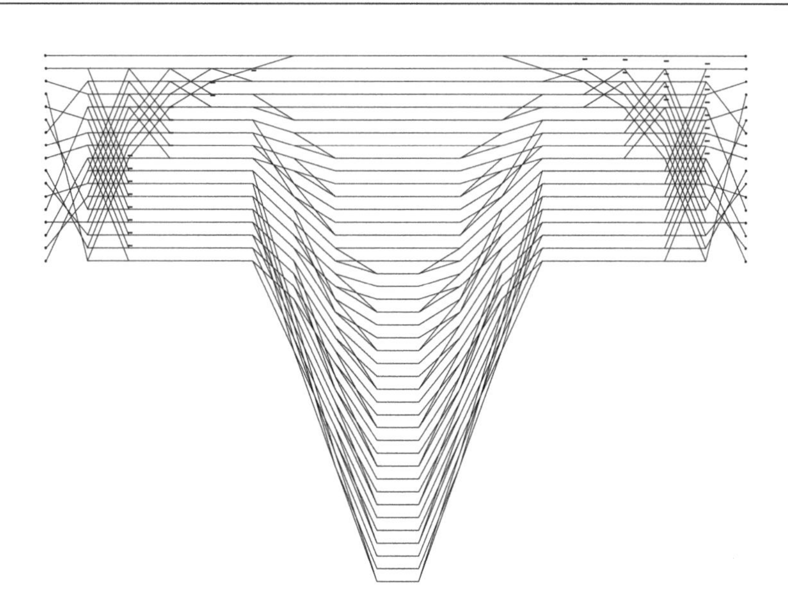

Figure 7.3: Flowgraph of length-17 DFT

The programs that accomplish this process are written in Matlab and C. Those that compute the appropriate matrices are written in Matlab. These matrices are then stored as two ASCII files, with the dimensions in one and the matrix elements in the second. A C program then reads the flies and compiles them to produce the final FFT program in C [335]

7.3.4 The Reduction Stage

The reduction of an Nth degree polynomial, $X(s)$, modulo the cyclotomic polynomial factors of $\left(s^N-1\right)$ requires only additions for many N, however, the actual number of additions depends upon the way in which the reduction proceeds. The reduction is most efficiently performed in steps. For example, if $N=4$ and $\left((X(s))_{s-1},\left((X(s))_{s+1}\right.\right.$ and $\left((X(s))_{s^2+1}\right.$ where the double parenthesis denote polynomial reduction modulo $(s-1)$, $s+1$, and s^2+1, then in the first step $((X(s)))_{s^2-1}$, and $((Xs))_{s^2+1}$ should be computed. In the second step, $((Xs))_{s-1}$ and $((Xs))_{s+1}$ can be found by reducing $((X(s)))_{s^2-1}$ This process is described by the diagram in Figure 7.4.

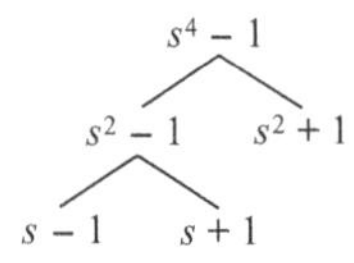

Figure 7.4: Factorization of s^4-1 in steps

When N is even, the appropriate first factorization is $\left(S^{N/2}-1\right)\left(s^{N/2}+1\right)$, however, the next appropriate factorization is frequently less obvious. The following procedure has been found to generate a factorization in steps that coincides with the factorization that minimizes the cumulative number of additions incurred by the steps. The prime factors of N are the basis of this procedure and their importance is clear from the useful well-known equation $s^N-1=\prod_{n|N}C_n(s)$ where $C_n(s)$ is the nth cyclotomic polynomial.

We first introduce the following two functions defined on the positive integers,

$$\psi(N)=\text{the smallest prime factor of } N \text{ for} N>1 \tag{7.40}$$

and $\psi(1)=1$.

Suppose $P(s)$ is equal to either $\left(s^N-1\right)$ or an intermediate noncyclotomic polynomial appearing in the factorization process, for example, (a^2-1), above. Write $P(s)$ in terms of its cyclotomic factors,

$$P(s)=C_{k_1}(s)\ C_{k_2}(s)\ \cdots C_{k_L} \tag{7.41}$$

define the two sets, G and G , by

$$G = \{k_1, \cdots, k_L\} \text{ and } G = \{k/gcd(G) : k \in G\} \tag{7.42}$$

and define the two integers, t and T, by

$$t = min\{\psi(k) : k \in G, k > 1\} \text{ and } T = maxnu(k,t) : k \in G\} \tag{7.43}$$

Then form two new sets,

$$A = \{k \in G : T \mid k\} \text{ and } B = \{k \in G : T|k\} \tag{7.44}$$

The factorization of $P(s)$,

$$P(s) = \left(\prod_{k \in A} C_k(s)\right)\left(\prod_{k \in B} C_k(s)\right) \tag{7.45}$$

has been found useful in the procedure for factoring $\left(s^N - 1\right)$. This is best illustrated with an example.

Example: $N = 36$

Step 1. Let $P(s) = s^{36} - 1$. Since $P = C_1 C_2 C_3 C_4 C_6 C_9 C_{12} C_{18} C_{36}$

$$G = G = \{1, 2, 3, 4, 6, 9, 12, 18, 36\} \tag{7.46}$$

$$t = min\{2, 3\} = 2 \tag{7.47}$$

$$A = \{k \in G : 4|k\} = \{1, 2, 3, 6, 9, 18\} \tag{7.48}$$

$$B = \{k \in G : 4|k\} = \{4, 12, 36\} \tag{7.49}$$

Hence the factorization of $s^{36} - 1$ into two intermediate polynomials is as expected,

$$\prod_{k \in A} C_k(s) = s^{18} - 1, \quad \prod_{k \in B} C_k(s) = s^{18} + 1 \tag{7.50}$$

If a 36th degree polynomial, $X(s)$, is represented by a vector of coefficients, $X = (x_{35}, \cdots, x_0)^{'}$, then $\left((X(s))_{s^{18}-1}\right.$ (represented by X') and $\left((X(s))_{s^{18}+1}\right.$ (represented by X") is given by

$$test \tag{7.51}$$

which entails 36 additions.

Step 2. This procedure is repeated with $P(s) = s^{18} - 1$ and $P(s) = s^{18} + 1$. We will just show it for the later. Let $P(s) = s^{18} + 1$. Since $P = C_4 C_{12} C_{36}$

$$G = \{4, 12, 36\}, \quad G' = \{l, 3, 9\} \tag{7.52}$$

$$t = min3 = 3 \tag{7.53}$$

$$T = max\nu(k, 3) : k \in G = maxl, 3, 9 = 9 \tag{7.54}$$

$$A = k \in G : 9|k\} = \{4, 12\} \tag{7.55}$$

$$B = k \in G : 9|k\} = \{36\} \tag{7.56}$$

This yields the two intermediate polynomials,

$$s^6 + 1, \quad \text{and} \quad s^{12} - s^6 + 1 \tag{7.57}$$

In the notation used above,

$$\begin{bmatrix} X' \\ X'' \end{bmatrix} = \begin{bmatrix} I_6 & -I_6 & I_6 \\ I_6 & I_6 & \\ -I_6 & & I_6 \end{bmatrix} X \tag{7.58}$$

entailing 24 additions. Continuing this process results in a factorization in steps

In order to see the number of additions this scheme uses for numbers of the form $N = P - 1$ (which is relevant to prime length FFT algorithms) figure 4 shows the number of additions the reduction process uses when the polynomial X(s) is real.

Figure 4: Number of Additions for Reduction Stage

7.3.5 The Polynomial Product Stage

The iterated convolution algorithm can be used to construct an N point linear convolution algorithm from shorter linear convolution algorithms [29]. Suppose the linear convolution y, of the n point vectors x and h (h known) is described by

$$y = E_n^T \; D \; E_n \; x \tag{7.59}$$

where E_n is an "expansion" matrix the elements of which are $\pm l$'s and 0's and D is an appropriate diagonal matrix. Because the only multiplications in this expression are by the elements of D, the number of multiplications required, $M(n)$, is equal to the number of rows of E_n. The number of additions is denoted by $A(n)$.

Given a matrix E_{n_1} and a matrix E_{n_2}, the iterated algorithm gives a method for combining E_{n_1} and E_{n_2} to construct a valid expansion matrix, E_n, for $N \leq n_1 n_2$. Specifically,

$$E_{n1,n2} = \left(I_{M(n_2)} \otimes E_{n_1}\right)\left(E_{n_2} \times I_{n_1}\right) \tag{7.60}$$

The product $n_1 n_2$ may be greater than N, for zeros can be (conceptually) appended to x. The operation count associated with $E_{n1,n2}$ is

$$A\left(n_1, n_2\right) = n!A\left(n_2\right) + A\left(n_1\right) M n_2 \tag{7.61}$$

$$M\left(n_1, n_2\right) = M\left(n_1\right)\ M\left(n_2\right) \tag{7.62}$$

Although they are both valid expansion matrices, $E_{n1,n2} \neq E_{n2,n1}$ and $A_{n1,n2} \neq A_{n2,n1}$ Because $M_{n_1,n_2} \neq M_{n_2,n_1}$ it is desirable to chose an ordering of factors to minimize the additions incurred by the expansion matrix. The following [7], [263] follows from above.

7.3.5.1 Multiple Factors

Note that a valid expansion matrix, E_N, can be constructed from $E_{n1,n2}$ and E_{n3}, for $N \leq n_1 n_2 n_3$. In general, any number of factors can be used to create larger expansion matrices. The operation count associated with $E_{n1,n2,n3}$ is

$$A\left(n_1, n_2, n_3\right) = n_1 n_2 A\left(n_3\right) + n_1 A\left(n_2\right) M\left(n_3\right) + A\left(n_1\right) M\left(n_2\right) M\left(n_3\right) \tag{7.63}$$

$$M\left(n_1, n_2, n_3\right) = M\left(n_1\right) M\left(n_2\right) M\left(n_3\right) \tag{7.64}$$

These equations generalize in the predicted way when more factors are considered. Because the ordering of the factors is relevant in the equation for $A\left(.\right)$ but not for $M\left(.\right)$, it is again desirable to order the factors to minimize the number of additions. By exploiting the following property of the expressions for $A\left(.\right)$ and $M\left(.\right)$, the optimal ordering can be found [7].

reservation of Optimal Ordering. Suppose $A\left(n_1, n_2, n_3\right) \leq min\{A\left(n_{k_1}, n_{k_2}, n_{k_3}\right) : k_1, k_2, k_3 \in \{1, 2, 3\}$ and distinct$\}$, then

1. $$A\left(n_1, n_2\right) \leq A\left(n_2, n_1\right) \tag{7.65}$$

2. $$A\left(n_2, n_3\right) \leq A\left(n_3, n_2\right) \tag{7.66}$$

3. $$A\left(n_1, n_3\right) \leq A\left(n_3, n_1\right) \tag{7.67}$$

The generalization of this property to more than two factors reveals that an optimal ordering of $\{n_1, \cdots, n_{L-i}\}$ is preserved in an optimal ordering of $\{n_1, \cdots n_L\}$. Therefore, if $(n_1, \cdots n_L)$ is an optimal ordering of $\{n_1, \cdots n_L\}$, then (n_k, n_{k+1}) is an optimal ordering of $\{n_k, n_{k+1}\}$ and consequently

$$\frac{A(n_k)}{M(n_k) - n_k} \leq \frac{A(n_{k+1})}{M(n_{k+1}) - n_{k+1}} \tag{7.68}$$

for all $k = 1, 2, \cdots, L-1$.

This immediately suggests that an optimal ordering of $\{n_1, \cdots n_L\}$ is one for which

$$\frac{A(n_1)}{M(n_1) - n_1} \cdots \frac{A(n_L)}{M(n_L) - n_L} \tag{7.69}$$

is nondecreasing. Hence, ordering the factors, $\{n_1, \cdots n_L\}$, to minimize the number of additions incurred by $E_{n_1, \cdots, n_L}$ simply involves computing the appropriate ratios.

7.3.6 Discussion and Conclusion

We have designed prime length FFTs up to length 53 that are as good as the previous designs that only went up to 19. Table 1 gives the operation counts for the new and previously designed modules, assuming complex inputs.

It is interesting to note that the operation counts depend on the factorability of $P-1$. The primes 11, 23, and 47 are all of the form $1 + 2P_1$ making the design of efficient FFTs for these lengths more difficult.

Further deviations from the original Winograd approach than we have made could prove useful for longer lengths. We investigated, for example, the use of twiddle factors at appropriate points in the decomposition stage; these can sometimes be used to divide the cyclic convolution into smaller convolutions. Their use means, however, that the 'center* multiplications would no longer be by purely real or imaginary numbers.

N	Mult	Adds
7	16	72
11	40	168
13	40	188
17	82	274
19	88	360
23	174	672
29	190	766
31	160	984
37	220	920
41	282	1140
43	304	1416
47	640	2088
53	556	2038

Table 7.2: Operation counts for prime length DFTs

The approach in writing a program that writes another program is a valuable one for several reasons. Programming the design process for the design of prime length FFTs has the advantages of being practical, error-free, and flexible. The flexibility is important because it allows for modification and experimentation with different algorithmic ideas. Above all, it has allowed longer DFTs to be reliably designed.

More details on the generation of programs for prime length FFTs can be found in the 1993 Technical Report.

Chapter 8

DFT and FFT: An Algebraic View[1]

by Markus Pueschel, Carnegie Mellon University

In infinite, or non-periodic, discrete-time signal processing, there is a strong connection between the z-transform, Laurent series, convolution, and the discrete-time Fourier transform (DTFT) [277]. As one may expect, a similar connection exists for the DFT but bears surprises. Namely, it turns out that the proper framework for the DFT requires modulo operations of polynomials, which means working with so-called polynomial algebras [138]. Associated with polynomial algebras is the Chinese remainder theorem, which describes the DFT algebraically and can be used as a tool to concisely derive various FFTs as well as convolution algorithms [268], [409], [414], [12] (see also Winograd's Short DFT Algorithms (Chapter 7)). The polynomial algebra framework was fully developed for signal processing as part of the algebraic signal processing theory (ASP). ASP identifies the structure underlying many transforms used in signal processing, provides deep insight into their properties, and enables the derivation of their fast algorithms [295], [293], [291], [294]. Here we focus on the algebraic description of the DFT and on the algebraic derivation of the general-radix Cooley-Tukey FFT from Factoring the Signal Processing Operators (Chapter 6). The derivation will make use of and extend the Polynomial Description of Signals (Chapter 4). We start with motivating the appearance of modulo operations.

The z-transform associates with infinite discrete signals $X = (\cdots, x(-1), x(0), x(1), \cdots)$ a Laurent series:

$$X \mapsto X(s) = \sum_{n \in \mathbb{Z}} x(n) s^n. \tag{8.1}$$

Here we used $s = z^{-1}$ to simplify the notation in the following. The DTFT of X is the evaluation of $X(s)$ on the unit circle

$$X\left(e^{-j\omega}\right), \quad -\pi < \omega \leq \pi. \tag{8.2}$$

[1]This content is available online at <http://cnx.org/content/m16331/1.14/>.

Finally, filtering or (linear) convolution is simply the multiplication of Laurent series,

$$H * X \leftrightarrow H(s) X(s). \tag{8.3}$$

For finite signals $X = (x(0), \cdots, x(N-1))$ one expects that the equivalent of (8.1) becomes a mapping to polynomials of degree $N-1$,

$$X \mapsto X(s) = \sum_{n=0}^{N-1} x(n) s^n, \tag{8.4}$$

and that the DFT is an evaluation of these polynomials. Indeed, the definition of the DFT in Winograd's Short DFT Algorithms (Chapter 7) shows that

$$C(k) = X\left(W_N^k\right) = X\left(e^{-j\frac{2\pi k}{N}}\right), \quad 0 \leq k < N, \tag{8.5}$$

i.e., the DFT computes the evaluations of the polynomial $X(s)$ at the nth roots of unity.

The problem arises with the equivalent of (8.3), since the multiplication $H(s)X(s)$ of two polynomials of degree $N-1$ yields one of degree $2N-2$. Also, it does not coincide with the circular convolution known to be associated with the DFT. The solution to both problems is to reduce the product modulo $s^n - 1$:

$$H *_{\text{circ}} X \leftrightarrow H(s) X(s) \mod (s^n - 1). \tag{8.6}$$

Concept	Infinite Time	Finite Time
Signal	$X(s) = \sum_{n \in \mathbb{Z}} x(n) s^n$	$\sum_{n=0}^{N-1} x(n) s^n$
Filter	$H(s) = \sum_{n \in \mathbb{Z}} h(n) s^n$	$\sum_{n=0}^{N-1} h(n) s^n$
Convolution	$H(s) X(s)$	$H(s) X(s) \bmod (s^n - 1)$
Fourier transform	DTFT: $X(e^{-j\omega}), \quad -\pi < \omega \leq \pi$	DFT: $X\left(e^{-j\frac{2\pi k}{n}}\right), \quad 0 \leq k < n$

Table 8.1: Infinite and finite discrete time signal processing.

The resulting polynomial then has again degree $N-1$ and this form of convolution becomes equivalent to circular convolution of the polynomial coefficients. We also observe that the evaluation points in (8.5) are precisely the roots of $s^n - 1$. This connection will become clear in this chapter.

The discussion is summarized in Table 8.1.

The proper framework to describe the multiplication of polynomials modulo a fixed polynomial are polynomial algebras. Together with the Chinese remainder theorem, they provide the theoretical underpinning for the DFT and the Cooley-Tukey FFT.

In this chapter, the DFT will naturally arise as a linear mapping with respect to chosen bases, i.e., as a matrix. Indeed, the definition shows that if all input and outputs are collected into vectors $X = (X(0), \cdots, X(N-1))$ and $C = (C(0), \cdots C(N-1))$, then Winograd's Short DFT Algorithms (Chapter 7) is equivalent to

$$C = DFT_N X, \tag{8.7}$$

where

$$DFT_N = \left[W_N^{kn}\right]_{0 \leq k,n < N}. \tag{8.8}$$

The matrix point of view is adopted in the FFT books [388], [381].

8.1 Polynomial Algebras and the DFT

In this section we introduce polynomial algebras and explain how they are associated to transforms. Then we identify this connection for the DFT. Later we use polynomial algebras to derive the Cooley-Tukey FFT.

For further background on the mathematics in this section and polynomial algebras in particular, we refer to [138].

8.1.1 Polynomial Algebra

An algebra $\mathcal{A}$ is a vector space that also provides a multiplication of its elements such that the distributivity law holds (see [138] for a complete definition). Examples include the sets of complex or real numbers $\mathbb{C}$ or $\mathbb{R}$, and the sets of complex or real polynomials in the variable s: $\mathbb{C}[s]$ or $\mathbb{R}[s]$.

The key player in this chapter is the **polynomial algebra**. Given a fixed polynomial $P(s)$ of degree $deg(P) = N$, we define a polynomial algebra as the set

$$\mathbb{C}[s]/P(s) = \{X(s) \mid deg(X) < deg(P)\} \tag{8.9}$$

of polynomials of degree smaller than N with addition and multiplication modulo P. Viewed as a vector space, $\mathbb{C}[s]/P(s)$ hence has dimension N.

Every polynomial $X(s) \in \mathbb{C}[s]$ is reduced to a unique polynomial $R(s)$ modulo $P(s)$ of degree smaller than N. $R(s)$ is computed using division with rest, namely

$$X(s) = Q(s)P(s) + R(s), \quad deg(R) < deg(P). \tag{8.10}$$

Regarding this equation modulo P, $P(s)$ becomes zero, and we get

$$X(s) \equiv R(s) \mod P(s). \tag{8.11}$$

We read this equation as "$X(s)$ is congruent (or equal) $R(s)$ modulo $P(s)$." We will also write $X(s) \bmod P(s)$ to denote that $X(s)$ is reduced modulo $P(s)$. Obviously,

$$P(s) \equiv 0 \bmod P(s). \tag{8.12}$$

As a simple example we consider $\mathcal{A} = \mathbb{C}[s]/(s^2-1)$, which has dimension 2. A possible basis is $b = (1, s)$. In $\mathcal{A}$, for example, $s \cdot (s+1) = s^2 + s \equiv s + 1 \bmod (s^2-1)$, obtained through division with rest

$$s^2 + s = 1 \cdot (s^2 - 1) + (s+1) \tag{8.13}$$

or simply by replacing s^2 with 1 (since $s^2 - 1 = 0$ implies $s^2 = 1$).

8.1.2 Chinese Remainder Theorem (CRT)

Assume $P(s) = Q(s)R(s)$ factors into two coprime (no common factors) polynomials Q and R. Then the Chinese remainder theorem (CRT) for polynomials is the linear mapping[2]

$$\begin{aligned}\Delta:\ \mathbb{C}[s]/P(s) &\rightarrow \mathbb{C}[s]/Q(s) \oplus \mathbb{C}[s]/R(s),\\ X(s) &\mapsto (X(s) \bmod Q(s), X(s) \bmod R(s)).\end{aligned} \tag{8.14}$$

Here, $\oplus$ is the Cartesian product of vector spaces with elementwise operation (also called outer direct sum). In words, the CRT asserts that computing (addition, multiplication, scalar multiplication) in $\mathbb{C}[s]/P(s)$ is equivalent to computing in parallel in $\mathbb{C}[s]/Q(s)$ and $\mathbb{C}[s]/R(s)$.

If we choose bases b, c, d in the three polynomial algebras, then Δ can be expressed as a matrix. As usual with linear mappings, this matrix is obtained by mapping every element of b with Δ, expressing it in the concatenation $c \cup d$ of the bases c and d, and writing the results into the columns of the matrix.

As an example, we consider again the polynomial $P(s) = s^2 - 1 = (s-1)(s+1)$ and the CRT decomposition

$$\Delta:\ \mathbb{C}[s]/(s^2-1) \rightarrow \mathbb{C}[s]/(x-1) \oplus \mathbb{C}[s]/(x+1). \tag{8.15}$$

As bases, we choose $b = (1, x)$, $c = (1)$, $d = (1)$. $\Delta(1) = (1, 1)$ with the same coordinate vector in $c \cup d = (1, 1)$. Further, because of $x \equiv 1 \bmod (x-1)$ and $x \equiv -1 \bmod (x+1)$, $\Delta(x) = (x, x) \equiv (1, -1)$ with the same coordinate vector. Thus, Δ in matrix form is the so-called butterfly matrix, which is a DFT of size 2: $DFT_2 = \begin{bmatrix} 1 & 1 \\ 1 & -1 \end{bmatrix}$.

[2]More precisely, isomorphism of algebras or isomorphism of $\mathcal{A}$-modules.

8.1.3 Polynomial Transforms

Assume $P(s) \in \mathbb{C}[s]$ has pairwise distinct zeros $\alpha = (\alpha_0, \cdots, \alpha_{N-1})$. Then the CRT can be used to completely decompose $\mathbb{C}[s]/P(s)$ into its **spectrum**:

$$\begin{array}{rcl} \Delta : \mathbb{C}[s]/P(s) & \to & \mathbb{C}[s]/(s-\alpha_0) \oplus \cdots \oplus \mathbb{C}[s]/(s-\alpha_{N-1}), \\ X(s) & \mapsto & (X(s) \bmod (s-\alpha_0), \cdots, X(s) \bmod (s-\alpha_{N-1})) \\ & & = (s(\alpha_0), \cdots, s(\alpha_{N-1})). \end{array} \quad (8.16)$$

If we choose a basis $b = (P_0(s), \cdots, P_{N-1}(s))$ in $\mathbb{C}[s]/P(s)$ and bases $b_i = (1)$ in each $\mathbb{C}[s]/(s-\alpha_i)$, then Δ, as a linear mapping, is represented by a matrix. The matrix is obtained by mapping every basis element P_n, $0 \leq n < N$, and collecting the results in the columns of the matrix. The result is

$$\mathcal{P}_{b,\alpha} = [P_n(\alpha_k)]_{0 \leq k,n < N} \quad (8.17)$$

and is called the **polynomial transform** for $\mathcal{A} = \mathbb{C}[s]/P(s)$ with basis b.

If, in general, we choose $b_i = (\beta_i)$ as spectral basis, then the matrix corresponding to the decomposition (8.16) is the **scaled polynomial transform**

$$diag_{0 \leq k < N}(1/\beta_n)\mathcal{P}_{b,\alpha}, \quad (8.18)$$

where $diag_{0 \leq n < N}(\gamma_n)$ denotes a diagonal matrix with diagonal entries γ_n.

We jointly refer to polynomial transforms, scaled or not, as Fourier transforms.

8.1.4 DFT as a Polynomial Transform

We show that the DFT_N is a polynomial transform for $\mathcal{A} = \mathbb{C}[s]/(s^N - 1)$ with basis $b = (1, s, \cdots, s^{N-1})$. Namely,

$$s^N - 1 = \prod_{0 \leq k < N} (x - W_N^k), \quad (8.19)$$

which means that Δ takes the form

$$\begin{array}{rcl} \Delta : \mathbb{C}[s]/(s^N-1) & \to & \mathbb{C}[s]/(s-W_N^0) \oplus \cdots \oplus \mathbb{C}[s]/(s-W_N^{N-1}), \\ X(s) & \mapsto & (X(s) \bmod (s-W_N^0), \cdots, X(s) \bmod (s-W_N^{N-1})) \\ & & = (X(W_N^0), \cdots, X(W_N^{N-1})). \end{array} \quad (8.20)$$

The associated polynomial transform hence becomes

$$\mathcal{P}_{b,\alpha} = [W_N^{kn}]_{0 \leq k,n < N} = DFT_N. \quad (8.21)$$

This interpretation of the DFT has been known at least since [409], [268] and clarifies the connection between the evaluation points in (8.5) and the circular convolution in (8.6).

In [40], DFTs of types 1–4 are defined, with type 1 being the standard DFT. In the algebraic framework, type 3 is obtained by choosing $\mathcal{A} = \mathbb{C}[s] / (s^N + 1)$ as algebra with the same basis as before:

$$\mathcal{P}_{b,\alpha} = \left[W_N^{(k+1/2)n}\right]_{0 \leq k,n < N} = DFT\text{-}3_N, \tag{8.22}$$

The DFTs of type 2 and 4 are scaled polynomial transforms [295].

8.2 Algebraic Derivation of the Cooley-Tukey FFT

Knowing the polynomial algebra underlying the DFT enables us to derive the Cooley-Tukey FFT **algebraically**. This means that instead of manipulating the DFT definition, we manipulate the polynomial algebra $\mathbb{C}[s] / (s^N - 1)$. The basic idea is intuitive. We showed that the DFT is the matrix representation of the complete decomposition (8.20). The Cooley-Tukey FFT is now derived by performing this decomposition **in steps** as shown in Figure 8.1. Each step yields a sparse matrix; hence, the DFT_N is factorized into a product of sparse matrices, which will be the matrix representation of the Cooley-Tukey FFT.

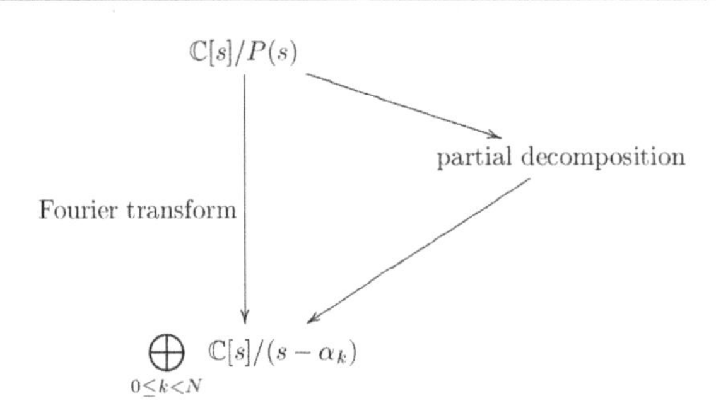

Figure 8.1: Basic idea behind the algebraic derivation of Cooley-Tukey type algorithms

This stepwise decomposition can be formulated generically for polynomial transforms [292], [294]. Here, we consider only the DFT.

We first introduce the matrix notation we will use and in particular the Kronecker product formalism that became mainstream for FFTs in [388], [381].

Then we first derive the radix-2 FFT using a **factorization** of $s^N - 1$. Subsequently, we obtain the general-radix FFT using a **decomposition** of $s^N - 1$.

8.2.1 Matrix Notation

We denote the $N \times N$ identity matrix with I_N, and diagonal matrices with

$$diag_{0 \leq k < N}(\gamma_k) = \begin{bmatrix} \gamma_0 & & \\ & \ddots & \\ & & \gamma_{N-1} \end{bmatrix}. \tag{8.23}$$

The $N \times N$ **stride permutation** matrix is defined for $N = KM$ by the permutation

$$L_M^N : iK + j \mapsto jM + i \tag{8.24}$$

for $0 \leq i < K$, $0 \leq j < M$. This definition shows that L_M^N transposes a $K \times M$ matrix stored in row-major order. Alternatively, we can write

$$L_M^N : i \mapsto \text{iM mod } N-1, \ \text{for } 0 \leq i < N-1, \ N-1 \mapsto N-1. \tag{8.25}$$

For example ($\cdot$ means 0),

$$L_2^6 = \begin{bmatrix} 1 & \cdot & \cdot & \cdot & \cdot & \cdot \\ \cdot & \cdot & 1 & \cdot & \cdot & \cdot \\ \cdot & \cdot & \cdot & \cdot & 1 & \cdot \\ \cdot & 1 & \cdot & \cdot & \cdot & \cdot \\ \cdot & \cdot & \cdot & 1 & \cdot & \cdot \\ \cdot & \cdot & \cdot & \cdot & \cdot & 1 \end{bmatrix}. \tag{8.26}$$

$L_{N/2}^N$ is sometimes called the perfect shuffle.

Further, we use matrix operators; namely the direct sum

$$A \oplus B = \begin{bmatrix} A & \\ & B \end{bmatrix} \tag{8.27}$$

and the Kronecker or tensor product

$$A \otimes B = [a_{k,\ell} B]_{k,\ell}, \quad \text{for } A = [a_{k,\ell}]. \tag{8.28}$$

In particular,

$$I_n \otimes A = A \oplus \cdots \oplus A = \begin{bmatrix} A & & \\ & \ddots & \\ & & A \end{bmatrix} \tag{8.29}$$

is block-diagonal.

We may also construct a larger matrix as a matrix of matrices, e.g.,

$$\begin{bmatrix} A & B \\ B & A \end{bmatrix}. \tag{8.30}$$

If an algorithm for a transform is given as a product of sparse matrices built from the constructs above, then an algorithm for the transpose or inverse of the transform can be readily derived using mathematical properties including

$$\begin{aligned} (AB)^T &= B^T A^T, & (AB)^{-1} &= B^{-1} A^{-1}, \\ (A \oplus B)^T &= A^T \oplus B^T, & (A \oplus B)^{-1} &= A^{-1} \oplus B^{-1}, \\ (A \otimes B)^T &= A^T \otimes B^T, & (A \otimes B)^{-1} &= A^{-1} \otimes B^{-1}. \end{aligned} \tag{8.31}$$

Permutation matrices are orthogonal, i.e., $P^T = P^{-1}$. The transposition or inversion of diagonal matrices is obvious.

8.2.2 Radix-2 FFT

The DFT decomposes $\mathcal{A} = \mathbb{C}[s] / \left(s^N - 1\right)$ with basis $b = \left(1, s, \cdots, s^{N-1}\right)$ as shown in (8.20). We assume $N = 2M$. Then

$$s^{2M} - 1 = \left(s^M - 1\right)\left(s^M + 1\right) \tag{8.32}$$

factors and we can apply the CRT in the following steps:

$$\begin{aligned} & \mathbb{C}[s] / \left(s^N - 1\right) \\ \rightarrow\ & \mathbb{C}[s] / \left(s^M - 1\right) \oplus \mathbb{C}[s] / \left(s^M + 1\right) \end{aligned} \tag{8.33}$$

$$\rightarrow \bigoplus_{0 \le i < M} \mathbb{C}[s] / \left(x - W_N^{2i}\right) \oplus \bigoplus_{0 \le i < M} \mathbb{C}[s] / \left(x - W_M^{2i+1}\right) \tag{8.34}$$

$$\rightarrow \underset{0 \leq i < N}{\oplus} \mathbb{C}[s] / \left(x - W_N^i\right). \tag{8.35}$$

As bases in the smaller algebras $\mathbb{C}[s] / \left(s^M - 1\right)$ and $\mathbb{C}[s] / \left(s^M + 1\right)$, we choose $c = d = \left(1, s, \cdots, s^{M-1}\right)$. The derivation of an algorithm for DFT_N based on (8.33)-(8.35) is now completely mechanical by reading off the matrix for each of the three decomposition steps. The product of these matrices is equal to the DFT_N.

First, we derive the base change matrix B corresponding to (8.33). To do so, we have to express the base elements $s^n \in b$ in the basis $c \cup d$; the coordinate vectors are the columns of B. For $0 \leq n < M$, s^n is actually contained in c and d, so the first M columns of B are

$$B = \begin{bmatrix} I_M & * \\ I_M & * \end{bmatrix}, \tag{8.36}$$

where the entries $*$ are determined next. For the base elements s^{M+n}, $0 \leq n < M$, we have

$$\begin{aligned} s^{M+n} &\equiv s^n \bmod \left(s^M - 1\right), \\ s^{M+n} &\equiv -s^n \bmod \left(s^M + 1\right), \end{aligned} \tag{8.37}$$

which yields the final result

$$B = \begin{bmatrix} I_M & I_M \\ I_M & -I_M \end{bmatrix} = DFT_2 \otimes I_M. \tag{8.38}$$

Next, we consider step (8.34). $\mathbb{C}[s] / \left(s^M - 1\right)$ is decomposed by DFT_M and $\mathbb{C}[s] / \left(s^M + 1\right)$ by $DFT\text{-}3_M$ in (8.22).

Finally, the permutation in step (8.35) is the perfect shuffle L_M^N, which interleaves the even and odd spectral components (even and odd exponents of W_N).

The final algorithm obtained is

$$DFT_{2M} = L_M^N \left(DFT_M \oplus DFT\text{-}3_M\right) \left(DFT_2 \otimes I_M\right). \tag{8.39}$$

To obtain a better known form, we use $DFT\text{-}3_M = DFT_M D_M$, with $D_M = diag_{0 \leq i < M}\left(W_N^i\right)$, which is evident from (8.22). It yields

$$\begin{aligned} DFT_{2M} &= L_M^N \left(DFT_M \oplus DFT_M D_M\right) \left(DFT_2 \otimes I_M\right) \\ &= L_M^N \left(I_2 \otimes DFT_M\right) \left(I_M \oplus D_M\right) \left(DFT_2 \otimes I_M\right). \end{aligned} \tag{8.40}$$

The last expression is the radix-2 decimation-in-frequency Cooley-Tukey FFT. The corresponding decimation-in-time version is obtained by transposition using (8.31) and the symmetry of the DFT:

$$DFT_{2M} = (DFT_2 \otimes I_M)(I_M \oplus D_M)(I_2 \otimes DFT_M) L_2^N. \tag{8.41}$$

The entries of the diagonal matrix $I_M \oplus D_M$ are commonly called **twiddle factors**.

The above method for deriving DFT algorithms is used extensively in [268].

8.2.3 General-radix FFT

To algebraically derive the general-radix FFT, we use the **decomposition property** of $s^N - 1$. Namely, if $N = KM$ then

$$s^N - 1 = \left(s^M\right)^K - 1. \tag{8.42}$$

Decomposition means that the polynomial is written as the composition of two polynomials: here, s^M is inserted into $s^K - 1$. Note that this is a special property: most polynomials do not decompose.

Based on this polynomial decomposition, we obtain the following stepwise decomposition of $\mathbb{C}[s]/\left(s^N - 1\right)$, which is more general than the previous one in (8.33)–(8.35). The basic idea is to first decompose with respect to the outer polynomial $t^K - 1$, $t = s^M$, and then completely [292]:

$$\begin{aligned} \mathbb{C}[s]/\left(s^N - 1\right) &= \mathbb{C}[x]/\left(\left(s^M\right)^K - 1\right) \\ \rightarrow & \bigoplus_{0 \le i < K} \mathbb{C}[s]/\left(s^M - W_K^i\right) \end{aligned} \tag{8.43}$$

$$\rightarrow \bigoplus_{0 \le i < K} \bigoplus_{0 \le j < M} \mathbb{C}[s]/\left(x - W_N^{jK+i}\right) \tag{8.44}$$

$$\rightarrow \bigoplus_{0 \le i < N} \mathbb{C}[s]/\left(x - W_N^i\right). \tag{8.45}$$

As bases in the smaller algebras $\mathbb{C}[s]/\left(s^M - W_K^i\right)$ we choose $c_i = \left(1, s, \cdots, s^{M-1}\right)$. As before, the derivation is completely mechanical from here: only the three matrices corresponding to (8.43)–(8.45) have to be read off.

The first decomposition step requires us to compute $s^n \bmod \left(s^M - W_K^i\right)$, $0 \le n < N$. To do so, we decompose the index n as $n = \ell M + m$ and compute

$$s^n = s^{\ell M + m} = \left(s^M\right)^\ell s^m \equiv W_k^{\ell m} s^m \bmod \left(s^M - W_K^i\right). \tag{8.46}$$

This shows that the matrix for (8.43) is given by $DFT_K \otimes I_M$.

In step (8.44), each $\mathbb{C}[s]/\left(s^M - W_K^i\right)$ is completely decomposed by its polynomial transform

$$DFT_M(i,K) = DFT_M \cdot diag_{0\le i<M}\left(W_N^{ij}\right). \tag{8.47}$$

At this point, $\mathbb{C}[s]/\left(s^N - 1\right)$ is completely decomposed, but the spectrum is ordered according to $jK+i$, $0 \le i < M$, $0 \le j < K$ (j runs faster). The desired order is $iM+j$.

Thus, in step (8.45), we need to apply the permutation $jK+i \mapsto iM+j$, which is exactly the stride permutation L_M^N in (8.24).

In summary, we obtain the Cooley-Tukey decimation-in-frequency FFT with arbitrary radix:

$$\begin{aligned} & L_M^N\left(\bigoplus_{0\le i<K} DFT_M \cdot diag_{j=0}^{M-1}\left(W_N^{ij}\right)\right)\left(DFT_k \otimes I_M\right) \\ = \quad & L_M^N\left(I_K \otimes DFT_M\right) T_M^N \left(DFT_k \otimes I_M\right). \end{aligned} \tag{8.48}$$

The matrix T_M^N is diagonal and usually called the **twiddle matrix**. Transposition using (8.31) yields the corresponding decimation in time version:

$$\left(DFT_k \otimes I_M\right) T_M^N \left(I_K \otimes DFT_M\right) L_K^N. \tag{8.49}$$

8.3 Discussion and Further Reading

This chapter only scratches the surface of the connection between algebra and the DFT or signal processing in general. We provide a few references for further reading.

8.3.1 Algebraic Derivation of Transform Algorithms

As mentioned before, the use of polynomial algebras and the CRT underlies much of the early work on FFTs and convolution algorithms [409], [268], [12]. For example, Winograd's work on FFTs minimizes the number of non-rational multiplications. This and his work on complexity theory in general makes heavy use of polynomial algebras [409], [414], [417] (see Chapter Winograd's Short DFT Algorithms (Chapter 7) for more information and references). See [72] for a broad treatment of algebraic complexity theory.

Since $\mathbb{C}[x]/\left(s^N - 1\right) = \mathbb{C}[C_N]$ can be viewed a group algebra for the cyclic group, the methods shown in this chapter can be translated into the context of group representation theory. For example, [256] derives the general-radix FFT using group theory and also uses already the Kronecker product formalism. So does Beth and started the area of FFTs for more general groups [23], [231]. However, Fourier transforms for groups have found only sporadic applications [317]. Along a related line of work, [117] shows that using group theory it is possible that to discover and generate certain algorithms for trigonometric transforms, such as discrete cosine transforms (DCTs), automatically using a computer program.

More recently, the polynomial algebra framework was extended to include most trigonometric transforms used in signal processing [293], [295], namely, besides the DFT, the discrete cosine and sine transforms and various real DFTs including the discrete Hartley transform. It turns out that the same techniques shown in this chapter can then be applied to derive, explain, and classify most of the known algorithms for these transforms and even obtain a large class of new algorithms including general-radix algorithms for the discrete cosine and sine transforms (DCTs/DSTs) [292], [294], [398], [397].

This latter line of work is part of the algebraic signal processing theory briefly discussed next.

8.3.2 Algebraic Signal Processing Theory

The algebraic properties of transforms used in the above work on algorithm derivation hints at a connection between algebra and (linear) signal processing itself. This is indeed the case and was fully developed in a recent body of work called algebraic signal processing theory (ASP). The foundation of ASP is developed in [295], [293], [291].

ASP first identifies the algebraic structure of (linear) signal processing: the common assumptions on available operations for filters and signals make the set of filters an **algebra**$\mathcal{A}$ and the set of signals an associated $\mathcal{A}$-module $\mathcal{M}$. ASP then builds a signal processing theory formally from the axiomatic definition of a **signal model**: a triple $(\mathcal{A}, \mathcal{M}, \Phi)$, where Φ generalizes the idea of the z-transform to mappings from vector spaces of signal values to $\mathcal{M}$. If a signal model is given, other concepts, such as spectrum, Fourier transform, frequency response are automatically defined but take different forms for different models. For example, infinite and finite time as discussed in Table 8.1 are two examples of signal models. Their complete definition is provided in Table 8.2 and identifies the proper notion of a finite z-transform as a mapping $\mathbb{C}^n \rightarrow \mathbb{C}[s]/(s^n-1)$.

Signal model	Infinite time	Finite time
$\mathcal{A}$	$\{\sum_{n\in\mathbb{Z}} H(n)\, s^n \mid (\cdots, H(-1), H(0), H(1), \cdots) \in \ell^1(\mathbb{Z})\}$	$\mathbb{C}[x]/(s^n-1)$
		continued on next page

$\mathcal{M}$	$\{\sum_{n\in\mathbb{Z}} X(n)\, s^n \mid (\cdots, X(-1), X(0), X(1), \cdots) \in \ell^2(\mathbb{Z})\}$	$\mathbb{C}[s]/(s^n-1)$
Φ	$\Phi:\ \ell^2(\mathbb{Z}) \to \mathcal{M}$	$\Phi:\ \mathbb{C}^n \to \mathcal{M}$
	defined in (8.1)	defined in (8.4)

Table 8.2: Infinite and finite time models as defined in ASP.

ASP shows that many signal models are in principle possible, each with its own notion of filtering and Fourier transform. Those that support shift-invariance have commutative algebras. Since finite-dimensional commutative algebras are precisely polynomial algebras, their appearance in signal processing is explained. For example, ASP identifies the polynomial algebras underlying the DCTs and DSTs, which hence become Fourier transforms in the ASP sense. The signal models are called finite **space** models since they support signal processing based on an undirected shift operator, different from the directed time shift. Many more insights are provided by ASP including the need for and choices in choosing boundary conditions, properties of transforms, techniques for deriving new signal models, and the concise derivation of algorithms mentioned before.

Chapter 9

The Cooley-Tukey Fast Fourier Transform Algorithm[1]

The publication by Cooley and Tukey [90] in 1965 of an efficient algorithm for the calculation of the DFT was a major turning point in the development of digital signal processing. During the five or so years that followed, various extensions and modifications were made to the original algorithm [95]. By the early 1970's the practical programs were basically in the form used today. The standard development presented in [274], [299], [38] shows how the DFT of a length-N sequence can be simply calculated from the two length-N/2 DFT's of the even index terms and the odd index terms. This is then applied to the two half-length DFT's to give four quarter-length DFT's, and repeated until N scalars are left which are the DFT values. Because of alternately taking the even and odd index terms, two forms of the resulting programs are called decimation-in-time and decimation-in-frequency. For a length of 2^M, the dividing process is repeated $M = log_2 N$ times and requires N multiplications each time. This gives the famous formula for the computational complexity of the FFT of $N log_2 N$ which was derived in Multidimensional Index Mapping: Equation 34 (3.34).

Although the decimation methods are straightforward and easy to understand, they do not generalize well. For that reason it will be assumed that the reader is familiar with that description and this chapter will develop the FFT using the index map from Multidimensional Index Mapping (Chapter 3).

The Cooley-Tukey FFT always uses the Type 2 index map from Multidimensional Index Mapping: Equation 11 (3.11). This is necessary for the most popular forms that have $N = R^M$, but is also used even when the factors are relatively prime and a Type 1 map could be used. The time and frequency maps from Multidimensional Index Mapping: Equation 6 (3.6) and Multidimensional Index Mapping: Equation 12 (3.12) are

$$n = ((K_1 n_1 + K_2 n_2))_N \tag{9.1}$$

[1]This content is available online at <http://cnx.org/content/m16334/1.13/>.

$$k = ((K_3 k_1 + K_4 k_2))_N \tag{9.2}$$

Type-2 conditions Multidimensional Index Mapping: Equation 8 (3.8) and Multidimensional Index Mapping: Equation 11 (3.11) become

$$K_1 = aN_2 \quad \text{or} \quad K_2 = bN_1 \quad \text{but not both} \tag{9.3}$$

and

$$K_3 = cN_2 \quad \text{or} \quad K_4 = dN_1 \quad \text{but not both} \tag{9.4}$$

The row and column calculations in Multidimensional Index Mapping: Equation 15 (3.15) are uncoupled by Multidimensional Index Mapping: Equation 16 (3.16) which for this case are

$$((K_1 K_4))_N = 0 \quad \text{or} \quad ((K_2 K_3))_N = 0 \quad \text{but not both} \tag{9.5}$$

To make each short sum a DFT, the K_i must satisfy

$$((K_1 K_3))_N = N_2 \quad \text{and} \quad ((K_2 K_4))_N = N_1 \tag{9.6}$$

In order to have the smallest values for K_i the constants in (9.3) are chosen to be

$$a = d = K_2 = K_3 = 1 \tag{9.7}$$

which makes the index maps of (9.1) become

$$n = N_2 n_1 + n_2 \tag{9.8}$$

$$k = k_1 + N_1 k_2 \tag{9.9}$$

These index maps are all evaluated modulo N, but in (9.8), explicit reduction is not necessary since n never exceeds N. The reduction notation will be omitted for clarity. From Multidimensional Index Mapping: Equation 15 (3.15) and example Multidimensional Index Mapping: Equation 19 (3.19), the DFT is

$$X = \sum_{n_2=0}^{N_2-1} \sum_{n_1=0}^{N_1-1} x \, W_{N_1}^{n_1 k_1} \, W_N^{n_2 k_1} \, W_{N_2}^{n_2 k_2} \tag{9.10}$$

This map of (9.8) and the form of the DFT in (9.10) are the fundamentals of the Cooley-Tukey FFT.

The order of the summations using the Type 2 map in (9.10) cannot be reversed as it can with the Type-1 map. This is because of the W_N terms, the twiddle factors.

Turning (9.10) into an efficient program requires some care. From Multidimensional Index Mapping: Efficiencies Resulting from Index Mapping with the DFT (Section 3.3: Efficiencies Resulting from Index Mapping with the DFT) we know that all the factors should be equal. If $N = R^M$, with R called the radix, N_1 is first set equal to R and N_2 is then necessarily R^{M-1}. Consider n_1 to be the index along the rows and n_2 along the columns. The inner sum of (9.10) over n_1 represents a length-N_1 DFT for each value of n_2. These N_2 length-N_1 DFT's are the DFT's of the rows of the $x(n_1, n_2)$ array. The resulting array of row DFT's is multiplied by an array of twiddle factors which are the W_N terms in (9.10). The twiddle-factor array for a length-8 radix-2 FFT is

$$TF: \qquad W_8^{n_2 k_1} = \begin{bmatrix} W^0 & W^0 \\ W^0 & W^1 \\ W^0 & W^2 \\ W^0 & W^3 \end{bmatrix} = \begin{bmatrix} 1 & 1 \\ 1 & W \\ 1 & -j \\ 1 & -jW \end{bmatrix} \tag{9.11}$$

The twiddle factor array will always have unity in the first row and first column.

To complete (9.10) at this point, after the row DFT's are multiplied by the TF array, the N_1 length-N_2 DFT's of the columns are calculated. However, since the columns DFT's are of length R^{M-1}, they can be posed as a R^{M-2} by R array and the process repeated, again using length-R DFT's. After M stages of length-R DFT's with TF multiplications interleaved, the DFT is complete. The flow graph of a length-2 DFT is given in Figure 1 (7.18) and is called a butterfly because of its shape. The flow graph of the complete length-8 radix-2 FFT is shown in Figure 2 (7.19) .

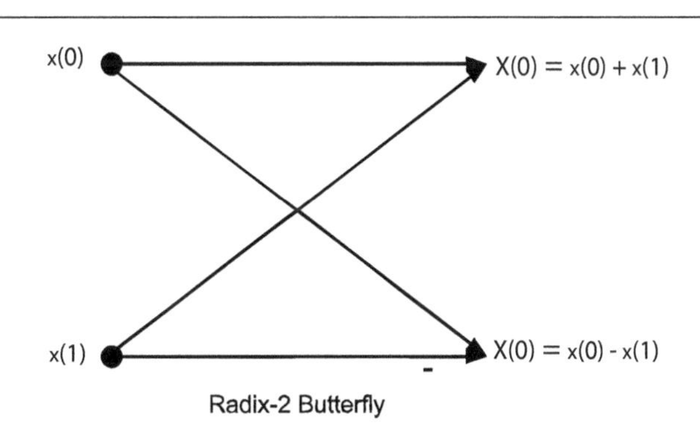

Figure 9.1: A Radix-2 Butterfly

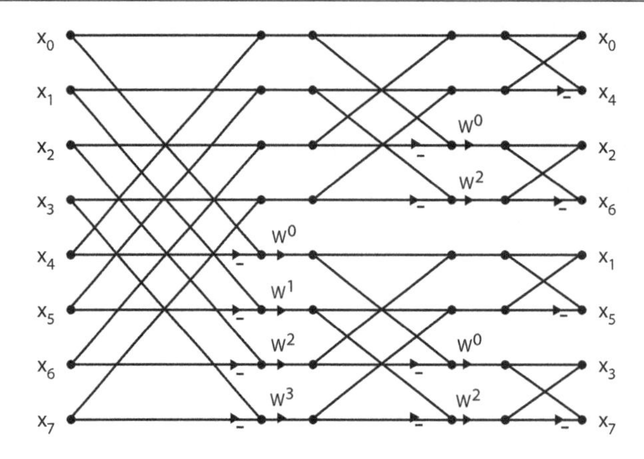

Figure 9.2: Length-8 Radix-2 FFT Flow Graph

This flow-graph, the twiddle factor map of (9.11), and the basic equation (9.10) should be completely understood before going further.

A very efficient indexing scheme has evolved over the years that results in a compact and efficient computer program. A FORTRAN program is given below that implements the radix-2 FFT. It should be studied [64] to see how it implements (9.10) and the flow-graph representation.

```
      N2 = N
      DO 10 K = 1, M
          N1 = N2
          N2 = N2/2
          E  = 6.28318/N1
          A  = 0
          DO 20 J = 1, N2
              C = COS (A)
              S =-SIN (A)
              A = J*E
              DO 30 I = J, N, N1
                  L = I + N2
                  XT   = X(I) - X(L)
                  X(I) = X(I) + X(L)
                  YT   = Y(I) - Y(L)
                  Y(I) = Y(I) + Y(L)
                  X(L) = XT*C - YT*S
                  Y(L) = XT*S + YT*C
30            CONTINUE
20        CONTINUE
10    CONTINUE
```

Listing 9.1: A Radix-2 Cooley-Tukey FFT Program

This discussion, the flow graph of Winograd's Short DFT Algorithms: Figure 2 (Figure 7.2) and the program of p. ?? are all based on the input index map of Multidimensional Index Mapping: Equation 6 (3.6) and (9.1) and the calculations are performed in-place. According to Multidimensional Index Mapping: In-Place Calculation of the DFT and Scrambling (Section 3.2: In-Place Calculation of the DFT and Scrambling), this means the output is scrambled in bit-reversed order and should be followed by an unscrambler to give the DFT in proper order. This formulation is called a decimation-in-frequency FFT [274], [299], [38]. A very similar algorithm based on the output index map can be derived which is called a decimation-in-time FFT. Examples of FFT programs are found in [64] and in the Appendix of this book.

9.1 Modifications to the Basic Cooley-Tukey FFT

Soon after the paper by Cooley and Tukey, there were improvements and extensions made. One very important discovery was the improvement in efficiency by using a larger radix of 4, 8 or even 16. For example, just as for the radix-2 butterfly, there are no multiplications required for a length-4 DFT, and therefore, a radix-4 FFT would have only twiddle factor multiplications. Because there are half as many stages in a radix-4 FFT, there would be half as many multiplications as in a radix-2 FFT. In practice, because some of the multiplications are by unity, the improvement is not by a factor of two, but it is significant. A radix-4 FFT is easily developed from the basic radix-2 structure by replacing the length-2 butterfly by a length-4 butterfly and making a few other modifications. Programs can be found in [64] and operation counts will be given in "Evaluation of the Cooley-Tukey FFT Algorithms" (Section 9.3: Evaluation of the Cooley-Tukey FFT Algorithms).

Increasing the radix to 8 gives some improvement but not as much as from 2 to 4. Increasing it to 16 is theoretically promising but the small decrease in multiplications is somewhat offset by an increase in additions and the program becomes rather long. Other radices are not attractive because they generally require a substantial number of multiplications and additions in the butterflies.

The second method of reducing arithmetic is to remove the unnecessary TF multiplications by plus or minus unity or by plus or minus the square root of minus one. This occurs when the exponent of W_N is zero or a multiple of $N/4$. A reduction of additions as well as multiplications is achieved by removing these extraneous complex multiplications since a complex multiplication requires at least two real additions. In a program, this reduction is usually achieved by having special butterflies for the cases where the TF is one or j. As many as four special butterflies may be necessary to remove all unnecessary arithmetic, but in many cases there will be no practical improvement above two or three.

In addition to removing multiplications by one or j, there can be a reduction in multiplications by using a special butterfly for TFs with $W_{N/8}$, which have equal real and imaginary parts. Also, for computers or hardware with multiplication considerably slower than addition, it is desirable to use an algorithm for complex multiplication that requires three multiplications and three additions rather than the conventional four multiplications and two additions. Note that this gives no reduction in the total number of arithmetic operations, but does give a trade of multiplications for additions. This is one reason not to use complex data types in programs but to explicitly program complex arithmetic.

A time-consuming and unnecessary part of the execution of a FFT program is the calculation of the sine and cosine terms which are the real and imaginary parts of the TFs. There are basically three approaches to obtaining the sine and cosine values. They can be calculated as needed which is what is done in the sample program above. One value per stage can be calculated and the others recursively calculated from those. That method is fast but suffers from accumulated round-off errors. The fastest method is to fetch precalculated values from a stored table. This has the disadvantage of requiring considerable memory space.

If all the N DFT values are not needed, special forms of the FFT can be developed using a process called pruning [226] which removes the operations concerned with the unneeded outputs.

Special algorithms are possible for cases with real data or with symmetric data [82]. The decimation-in-time algorithm can be easily modified to transform real data and save half the arithmetic required for complex data [357]. There are numerous other modifications to deal with special hardware considerations such as an array processor or a special microprocessor such as the Texas Instruments TMS320. Examples of programs that deal with some of these items can be found in [299], [64], [82].

9.2 The Split-Radix FFT Algorithm

Recently several papers [228], [106], [393], [350], [102] have been published on algorithms to calculate a length-2^M DFT more efficiently than a Cooley-Tukey FFT of any radix. They all have the same computational complexity and are optimal for lengths up through 16 and until recently was thought to give the best total add-multiply count possible for any power-of-two length. Yavne published an algorithm with the same computational complexity in 1968 [421], but it went largely unnoticed. Johnson and Frigo have recently reported the first improvement in almost 40 years [201]. The reduction in total operations is only a few percent, but it is a reduction.

The basic idea behind the split-radix FFT (SRFFT) as derived by Duhamel and Hollmann [106], [102] is the application of a radix-2 index map to the even-indexed terms and a radix-4 map to the odd- indexed terms. The basic definition of the DFT

$$C_k = \sum_{n=0}^{N-1} x_n \, W^{nk} \tag{9.12}$$

with $W = e^{-j2\pi/N}$ gives

$$C_{2k} = \sum_{n=0}^{N/2-1} \left[x_n + x_{n+N/2}\right] \, W^{2nk} \tag{9.13}$$

for the even index terms, and

$$C_{4k+1} = \sum_{n=0}^{N/4-1} \left[\left(x_n - x_{n+N/2}\right) - j\left(x_{n+N/4} - x_{n+3N/4}\right)\right] \, W^n \, W^{4nk} \tag{9.14}$$

and

$$C_{4k+3} = \sum_{n=0}^{N/4-1} \left[\left(x_n - x_{n+N/2} \right) + j \left(x_{n+N/4} - x_{n+3N/4} \right) \right] W^{3n} W^{4nk} \qquad (9.15)$$

for the odd index terms. This results in an L-shaped "butterfly" shown in Figure 9.3 which relates a length-N DFT to one length-N/2 DFT and two length-N/4 DFT's with twiddle factors. Repeating this process for the half and quarter length DFT's until scalars result gives the SRFFT algorithm in much the same way the decimation-in-frequency radix-2 Cooley-Tukey FFT is derived [274], [299], [38]. The resulting flow graph for the algorithm calculated in place looks like a radix-2 FFT except for the location of the twiddle factors. Indeed, it is the location of the twiddle factors that makes this algorithm use less arithmetic. The L- shaped SRFFT butterfly Figure 9.3 advances the calculation of the top half by one of the M stages while the lower half, like a radix-4 butterfly, calculates two stages at once. This is illustrated for $N = 8$ in Figure 9.4.

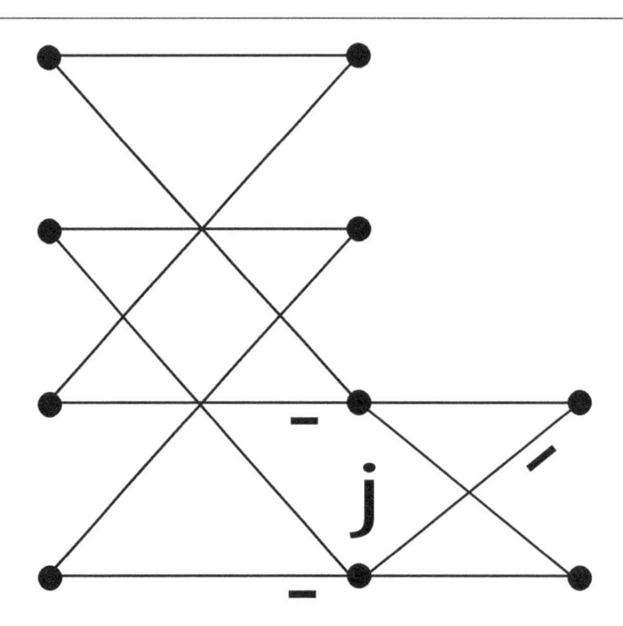

Figure 9.3: SRFFT Butterfly

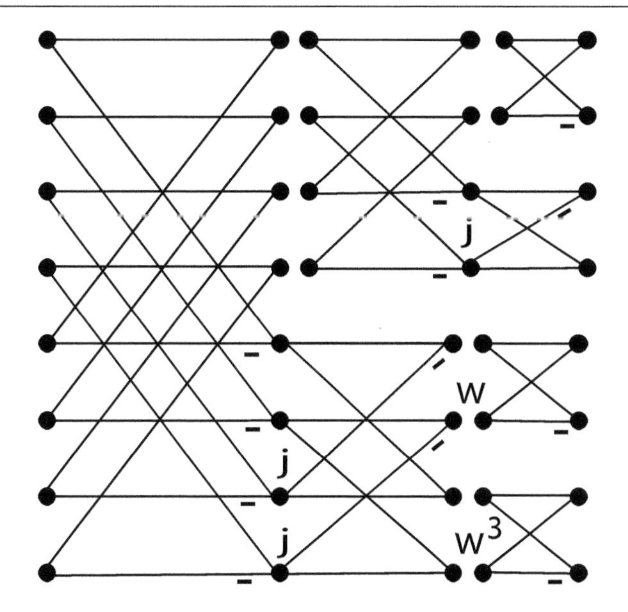

Figure 9.4: Length-8 SRFFT

Unlike the fixed radix, mixed radix or variable radix Cooley-Tukey FFT or even the prime factor algorithm or Winograd Fourier transform algorithm , the Split-Radix FFT does not progress completely stage by stage, or, in terms of indices, does not complete each nested sum in order. This is perhaps better seen from the polynomial formulation of Martens [228]. Because of this, the indexing is somewhat more complicated than the conventional Cooley-Tukey program.

A FORTRAN program is given below which implements the basic decimation-in-frequency split-radix FFT algorithm. The indexing scheme [350] of this program gives a structure very similar to the Cooley-Tukey programs in [64] and allows the same modifications and improvements such as decimation-in-time, multiple butterflies, table look-up of sine and cosine values, three real per complex multiply methods, and real data versions [102], [357].

```
      SUBROUTINE FFT(X,Y,N,M)
      N2 = 2*N
      DO  10 K = 1, M-1
          N2 = N2/2
          N4 = N2/4
          E  = 6.283185307179586/N2
          A = 0
          DO  20 J = 1, N4
              A3  = 3*A
              CC1 = COS(A)
              SS1 = SIN(A)
              CC3 = COS(A3)
              SS3 = SIN(A3)
              A   = J*E
              IS  = J
              ID  = 2*N2
40            DO 30 I0 = IS, N-1, ID
                  I1 = I0 + N4
                  I2 = I1 + N4
                  I3 = I2 + N4
                  R1     = X(I0) - X(I2)
                  X(I0) = X(I0) + X(I2)
                  R2     = X(I1) - X(I3)
                  X(I1) = X(I1) + X(I3)
                  S1     = Y(I0) - Y(I2)
                  Y(I0) = Y(I0) + Y(I2)
                  S2     = Y(I1) - Y(I3)
                  Y(I1) = Y(I1) + Y(I3)
                  S3     = R1 - S2
                  R1     = R1 + S2
                  S2     = R2 - S1
                  R2     = R2 + S1
                  X(I2) = R1*CC1 - S2*SS1
                  Y(I2) =-S2*CC1 - R1*SS1
                  X(I3) = S3*CC3 + R2*SS3
                  Y(I3) = R2*CC3 - S3*SS3
30            CONTINUE
              IS = 2*ID - N2 + J
              ID = 4*ID
              IF (IS.LT.N) GOTO 40
20        CONTINUE
10    CONTINUE
      IS = 1
      ID = 4
50    DO 60 I0 = IS, N, ID
          I1    = I0 + 1
```

As was done for the other decimation-in-frequency algorithms, the input index map is used and the calculations are done in place resulting in the output being in bit-reversed order. It is the three statements following label 30 that do the special indexing required by the SRFFT. The last stage is length- 2 and, therefore, inappropriate for the standard L-shaped butterfly, so it is calculated separately in the DO 60 loop. This program is considered a one-butterfly version. A second butterfly can be added just before statement 40 to remove the unnecessary multiplications by unity. A third butterfly can be added to reduce the number of real multiplications from four to two for the complex multiplication when W has equal real and imaginary parts. It is also possible to reduce the arithmetic for the two- butterfly case and to reduce the data transfers by directly programming a length-4 and length-8 butterfly to replace the last three stages. This is called a two-butterfly-plus version. Operation counts for the one, two, two-plus and three butterfly SRFFT programs are given in the next section. Some details can be found in [350].

The special case of a SRFFT for real data and symmetric data is discussed in [102]. An application of the decimation-in-time SRFFT to real data is given in [357]. Application to convolution is made in [110], to the discrete Hartley transform in [352], [110], to calculating the discrete cosine transform in [393], and could be made to calculating number theoretic transforms.

An improvement in operation count has been reported by Johnson and Frigo [201] which involves a scaling of multiplying factors. The improvement is small but until this result, it was generally thought the Split-Radix FFT was optimal for total floating point operation count.

9.3 Evaluation of the Cooley-Tukey FFT Algorithms

The evaluation of any FFT algorithm starts with a count of the real (or floating point) arithmetic. Table 9.1 gives the number of real multiplications and additions required to calculate a length-N FFT of complex data. Results of programs with one, two, three and five butterflies are given to show the improvement that can be expected from removing unnecessary multiplications and additions. Results of radices two, four, eight and sixteen for the Cooley-Tukey FFT as well as of the split-radix FFT are given to show the relative merits of the various structures. Comparisons of these data should be made with the table of counts for the PFA and WFTA programs in The Prime Factor and Winograd Fourier Transform Algorithms: Evaluation of the PFA and WFTA (Section 10.4: Evaluation of the PFA and WFTA). All programs use the four-multiply-two-add complex multiply algorithm. A similar table can be developed for the three-multiply-three-add algorithm, but the relative results are the same.

From the table it is seen that a greater improvement is obtained going from radix-2 to 4 than from 4 to 8 or 16. This is partly because length 2 and 4 butterflies have no multiplications while length 8, 16 and higher do. It is also seen that going from one to two butterflies gives more improvement than going from two to higher values. From an operation count point of view and from practical experience, a three butterfly radix-4 or a two butterfly radix-8 FFT is a good compromise. The radix-8 and 16 programs become long, especially with multiple butterflies, and they give a limited choice of transform length unless combined with some length 2 and 4 butterflies.

N	M1	M2	M3	M5	A1	A2	A3	A5
2	4	0	0	0	6	4	4	4
4	16	4	0	0	24	18	16	16
8	48	20	8	4	72	58	52	52
16	128	68	40	28	192	162	148	148
32	320	196	136	108	480	418	388	388
64	768	516	392	332	1152	1026	964	964
128	1792	1284	1032	908	2688	2434	2308	2308
256	4096	3076	2568	2316	6144	5634	5380	5380
512	9216	7172	6152	5644	13824	12802	12292	12292
1024	20480	16388	14344	13324	30720	28674	27652	27652
2048	45056	36868	32776	30732	67584	63490	61444	61444
4096	98304	81924	73736	69644	147456	139266	135172	135172
4	12	0	0	0	22	16	16	16
16	96	36	28	24	176	146	144	144
64	576	324	284	264	1056	930	920	920
256	3072	2052	1884	1800	5632	5122	5080	5080
1024	15360	11268	10588	10248	28160	26114	25944	25944

continued on next page

4096	73728	57348	54620	53256	135168	126978	126296	126296
8	32	4	4	4	66	52	52	52
64	512	260	252	248	1056	930	928	928
512	6144	4100	4028	3992	12672	11650	11632	11632
4096	65536	49156	48572	48280	135168	126978	126832	126832
16	80	20	20	20	178	148	148	148
256	2560	1540	1532	1528	5696	5186	5184	5184
4096	61440	45060	44924	44856	136704	128514	128480	128480
2	0	0	0	0	4	4	4	4
4	8	0	0	0	20	16	16	16
8	24	8	4	4	60	52	52	52
16	72	32	28	24	164	144	144	144
32	184	104	92	84	412	372	372	372
64	456	288	268	248	996	912	912	912
128	1080	744	700	660	2332	2164	2164	2164
256	2504	1824	1740	1656	5348	5008	5008	5008
512	5688	4328	4156	3988	12060	11380	11380	11380
1024	12744	10016	9676	9336	26852	25488	25488	25488
2048	28216	22760	22076	21396	59164	56436	56436	56436
4096	61896	50976	49612	48248	129252	123792	123792	123792

Table 9.1: Number of Real Multiplications and Additions for Complex Single Radix FFTs

In Table 9.1 Mi and Ai refer to the number of real multiplications and real additions used by an FFT with i separately written butterflies. The first block has the counts for Radix-2, the second for Radix-4, the third for Radix-8, the fourth for Radix-16, and the last for the Split-Radix FFT. For the split-radix FFT, M3 and A3 refer to the two- butterfly-plus program and M5 and A5 refer to the three-butterfly program.

The first evaluations of FFT algorithms were in terms of the number of real multiplications required as that was the slowest operation on the computer and, therefore, controlled the execution speed. Later with hardware arithmetic both the number of multiplications and additions became important. Modern systems have arithmetic speeds such that indexing and data transfer times become important factors. Morris [249] has looked at some of these problems and has developed a procedure called autogen to write partially straight-line program code to significantly reduce overhead and speed up FFT run times. Some hardware, such as the TMS320 signal processing chip, has the multiply and add operations combined. Some machines have vector instructions or have parallel processors. Because the execution speed of an FFT depends not only on the algorithm, but also on the hardware architecture and compiler, experiments must be run on the system to be used.

In many cases the unscrambler or bit-reverse-counter requires 10% of the execution time, therefore, if possible, it should be eliminated. In high-speed convolution where the convolution is done by multiplication of DFT's, a decimation-in-frequency FFT can be combined with a decimation-in-time inverse FFT to require no unscrambler. It is also possible for a radix-2 FFT to do the unscrambling inside the FFT but the structure is not very regular [299], [193]. Special structures can be found in [299] and programs for data that are real or have special symmetries are in [82], [102], [357].

Although there can be significant differences in the efficiencies of the various Cooley-Tukey and Split-Radix FFTs, the number of multiplications and additions for all of them is on the order of $NlogN$. That is fundamental to the class of algorithms.

9.4 The Quick Fourier Transform, An FFT based on Symmetries

The development of fast algorithms usually consists of using special properties of the algorithm of interest to remove redundant or unnecessary operations of a direct implementation. The discrete Fourier transform (DFT) defined by

$$C(k) = \sum_{n=0}^{N-1} x(n)\, W_N^{nk} \tag{9.16}$$

where

$$W_N = e^{-j2\pi/N} \tag{9.17}$$

has enormous capacity for improvement of its arithmetic efficiency. Most fast algorithms use the periodic and symmetric properties of its basis functions. The classical Cooley-Tukey FFT and prime factor FFT [64] exploit the periodic properties of the cosine and sine functions. Their use of the periodicities to share and, therefore, reduce arithmetic operations depends on the factorability of the length of the data to be transformed. For highly composite lengths, the number of floating-point operation is of order $N\,log(N)$ and for prime lengths it is of order N^2.

This section will look at an approach using the symmetric properties to remove redundancies. This possibility has long been recognized [176], [211], [344], [270] but has not been developed in any systematic way in the open literature. We will develop an algorithm, called the quick Fourier transform (QFT) [211], that will reduce the number of floating point operations necessary to compute the DFT by a factor of two to four over direct methods or Goertzel's method for prime lengths. Indeed, it seems the best general algorithm available for prime length DFTs. One can always do better by using Winograd type algorithms but they must be individually designed for each length. The Chirp Z-transform can be used for longer lengths.

9.4.1 Input and Output Symmetries

We use the fact that the cosine is an even function and the sine is an odd function. The kernel of the DFT or the basis functions of the expansion is given by

$$W_N^{nk} = e^{-j2\pi nk/N} = cos\left(2\pi nk/N\right) + j\, sin\left(2\pi nk/N\right) \tag{9.18}$$

which has an even real part and odd imaginary part. If the data $x(n)$ are decomposed into their real and imaginary parts and those into their even and odd parts, we have

$$x(n) = u(n) + j\, v(n) = \left[u_e(n) + u_o(n)\right] + j\,\left[v_e(n) + v_o(n)\right] \tag{9.19}$$

where the even part of the real part of $x(n)$ is given by

$$u_e(n) = \left(u(n) + u(-n)\right)/2 \tag{9.20}$$

and the odd part of the real part is

$$u_o(n) = \left(u(n) - u(-n)\right)/2 \tag{9.21}$$

with corresponding definitions of $v_e(n)$ and $v_o(n)$. Using Convolution Algorithms: Equation 32 (13.32) with a simpler notation, the DFT of Convolution Algorithms: Equation 29 (13.29) becomes

$$C(k) = \sum_{n=0}^{N-1} (u + j\, v)\,(cos - j sin)\,. \tag{9.22}$$

The sum over an integral number of periods of an odd function is zero and the sum of an even function over half of the period is one half the sum over the whole period. This causes (9.16) and (9.22) to become

$$C(k) = \sum_{n=0}^{N/2-1} \left[u_e\, cos + v_o\, sin\right] + j\,\left[v_e\, cos - v_o\, sin\right]. \tag{9.23}$$

for $k = 0, 1, 2, \cdots, N-1$.

The evaluation of the DFT using equation (9.23) requires half as many real multiplication and half as many real additions as evaluating it using (9.16) or (9.22). We have exploited the symmetries of the sine and cosine as functions of the time index n. This is independent of whether the length is composite or not. Another view of this formulation is that we have used the property of associatively of multiplication and addition. In other words, rather than multiply two data points by the same value of a sine or cosine then add the results, one should add the data points first then multiply the sum by the sine or cosine which requires one rather than two multiplications.

Next we take advantage of the symmetries of the sine and cosine as functions of the frequency index k. Using these symmetries on (9.23) gives

$$C(k) = \sum_{n=0}^{N/2-1} [u_e \, cos + v_o \, sin] + j \, [v_e \, cos - v_o \, sin] \tag{9.24}$$

$$C(N-k) = \sum_{n=0}^{N/2-1} [u_e \, cos - v_o \, sin] + j \, [v_e \, cos + v_o \, sin] \,. \tag{9.25}$$

for $k = 0, 1, 2, \cdots, N/2 - 1$. This again reduces the number of operations by a factor of two, this time because it calculates two output values at a time. The first reduction by a factor of two is always available. The second is possible only if both DFT values are needed. It is not available if you are calculating only one DFT value. The above development has not dealt with the details that arise with the difference between an even and an odd length. That is straightforward.

9.4.2 Further Reductions if the Length is Even

If the length of the sequence to be transformed is even, there are further symmetries that can be exploited. There will be four data values that are all multiplied by plus or minus the same sine or cosine value. This means a more complicated pre-addition process which is a generalization of the simple calculation of the even and odd parts in (9.20) and (9.21) will reduce the size of the order N^2 part of the algorithm by still another factor of two or four. It the length is divisible by 4, the process can be repeated. Indeed, it the length is a power of 2, one can show this process is equivalent to calculating the DFT in terms of discrete cosine and sine transforms [156], [159] with a resulting arithmetic complexity of order $N \, log \, (N)$ and with a structure that is well suited to real data calculations and pruning.

If the flow-graph of the Cooley-Tukey FFT is compared to the flow-graph of the QFT, one notices both similarities and differences. Both progress in stages as the length is continually divided by two. The Cooley-Tukey algorithm uses the periodic properties of the sine and cosine to give the familiar horizontal tree of butterflies. The parallel diagonal lines in this graph represent the parallel stepping through the data in synchronism with the periodic basis functions. The QFT has diagonal lines that connect the first data point with the last, then the second with the next to last, and so on to give a "star" like picture. This is interesting in that one can look at the flow graph of an algorithm developed by some completely different strategy and often find section with the parallel structures and other parts with the star structure. These must be using some underlying periodic and symmetric properties of the basis functions.

9.4.3 Arithmetic Complexity and Timings

A careful analysis of the QFT shows that $2N$ additions are necessary to compute the even and odd parts of the input data. This is followed by the length $N/2$ inner product that requires $4(N/2)^2 = N^2$ real multiplications and an equal number of additions. This is followed by the calculations necessary for the simultaneous calculations of the first half and last half of $C(k)$ which requires $4(N/2) = 2N$ real additions. This means the total QFT algorithm requires M^2 real multiplications and $N^2 + 4N$ real additions. These numbers along with those for the Goertzel algorithm [52], [64], [270] and the direct calculation of the DFT are included in the following table. Of the various order-N^2 DFT algorithms, the QFT seems to be the most efficient general method for an arbitrary length N.

Algorithm	Real Mults.	Real Adds	Trig Eval.
Direct DFT	$4\,N^2$	$4\,N^2$	$2\,N^2$
Mod. 2nd Order Goertzel	$N^2 + N$	$2\,N^2 + N$	N
QFT	N^2	$N^2 + 4N$	$2N$

Table 9.2

Timings of the algorithms on a PC in milliseconds are given in the following table.

Algorithm	$N = 125$	$N = 256$
Direct DFT	4.90	19.83
Mod. 2O. Goertzel	1.32	5.55
QFT	1.09	4.50
Chirp + FFT	1.70	3.52

Table 9.3

These timings track the floating point operation counts fairly well.

9.4.4 Conclusions

The QFT is a straight-forward DFT algorithm that uses all of the possible symmetries of the DFT basis function with no requirements on the length being composite. These ideas have been proposed before, but have not been published or clearly developed by [211], [344], [342], [168]. It seems that the basic QFT is practical and useful as a general algorithm for lengths up to a hundred or so. Above that, the chirp z-transform [64] or other filter based methods will be superior. For special cases and shorter lengths, methods based on Winograd's theories will always be superior. Nevertheless, the QFT has a definite place in the array of DFT algorithms and is not well known. A Fortran program is included in the appendix.

It is possible, but unlikely, that further arithmetic reduction could be achieved using the fact that W_N has unity magnitude as was done in second-order Goertzel algorithm. It is also possible that some way of combining the Goertzel and QFT algorithm would have some advantages. A development of a complete QFT decomposition of a DFT of length-2^M shows interesting structure [156], [159] and arithmetic complexity comparable to average Cooley-Tukey FFTs. It does seem better suited to real data calculations with pruning.

Chapter 10

The Prime Factor and Winograd Fourier Transform Algorithms[1]

The prime factor algorithm (PFA) and the Winograd Fourier transform algorithm (WFTA) are methods for efficiently calculating the DFT which use, and in fact, depend on the Type-1 index map from Multidimensional Index Mapping: Equation 10 (3.10) and Multidimensional Index Mapping: Equation 6 (3.6). The use of this index map preceded Cooley and Tukey's paper [150], [302] but its full potential was not realized until it was combined with Winograd's short DFT algorithms. The modern PFA was first presented in [213] and a program given in [57]. The WFTA was first presented in [407] and programs given in [236], [83].

The number theoretic basis for the indexing in these algorithms may, at first, seem more complicated than in the Cooley-Tukey FFT; however, if approached from the general index mapping point of view of Multidimensional Index Mapping (Chapter 3), it is straightforward, and part of a common approach to breaking large problems into smaller ones. The development in this section will parallel that in The Cooley-Tukey Fast Fourier Transform Algorithm (Chapter 9).

The general index maps of Multidimensional Index Mapping: Equation 6 (3.6) and Multidimensional Index Mapping: Equation 12 (3.12) must satisfy the Type-1 conditions of Multidimensional Index Mapping: Equation 7 (3.7) and Multidimensional Index Mapping: Equation 10 (3.10) which are

$$K_1 = aN_2 \quad \text{and} \quad K_2 = bN_1 \quad \text{with} \quad (K_1, N_1) = (K_2, N_2) = 1 \tag{10.1}$$

$$K_3 = cN_2 \quad \text{and} \quad K_4 = dN_1 \quad \text{with} \quad (K_3, N_1) = (K_4, N_2) = 1 \tag{10.2}$$

The row and column calculations in Multidimensional Index Mapping: Equation 15 (3.15) are uncoupled by Multidimensional Index Mapping: Equation 16 (3.16) which for this case

[1]This content is available online at <http://cnx.org/content/m16335/1.9/>.

are

$$((K_1K_4))_N = ((K_2K_3))_N = 0 \tag{10.3}$$

In addition, to make each short sum a DFT, the K_i must also satisfy

$$((K_1K_3))_N = N_2 \;\; and \;\; ((K_2K_4))_N = N_1 \tag{10.4}$$

In order to have the smallest values for K_i, the constants in (10.1) are chosen to be

$$a = b = 1, \quad c = \left((N_2^{-1})\right)_N, \quad d = \left((N_1^{-1})\right)_N \tag{10.5}$$

which gives for the index maps in (10.1)

$$n = ((N_2n_1 + N_1n_2))_N \tag{10.6}$$

$$k = ((K_3k_1 + K_4k_2))_N \tag{10.7}$$

The frequency index map is a form of the Chinese remainder theorem. Using these index maps, the DFT in Multidimensional Index Mapping: Equation 15 (3.15) becomes

$$X = \sum_{n_2=0}^{N_2-1} \sum_{n_1=0}^{N_1-1} x \, W_{N_1}^{n_1k_1} \, W_{N_2}^{n_2k_2} \tag{10.8}$$

which is a pure two-dimensional DFT with no twiddle factors and the summations can be done in either order. Choices other than (10.5) could be used. For example, $a = b = c = d = 1$ will cause the input and output index map to be the same and, therefore, there will be no scrambling of the output order. The short summations in (96), however, will no longer be short DFT's [57].

An important feature of the short Winograd DFT's described in Winograd's Short DFT Algorithms (Chapter 7) that is useful for both the PFA and WFTA is the fact that the multiplier constants in Winograd's Short DFT Algorithms: Equation 6 (7.6) or Winograd's Short DFT Algorithms: Equation 8 (7.8) are either real or imaginary, never a general complex number. For that reason, multiplication by complex data requires only two real multiplications, not four. That is a very significant feature. It is also true that the j multiplier can be commuted from the D operator to the last part of the A^T operator. This means the D operator has only real multipliers and the calculations on real data remains real until the last stage. This can be seen by examining the short DFT modules in [65], [198] and in the appendices.

10.1 The Prime Factor Algorithm

If the DFT is calculated directly using (10.8), the algorithm is called a prime factor algorithm [150], [302] and was discussed in Winograd's Short DFT Algorithms (Chapter 7) and Multidimensional Index Mapping: In-Place Calculation of the DFT and Scrambling (Section 3.2:

In-Place Calculation of the DFT and Scrambling). When the short DFT's are calculated by the very efficient algorithms of Winograd discussed in Factoring the Signal Processing Operators (Chapter 6), the PFA becomes a very powerful method that is as fast or faster than the best Cooley-Tukey FFT's [57], [213].

A flow graph is not as helpful with the PFA as it was with the Cooley-Tukey FFT, however, the following representation in Figure 10.1 which combines Figures Multidimensional Index Mapping: Figure 1 (Figure 3.1) and Winograd's Short DFT Algorithms: Figure 2 (Figure 7.2) gives a good picture of the algorithm with the example of Multidimensional Index Mapping: Equation 25 (3.25)

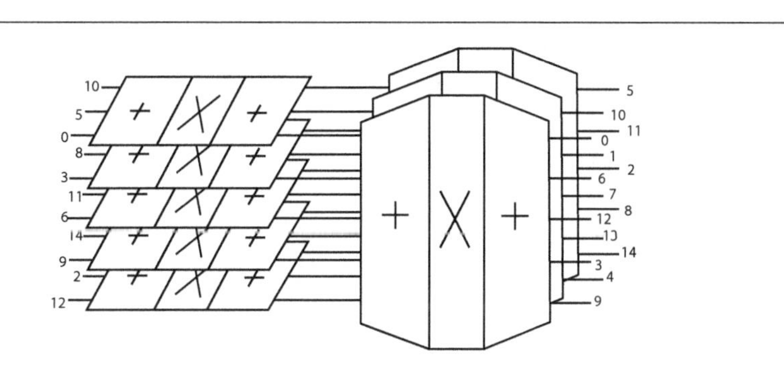

Figure 10.1: A Prime Factor FFT for N = 15

If N is factored into three factors, the DFT of (10.8) would have three nested summations and would be a three-dimensional DFT. This principle extends to any number of factors; however, recall that the Type-1 map requires that all the factors be relatively prime. A very simple three-loop indexing scheme has been developed [57] which gives a compact, efficient PFA program for any number of factors. The basic program structure is illustrated in p. **??** with the short DFT's being omitted for clarity. Complete programs are given in [65] and in the appendices.

```
C---------------PFA INDEXING LOOPS--------------
        DO 10 K = 1, M
           N1 = NI(K)
           N2 = N/N1
           I(1) = 1
           DO 20 J = 1, N2
              DO 30 L=2, N1
                 I(L) = I(L-1) + N2
                 IF (I(L .GT.N)  I(L) = I(L) - N
   30         CONTINUE
              GOTO (20,102,103,104,105), N1
              I(1) = I(1) + N1
   20      CONTINUE
   10   CONTINUE
        RETURN
C----------------MODULE FOR N=2-----------------
  102   R1        = X(I(1))
        X(I(1))   = R1 + X(I(2))
        X(I(2))   = R1 - X(I(2))
        R1        = Y(I(1))
        Y(I(1))   = R1 + Y(I(2))
        Y(I(2))   = R1 - Y(I(2))
        GOTO 20
C----------------OTHER MODULES------------------
  103   Length-3 DFT
  104   Length-4 DFT
  105   Length-5 DFT
        etc.
```

Listing 10.1: Part of a FORTRAN PFA Program

As in the Cooley-Tukey program, the DO 10 loop steps through the M stages (factors of N) and the DO 20 loop calculates the N/N1 length-N1 DFT's. The input index map of (10.6) is implemented in the DO 30 loop and the statement just before label 20. In the PFA, each stage or factor requires a separately programmed module or butterfly. This lengthens the PFA program but an efficient Cooley-Tukey program will also require three or more butterflies.

Because the PFA is calculated in-place using the input index map, the output is scrambled. There are five approaches to dealing with this scrambled output. First, there are some applications where the output does not have to be unscrambled as in the case of high-speed convolution. Second, an unscrambler can be added after the PFA to give the output in correct order just as the bit-reversed-counter is used for the Cooley-Tukey FFT. A simple unscrambler is given in [65], [57] but it is not in place. The third method does the unscrambling in the modules while they are being calculated. This is probably the fastest method but the program must be written for a specific length [65], [57]. A fourth method is similar and achieves the unscrambling by choosing the multiplier constants in the modules properly [198]. The fifth method uses a separate indexing method for the input and output of each module [65], [320].

10.2 The Winograd Fourier Transform Algorithm

The Winograd Fourier transform algorithm (WFTA) uses a very powerful property of the Type-1 index map and the DFT to give a further reduction of the number of multiplications in the PFA. Using an operator notation where F_1 represents taking row DFT's and F_2 represents column DFT's, the two-factor PFA of (10.8) is represented by

$$X = F_2\ F_1\ \ x \tag{10.9}$$

It has been shown [410], [190] that if each operator represents identical operations on each row or column, they commute. Since F_1 and F_2 represent length N_1 and N_2 DFT's, they commute and (10.9) can also be written

$$X = F_1\ F_2\ \ x \tag{10.10}$$

If each short DFT in F is expressed by three operators as in Winograd's Short DFT Algorithms: Equation 8 (7.8) and Winograd's Short DFT Algorithms: Figure 2 (Figure 7.2), F can be factored as

$$F = A^T D A \tag{10.11}$$

where A represents the set of additions done on each row or column that performs the residue reduction as Winograd's Short DFT Algorithms: Equation 30 (7.30). Because of the appearance of the flow graph of A and because it is the first operator on x, it is called a preweave operator [236]. D is the set of M multiplications and A^T (or B^T or C^T) from Winograd's Short DFT Algorithms: Equation 5 (7.5) or Winograd's Short DFT Algorithms: Equation 6 (7.6) is the reconstruction operator called the postweave. Applying (10.11) to (10.9) gives

$$X = A_2^T\ D_2\ A_2\ A_1^T\ D_1\ A_1\ \ x \tag{10.12}$$

This is the PFA of (10.8) and Figure 10.1 where $A_1 D_1 A_1$ represents the row DFT's on the array formed from x. Because these operators commute, (10.12) can also be written as

$$X = A_2^T \; A_1^T \; D_2 \; D_1 \; A_2 \; A_1 \quad x \tag{10.13}$$

or

$$X = A_1^T \; A_2^T \; D_2 \; D_1 \; A_2 \; A_1 \quad x \tag{10.14}$$

but the two adjacent multiplication operators can be premultiplied and the result represented by one operator $D = D_2 \; D_1$ which is no longer the same for each row or column. Equation (10.14) becomes

$$X = A_1^T \; A_2^T \; D \; A_2 \; A_1 \quad x \tag{10.15}$$

This is the basic idea of the Winograd Fourier transform algorithm. The commuting of the multiplication operators together in the center of the algorithm is called nesting and it results in a significant decrease in the number of multiplications that must be done at the execution of the algorithm. Pictorially, the PFA of Figure 10.1 becomes [213] the WFTA in Figure 10.2.

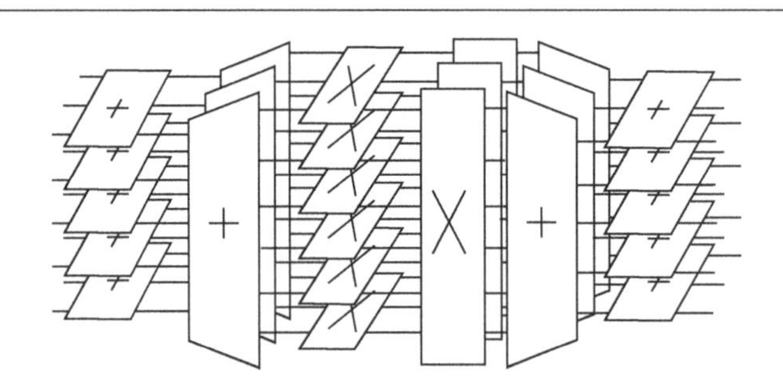

Figure 10.2: A Length-15 WFTA with Nested Multiplications

The rectangular structure of the preweave addition operators causes an expansion of the data in the center of the algorithm. The 15 data points in Figure 10.2 become 18 intermediate values. This expansion is a major problem in programming the WFTA because it prevents a straightforward in-place calculation and causes an increase in the number of required additions and in the number of multiplier constants that must be precalculated and stored.

From Figure 10.2 and the idea of premultiplying the individual multiplication operators, it can be seen why the multiplications by unity had to be considered in Winograd's Short DFT Algorithms: Table 1 (Table 7.1). Even if a multiplier in D_1 is unity, it may not be in $D_2 D_1$. In Figure 10.2 with factors of three and five, there appear to be 18 multiplications required because of the expansion of the length-5 preweave operator, A_2, however, one of multipliers in each of the length three and five operators is unity, so one of the 18 multipliers in the product is unity. This gives 17 required multiplications - a rather impressive reduction from the $15^2 = 225$ multiplications required by direct calculation. This number of 17 complex multiplications will require only 34 real multiplications because, as mentioned earlier, the multiplier constants are purely real or imaginary while the 225 complex multiplications are general and therefore will require four times as many real multiplications.

The number of additions depends on the order of the pre- and postweave operators. For example in the length-15 WFTA in Figure 10.2, if the length-5 had been done first and last, there would have been six row addition preweaves in the preweave operator rather than the five shown. It is difficult to illustrate the algorithm for three or more factors of N, but the ideas apply to any number of factors. Each length has an optimal ordering of the pre- and postweave operators that will minimize the number of additions.

A program for the WFTA is not as simple as for the FFT or PFA because of the very characteristic that reduces the number of multiplications, the nesting. A simple two-factor example program is given in [65] and a general program can be found in [236], [83]. The same lengths are possible with the PFA and WFTA and the same short DFT modules can be used, however, the multiplies in the modules must occur in one place for use in the WFTA.

10.3 Modifications of the PFA and WFTA Type Algorithms

In the previous section it was seen how using the permutation property of the elementary operators in the PFA allowed the nesting of the multiplications to reduce their number. It was also seen that a proper ordering of the operators could minimize the number of additions. These ideas have been extended in formulating a more general algorithm optimizing problem. If the DFT operator F in (10.11) is expressed in a still more factored form obtained from Winograd's Short DFT Algorithms: Equation 30 (7.30), a greater variety of ordering can be optimized. For example if the A operators have two factors

$$F_1 = A_1^T A_1^{'T} \; D_1 \; A_1^{'} A_1 \tag{10.16}$$

The DFT in (10.10) becomes

$$X = A_2^T A_2^{'T} D_2 A'_2 A_2 A_1^T A_1^{'T} D_1 A'_1 A_1 x \tag{10.17}$$

The operator notation is very helpful in understanding the central ideas, but may hide some important facts. It has been shown [410], [198] that operators in different F_i commute with each other, but the order of the operators within an F_i cannot be changed. They represent the matrix multiplications in Winograd's Short DFT Algorithms: Equation 30 (7.30) or Winograd's Short DFT Algorithms: Equation 8 (7.8) which do not commute.

This formulation allows a very large set of possible orderings, in fact, the number is so large that some automatic technique must be used to find the "best". It is possible to set up a criterion of optimality that not only includes the number of multiplications but the number of additions as well. The effects of relative multiply-add times, data transfer times, CPU register and memory sizes, and other hardware characteristics can be included in the criterion. Dynamic programming can then be applied to derive an optimal algorithm for a particular application [190]. This is a very interesting idea as there is no longer a single algorithm, but a class and an optimizing procedure. The challenge is to generate a broad enough class to result in a solution that is close to a global optimum and to have a practical scheme for finding the solution.

Results obtained applying the dynamic programming method to the design of fairly long DFT algorithms gave algorithms that had fewer multiplications and additions than either a pure PFA or WFTA [190]. It seems that some nesting is desirable but not total nesting for four or more factors. There are also some interesting possibilities in mixing the Cooley-Tukey with this formulation. Unfortunately, the twiddle factors are not the same for all rows and columns, therefore, operations cannot commute past a twiddle factor operator. There are ways of breaking the total algorithm into horizontal paths and using different orderings along the different paths [264], [198]. In a sense, this is what the split-radix FFT does with its twiddle factors when compared to a conventional Cooley-Tukey FFT.

There are other modifications of the basic structure of the Type-1 index map DFT algorithm. One is to use the same index structure and conversion of the short DFT's to convolution as the PFA but to use some other method for the high-speed convolution. Table look-up of partial products based on distributed arithmetic to eliminate all multiplications [78] looks promising for certain very specific applications, perhaps for specialized VLSI implementation. Another possibility is to calculate the short convolutions using number-theoretic transforms [30], [236], [264]. This would also require special hardware. Direct calculation of short convolutions is faster on certain pipelined processor such as the TMS-320 microprocessor [216].

10.4 Evaluation of the PFA and WFTA

As for the Cooley-Tukey FFT's, the first evaluation of these algorithms will be on the number of multiplications and additions required. The number of multiplications to compute the PFA in (10.8) is given by Multidimensional Index Mapping: Equation 3 (3.3). Using the notation that $T(N)$ is the number of multiplications or additions necessary to calculate a length-N DFT, the total number for a four-factor PFA of length-N, where $N = N_1 N_2 N_3 N_4$ is

$$T(N) = N_1 N_2 N_3 T(N_4) + N_2 N_3 N_4 T(N_1) + N_3 N_4 N_1 T(N_2) + N_4 N_1 N_2 T(N_3) \quad (10.18)$$

The count of multiplies and adds in Table 10.1 are calculated from (105) with the counts of the factors taken from Winograd's Short DFT Algorithms: Table 1 (Table 7.1). The list of lengths are those possible with modules in the program of length 2, 3, 4, 5, 7, 8, 9 and 16 as is true for the PFA in [65], [57] and the WFTA in [236], [83]. A maximum of four relatively prime lengths can be used from this group giving 59 different lengths over the range from 2 to 5040. The radix-2 or split-radix FFT allows 12 different lengths over the same range. If modules of length 11 and 13 from [188] are added, the maximum length becomes 720720 and the number of different lengths becomes 239. Adding modules for 17, 19 and 25 from [188] gives a maximum length of 1163962800 and a very large and dense number of possible lengths. The length of the code for the longer modules becomes excessive and should not be included unless needed.

The number of multiplications necessary for the WFTA is simply the product of those necessary for the required modules, including multiplications by unity. The total number may contain some unity multipliers but it is difficult to remove them in a practical program. Table 10.1 contains both the total number (MULTS) and the number with the unity multiplies removed (RMULTS).

Calculating the number of additions for the WFTA is more complicated than for the PFA because of the expansion of the data moving through the algorithm. For example the number of additions, TA, for the length-15 example in Figure 10.2 is given by

$$TA(N) = N_2 TA(N_1) + TM_1 TA(N_2) \tag{10.19}$$

where $N_1 = 3$, $N_2 = 5$, $TM_1 =$ the number of multiplies for the length-3 module and hence the expansion factor. As mentioned earlier there is an optimum ordering to minimize additions. The ordering used to calculate Table 10.1 is the ordering used in [236], [83] which is optimal in most cases and close to optimal in the others.

Length	PFA	PFA	WFTA	WFTA	WFTA
N	Mults	Adds	Mults	RMults	Adds
10	20	88	24	20	88
12	16	96	24	16	96
14	32	172	36	32	172
15	50	162	36	34	162

continued on next page

18	40	204	44	40	208
20	40	216	48	40	216
21	76	300	54	52	300
24	44	252	48	36	252
28	64	400	72	64	400
30	100	384	72	68	384
35	150	598	108	106	666
36	80	480	88	80	488
40	100	532	96	84	532
42	152	684	108	104	684
45	190	726	132	130	804
48	124	636	108	92	660
56	156	940	144	132	940
60	200	888	144	136	888
63	284	1236	198	196	1394
70	300	1336	216	212	1472
72	196	1140	176	164	1156
80	260	1284	216	200	1352
84	304	1536	216	208	1536
90	380	1632	264	260	1788
105	590	2214	324	322	2418
112	396	2188	324	308	2332
120	460	2076	288	276	2076

continued on next page

126	568	2724	396	392	3040
140	600	2952	432	424	3224
144	500	2676	396	380	2880
168	692	3492	432	420	3492
180	760	3624	528	520	3936
210	1180	4848	648	644	5256
240	1100	4812	648	632	5136
252	1136	5952	792	784	6584
280	1340	6604	864	852	7148
315	2050	8322	1188	1186	10336
336	1636	7908	972	956	8508
360	1700	8148	1056	1044	8772
420	2360	10536	1296	1288	11352
504	2524	13164	1584	1572	14428
560	3100	14748	1944	1928	17168
630	4100	17904	2376	2372	21932
720	3940	18276	2376	2360	21132
840	5140	23172	2592	2580	24804
1008	5804	29100	3564	3548	34416
1260	8200	38328	4752	4744	46384
1680	11540	50964	5832	5816	59064
2520	17660	82956	9504	9492	99068
5040	39100	179772	21384	21368	232668

Table 10.1: Number of Real Multiplications and Additions for Complex PFA and WFTA FFTs

from Table 10.1 we see that compared to the PFA or any of the Cooley-Tukey FFT's, the WFTA has significantly fewer multiplications. For the shorter lengths, the WFTA and the PFA have approximately the same number of additions; however for longer lengths, the PFA has fewer and the Cooley-Tukey FFT's always have the fewest. If the total arithmetic, the number of multiplications plus the number of additions, is compared, the split-radix FFT, PFA and WFTA all have about the same count. Special versions of the PFA and WFTA have been developed for real data [178], [358].

The size of the Cooley-Tukey program is the smallest, the PFA next and the WFTA largest. The PFA requires the smallest number of stored constants, the Cooley-Tukey or split-radix FFT next, and the WFTA requires the largest number. For a DFT of approximately 1000, the PFA stores 28 constants, the FFT 2048 and the WFTA 3564. Both the FFT and PFA can be calculated in-place and the WFTA cannot. The PFA can be calculated in-order without an unscrambler. The radix-2 FFT can also, but it requires additional indexing overhead [194]. The indexing and data transfer overhead is greatest for the WFTA because the separate preweave and postweave sections each require their indexing and pass through the complete data. The shorter modules in the PFA and WFTA and the butterflies in the radix 2 and 4 FFT's are more efficient than the longer ones because intermediate calculations can be kept in cpu registers rather general memory [250]. However, the shorter modules and radices require more passes through the data for a given approximate length. A proper comparison will require actual programs to be compiled and run on a particular machine. There are many open questions about the relationship of algorithms and hardware architecture.

Chapter 11

Implementing FFTs in Practice[1]

by Steven G. Johnson (Department of Mathematics, Massachusetts Institute of Technology) and Matteo Frigo (Cilk Arts, Inc.)

11.1 Introduction

Although there are a wide range of fast Fourier transform (FFT) algorithms, involving a wealth of mathematics from number theory to polynomial algebras, the vast majority of FFT implementations in practice employ some variation on the Cooley-Tukey algorithm [92]. The Cooley-Tukey algorithm can be derived in two or three lines of elementary algebra. It can be implemented almost as easily, especially if only power-of-two sizes are desired; numerous popular textbooks list short FFT subroutines for power-of-two sizes, written in the language *du jour*. The implementation of the Cooley-Tukey algorithm, at least, would therefore seem to be a long-solved problem. In this chapter, however, we will argue that matters are not as straightforward as they might appear.

For many years, the primary route to improving upon the Cooley-Tukey FFT seemed to be reductions in the count of arithmetic operations, which often dominated the execution time prior to the ubiquity of fast floating-point hardware (at least on non-embedded processors). Therefore, great effort was expended towards finding new algorithms with reduced arithmetic counts [114], from Winograd's method to achieve $\Theta(n)$ multiplications[2] (at the cost of many more additions) [411], [180], [116], [114] to the split-radix variant on Cooley-Tukey that long achieved the lowest known total count of additions and multiplications for power-of-two sizes [422], [107], [391], [230], [114] (but was recently improved upon [202], [225]). The question of the minimum possible arithmetic count continues to be of fundamental theoretical interest—it is not even known whether better than $\Theta(nlogn)$ complexity is possible, since $\Omega(nlogn)$ lower bounds on the count of additions have only been proven subject to

[1]This content is available online at <http://cnx.org/content/m16336/1.15/>.

[2]We employ the standard asymptotic notation of O for asymptotic upper bounds, Θ for asymptotic tight bounds, and Ω for asymptotic lower bounds [210].

restrictive assumptions about the algorithms [248], [280], [281]. Nevertheless, the difference in the number of arithmetic operations, for power-of-two sizes n, between the 1965 radix-2 Cooley-Tukey algorithm ($\sim 5nlog_2n$ [92]) and the currently lowest-known arithmetic count ($\sim \frac{34}{9}nlog_2n$ [202], [225]) remains only about 25%.

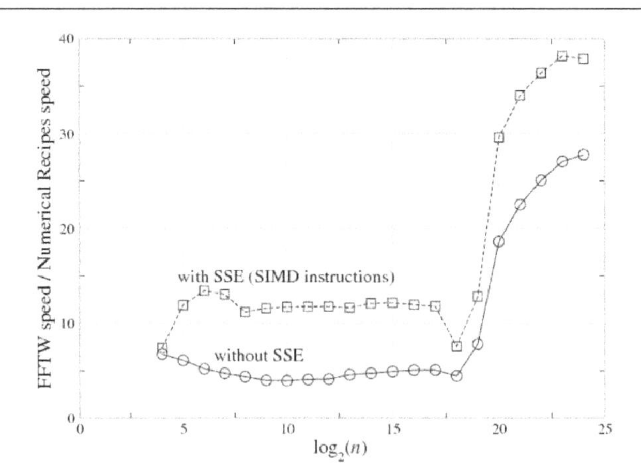

Figure 11.1: The ratio of speed (1/time) between a highly optimized FFT (FFTW 3.1.2 [133], [134]) and a typical textbook radix-2 implementation (*Numerical Recipes in C* [290]) on a 3 GHz Intel Core Duo with the Intel C compiler 9.1.043, for single-precision complex-data DFTs of size n, plotted versus log_2n. Top line (squares) shows FFTW with SSE SIMD instructions enabled, which perform multiple arithmetic operations at once (see section); bottom line (circles) shows FFTW with SSE disabled, which thus requires a similar number of arithmetic instructions to the textbook code. (This is not intended as a criticism of *Numerical Recipes*—simple radix-2 implementations are reasonable for pedagogy—but it illustrates the radical differences between straightforward and optimized implementations of FFT algorithms, even with similar arithmetic costs.) For $n \gtrsim 2^{19}$, the ratio increases because the textbook code becomes much slower (this happens when the DFT size exceeds the level-2 cache).

And yet there is a vast gap between this basic mathematical theory and the actual practice—highly optimized FFT packages are often an order of magnitude faster than the textbook subroutines, and the internal structure to achieve this performance is radically different from the typical textbook presentation of the "same" Cooley-Tukey algorithm. For example, Figure 11.1 plots the ratio of benchmark speeds between a highly optimized FFT [133], [134] and a typical textbook radix-2 implementation [290], and the former is faster by a factor of 5–40 (with a larger ratio as n grows). Here, we will consider some of the reasons for this discrepancy, and some techniques that can be used to address the difficulties faced by a practical high-performance FFT implementation.[3]

In particular, in this chapter we will discuss some of the lessons learned and the strategies adopted in the FFTW library. FFTW [133], [134] is a widely used free-software library that computes the discrete Fourier transform (DFT) and its various special cases. Its performance is competitive even with manufacturer-optimized programs [134], and this performance is **portable** thanks the structure of the algorithms employed, self-optimization techniques, and highly optimized kernels (FFTW's **codelets**) generated by a special-purpose compiler.

This chapter is structured as follows. First "Review of the Cooley-Tukey FFT" (Section 11.2: Review of the Cooley-Tukey FFT), we briefly review the basic ideas behind the Cooley-Tukey algorithm and define some common terminology, especially focusing on the many degrees of freedom that the abstract algorithm allows to implementations. Next, in "Goals and Background of the FFTW Project" (Section 11.3: Goals and Background of the FFTW Project), we provide some context for FFTW's development and stress that performance, while it receives the most publicity, is not necessarily the most important consideration in the implementation of a library of this sort. Third, in "FFTs and the Memory Hierarchy" (Section 11.4: FFTs and the Memory Hierarchy), we consider a basic theoretical model of the computer memory hierarchy and its impact on FFT algorithm choices: quite general considerations push implementations towards large radices and explicitly recursive structure. Unfortunately, general considerations are not sufficient in themselves, so we will explain in "Adaptive Composition of FFT Algorithms" (Section 11.5: Adaptive Composition of FFT Algorithms) how FFTW self-optimizes for particular machines by selecting its algorithm at runtime from a composition of simple algorithmic steps. Furthermore, "Generating Small FFT Kernels" (Section 11.6: Generating Small FFT Kernels) describes the utility and the principles of automatic code generation used to produce the highly optimized building blocks of this composition, FFTW's codelets. Finally, we will briefly consider an important non-performance issue, in "Numerical Accuracy in FFTs" (Section 11.7: Numerical Accuracy in FFTs).

11.2 Review of the Cooley-Tukey FFT

The (forward, one-dimensional) discrete Fourier transform (DFT) of an array $\mathbf{X}$ of n complex numbers is the array $\mathbf{Y}$ given by

$$\mathbf{Y}[k] = \sum_{\ell=0}^{n-1} \mathbf{X}[\ell]\, \omega_n^{\ell k}, \tag{11.1}$$

where $0 \le k < n$ and $\omega_n = exp(-2\pi i/n)$ is a primitive root of unity. Implemented directly, (11.1) would require $\Theta(n^2)$ operations; fast Fourier transforms are $O(nlogn)$ algorithms

[3]We won't address the question of parallelization on multi-processor machines, which adds even greater difficulty to FFT implementation—although multi-processors are increasingly important, achieving good serial performance is a basic prerequisite for optimized parallel code, and is already hard enough!

to compute the same result. The most important FFT (and the one primarily used in FFTW) is known as the "Cooley-Tukey" algorithm, after the two authors who rediscovered and popularized it in 1965 [92], although it had been previously known as early as 1805 by Gauss as well as by later re-inventors [173]. The basic idea behind this FFT is that a DFT of a composite size $n = n_1 n_2$ can be re-expressed in terms of smaller DFTs of sizes n_1 and n_2—essentially, as a two-dimensional DFT of size $n_1 \times n_2$ where the output is **transposed**. The choices of factorizations of n, combined with the many different ways to implement the data re-orderings of the transpositions, have led to numerous implementation strategies for the Cooley-Tukey FFT, with many variants distinguished by their own names [114], [389]. FFTW implements a space of **many** such variants, as described in "Adaptive Composition of FFT Algorithms" (Section 11.5: Adaptive Composition of FFT Algorithms), but here we derive the basic algorithm, identify its key features, and outline some important historical variations and their relation to FFTW.

The Cooley-Tukey algorithm can be derived as follows. If n can be factored into $n = n_1 n_2$, (11.1) can be rewritten by letting $\ell = \ell_1 n_2 + \ell_2$ and $k = k_1 + k_2 n_1$. We then have:

$$\mathbf{Y}\left[k_1 + k_2 n_1\right] = \sum_{\ell_2=0}^{n_2-1} \left[\left(\sum_{\ell_1=0}^{n_1-1} \mathbf{X}\left[\ell_1 n_2 + \ell_2\right] \omega_{n_1}^{\ell_1 k_1} \right) \omega_n^{\ell_2 k_1} \right] \omega_{n_2}^{\ell_2 k_2}, \tag{11.2}$$

where $k_{1,2} = 0, ..., n_{1,2} - 1$. Thus, the algorithm computes n_2 DFTs of size n_1 (the inner sum), multiplies the result by the so-called [139] **twiddle factors** $\omega_n^{\ell_2 k_1}$, and finally computes n_1 DFTs of size n_2 (the outer sum). This decomposition is then continued recursively. The literature uses the term **radix** to describe an n_1 or n_2 that is bounded (often constant); the small DFT of the radix is traditionally called a **butterfly**.

Many well-known variations are distinguished by the radix alone. A **decimation in time** (**DIT**) algorithm uses n_2 as the radix, while a **decimation in frequency** (**DIF**) algorithm uses n_1 as the radix. If multiple radices are used, e.g. for n composite but not a prime power, the algorithm is called **mixed radix**. A peculiar blending of radix 2 and 4 is called **split radix**, which was proposed to minimize the count of arithmetic operations [422], [107], [391], [230], [114] although it has been superseded in this regard [202], [225]. FFTW implements both DIT and DIF, is mixed-radix with radices that are **adapted** to the hardware, and often uses much larger radices (e.g. radix 32) than were once common. On the other end of the scale, a "radix" of roughly $\sqrt{n}$ has been called a **four-step** FFT algorithm (or **six-step**, depending on how many transposes one performs) [14]; see "FFTs and the Memory Hierarchy" (Section 11.4: FFTs and the Memory Hierarchy) for some theoretical and practical discussion of this algorithm.

A key difficulty in implementing the Cooley-Tukey FFT is that the n_1 dimension corresponds to discontiguous inputs ℓ_1 in $\mathbf{X}$ but contiguous outputs k_1 in $\mathbf{Y}$, and vice-versa for n_2. This is a matrix transpose for a single decomposition stage, and the composition of all such transpositions is a (mixed-base) digit-reversal permutation (or **bit-reversal**, for radix 2). The resulting necessity of discontiguous memory access and data re-ordering hinders efficient use of hierarchical memory architectures (e.g., caches), so that the optimal execution order of an FFT for given hardware is non-obvious, and various approaches have been proposed.

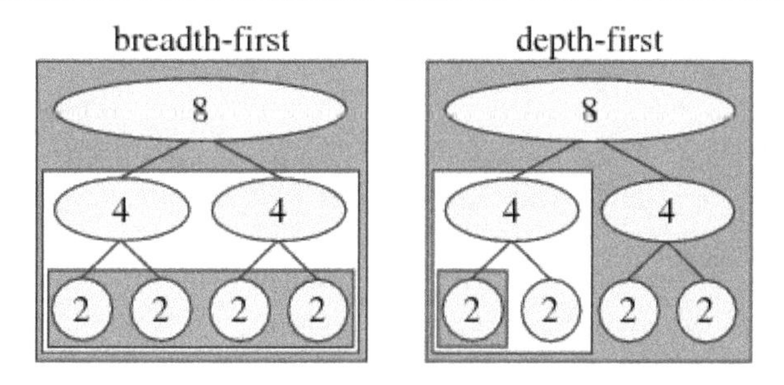

Figure 11.2: Schematic of traditional breadth-first (left) vs. recursive depth-first (right) ordering for radix-2 FFT of size 8: the computations for each nested box are completed before doing anything else in the surrounding box. Breadth-first computation performs all butterflies of a given size at once, while depth-first computation completes one sub-transform entirely before moving on to the next (as in the algorithm below).

One ordering distinction is between recursion and iteration. As expressed above, the Cooley-Tukey algorithm could be thought of as defining a tree of smaller and smaller DFTs, as depicted in Figure 11.2; for example, a textbook radix-2 algorithm would divide size n into two transforms of size $n/2$, which are divided into four transforms of size $n/4$, and so on until a base case is reached (in principle, size 1). This might naturally suggest a recursive implementation in which the tree is traversed "depth-first" as in Figure 11.2(right) and the algorithm of p. **??**—one size $n/2$ transform is solved completely before processing the other one, and so on. However, most traditional FFT implementations are non-recursive (with rare exceptions [341]) and traverse the tree "breadth-first" [389] as in Figure 11.2(left)—in the radix-2 example, they would perform n (trivial) size-1 transforms, then $n/2$ combinations into size-2 transforms, then $n/4$ combinations into size-4 transforms, and so on, thus making $log_2 n$ passes over the whole array. In contrast, as we discuss in "Discussion" (Section 11.5.2.6: Discussion), FFTW employs an explicitly recursive strategy that encompasses **both** depth-first and breadth-first styles, favoring the former since it has some theoretical and practical advantages as discussed in "FFTs and the Memory Hierarchy" (Section 11.4: FFTs and the Memory Hierarchy).

$\mathbf{Y}[0,...,n-1] \leftarrow recfft\ 2(n,\mathbf{X},\iota\ )$:
```
IF n=1 THEN
```
$Y[0] \leftarrow X[0]$
```
ELSE
```
$\mathbf{Y}[0,...,n/2-1] \leftarrow recfft2(n/2,\mathbf{X},2\iota\ )$

$\mathbf{Y}[n/2,...,n-1] \leftarrow recfft2(n/2,\mathbf{X}+\iota\ ,2\iota\ )$

FOR $k\_1 = 0$ TO $(n/2)-1$ DO

$t \leftarrow \mathbf{Y}[k\_1]$

$\mathbf{Y}[k\_1] \leftarrow t+\omega\ \_n\hat{\ }k\_1\mathbf{Y}[k\_1+n/2]$

$\mathbf{Y}[k\_1+n/2] \leftarrow t-\omega\ \_n\hat{\ }k\_1\mathbf{Y}[k\_1+n/2]$
```
    END FOR
END IF
```

Listing 11.1: A depth-first recursive radix-2 DIT Cooley-Tukey FFT to compute a DFT of a power-of-two size $n = 2^m$. The input is an array $\mathbf{X}$ of length n with stride ι (i.e., the inputs are $\mathbf{X}[\ell\iota]$ for $\ell = 0,...,n-1$) and the output is an array $\mathbf{Y}$ of length n (with stride 1), containing the DFT of $\mathbf{X}$ [Equation 1]. $\mathbf{X}+\iota$ denotes the array beginning with $\mathbf{X}[\iota]$. This algorithm operates out-of-place, produces in-order output, and does not require a separate bit-reversal stage.

A second ordering distinction lies in how the digit-reversal is performed. The classic approach is a single, separate digit-reversal pass following or preceding the arithmetic computations; this approach is so common and so deeply embedded into FFT lore that many practitioners find it difficult to imagine an FFT without an explicit bit-reversal stage. Although this pass requires only $O(n)$ time [207], it can still be non-negligible, especially if the data is out-of-cache; moreover, it neglects the possibility that data reordering during the transform may improve memory locality. Perhaps the oldest alternative is the Stockham **auto-sort** FFT [367], [389], which transforms back and forth between two arrays with each butterfly, transposing one digit each time, and was popular to improve contiguity of access for vector computers [372]. Alternatively, an explicitly recursive style, as in FFTW, performs the digit-reversal implicitly at the "leaves" of its computation when operating out-of-place (see section "Discussion" (Section 11.5.2.6: Discussion)). A simple example of this style, which computes in-order output using an out-of-place radix-2 FFT without explicit bit-reversal, is shown in the algorithm of p. **??** [corresponding to Figure 11.2(right)]. To operate in-place with $O(1)$ scratch storage, one can interleave small matrix transpositions with the butterflies [195], [375], [297], [166], and a related strategy in FFTW [134] is briefly described by "Discussion" (Section 11.5.2.6: Discussion).

Finally, we should mention that there are many FFTs entirely distinct from Cooley-Tukey.

Three notable such algorithms are the **prime-factor algorithm** for $gcd(n_1, n_2) = 1$ [278], along with Rader's [309] and Bluestein's [35], [305], [278] algorithms for prime n. FFTW implements the first two in its codelet generator for hard-coded n "Generating Small FFT Kernels" (Section 11.6: Generating Small FFT Kernels) and the latter two for general prime n (sections "Plans for prime sizes" (Section 11.5.2.5: Plans for prime sizes) and "Goals and Background of the FFTW Project" (Section 11.3: Goals and Background of the FFTW Project)). There is also the Winograd FFT [411], [180], [116], [114], which minimizes the number of multiplications at the expense of a large number of additions; this trade-off is not beneficial on current processors that have specialized hardware multipliers.

11.3 Goals and Background of the FFTW Project

The FFTW project, begun in 1997 as a side project of the authors Frigo and Johnson as graduate students at MIT, has gone through several major revisions, and as of 2008 consists of more than 40,000 lines of code. It is difficult to measure the popularity of a free-software package, but (as of 2008) FFTW has been cited in over 500 academic papers, is used in hundreds of shipping free and proprietary software packages, and the authors have received over 10,000 emails from users of the software. Most of this chapter focuses on performance of FFT implementations, but FFTW would probably not be where it is today if that were the only consideration in its design. One of the key factors in FFTW's success seems to have been its flexibility in addition to its performance. In fact, FFTW is probably the most flexible DFT library available:

- FFTW is written in portable C and runs well on many architectures and operating systems.
- FFTW computes DFTs in $O(nlogn)$ time for any length n. (Most other DFT implementations are either restricted to a subset of sizes or they become $\Theta(n^2)$ for certain values of n, for example when n is prime.)
- FFTW imposes no restrictions on the rank (dimensionality) of multi-dimensional transforms. (Most other implementations are limited to one-dimensional, or at most two- and three-dimensional data.)
- FFTW supports multiple and/or strided DFTs; for example, to transform a 3-component vector field or a portion of a multi-dimensional array. (Most implementations support only a single DFT of contiguous data.)
- FFTW supports DFTs of real data, as well as of real symmetric/anti-symmetric data (also called discrete cosine/sine transforms).

Our design philosophy has been to first define the most general reasonable functionality, and then to obtain the highest possible performance without sacrificing this generality. In this section, we offer a few thoughts about why such flexibility has proved important, and how it came about that FFTW was designed in this way.

FFTW's generality is partly a consequence of the fact the FFTW project was started in response to the needs of a real application for one of the authors (a spectral solver for Maxwell's equations [204]), which from the beginning had to run on heterogeneous hardware. Our initial application required multi-dimensional DFTs of three-component vector fields (magnetic fields in electromagnetism), and so right away this meant: (i) multi-dimensional FFTs; (ii) user-accessible loops of FFTs of discontiguous data; (iii) efficient support for non-power-of-two sizes (the factor of eight difference between $n \times n \times n$ and $2n \times 2n \times 2n$ was too much to tolerate); and (iv) saving a factor of two for the common real-input case was desirable. That is, the initial requirements already encompassed most of the features above, and nothing about this application is particularly unusual.

Even for one-dimensional DFTs, there is a common misperception that one should always choose power-of-two sizes if one cares about efficiency. Thanks to FFTW's code generator (described in "Generating Small FFT Kernels" (Section 11.6: Generating Small FFT Kernels)), we could afford to devote equal optimization effort to any n with small factors (2, 3, 5, and 7 are good), instead of mostly optimizing powers of two like many high-performance FFTs. As a result, to pick a typical example on the 3 GHz Core Duo processor of Figure 11.1, $n = 3600 = 2^4 \cdot 3^2 \cdot 5^2$ and $n = 3840 = 2^8 \cdot 3 \cdot 5$ both execute faster than $n = 4096 = 2^{12}$. (And if there are factors one particularly cares about, one can generate code for them too.)

One initially missing feature was efficient support for large prime sizes; the conventional wisdom was that large-prime algorithms were mainly of academic interest, since in real applications (including ours) one has enough freedom to choose a highly composite transform size. However, the prime-size algorithms are fascinating, so we implemented Rader's $O(nlogn)$ prime-n algorithm [309] purely for fun, including it in FFTW 2.0 (released in 1998) as a bonus feature. The response was astonishingly positive—even though users are (probably) never **forced** by their application to compute a prime-size DFT, it is rather inconvenient to always worry that collecting an unlucky number of data points will slow down one's analysis by a factor of a million. The prime-size algorithms are certainly slower than algorithms for nearby composite sizes, but in interactive data-analysis situations the difference between 1 ms and 10 ms means little, while educating users to avoid large prime factors is hard.

Another form of flexibility that deserves comment has to do with a purely technical aspect of computer software. FFTW's implementation involves some unusual language choices internally (the FFT-kernel generator, described in "Generating Small FFT Kernels" (Section 11.6: Generating Small FFT Kernels), is written in Objective Caml, a functional language especially suited for compiler-like programs), but its user-callable interface is purely in C with lowest-common-denominator datatypes (arrays of floating-point values). The advantage of this is that FFTW can be (and has been) called from almost any other programming language, from Java to Perl to Fortran 77. Similar lowest-common-denominator interfaces are apparent in many other popular numerical libraries, such as LAPACK [10]. Language preferences arouse strong feelings, but this technical constraint means that modern programming dialects are best hidden from view for a numerical library.

Ultimately, very few scientific-computing applications should have performance as their top priority. Flexibility is often far more important, because one wants to be limited only by one's imagination, rather than by one's software, in the kinds of problems that can be studied.

11.4 FFTs and the Memory Hierarchy

There are many complexities of computer architectures that impact the optimization of FFT implementations, but one of the most pervasive is the memory hierarchy. On any modern general-purpose computer, memory is arranged into a hierarchy of storage devices with increasing size and decreasing speed: the fastest and smallest memory being the CPU registers, then two or three levels of cache, then the main-memory RAM, then external storage such as hard disks.[4] Most of these levels are managed automatically by the hardware to hold the most-recently-used data from the next level in the hierarchy.[5] There are many complications, however, such as limited cache associativity (which means that certain locations in memory cannot be cached simultaneously) and cache lines (which optimize the cache for contiguous memory access), which are reviewed in numerous textbooks on computer architectures. In this section, we focus on the simplest abstract principles of memory hierarchies in order to grasp their fundamental impact on FFTs.

Because access to memory is in many cases the slowest part of the computer, especially compared to arithmetic, one wishes to load as much data as possible in to the faster levels of the hierarchy, and then perform as much computation as possible before going back to the slower memory devices. This is called **temporal locality**: if a given datum is used more than once, we arrange the computation so that these usages occur as close together as possible in time.

11.4.1 Understanding FFTs with an ideal cache

To understand temporal-locality strategies at a basic level, in this section we will employ an idealized model of a cache in a two-level memory hierarchy, as defined in [137]. This **ideal cache** stores Z data items from main memory (e.g. complex numbers for our purposes): when the processor loads a datum from memory, the access is quick if the datum is already in the cache (a **cache hit**) and slow otherwise (a **cache miss**, which requires the datum to be fetched into the cache). When a datum is loaded into the cache,[6] it must replace some

[4] A hard disk is utilized by "out-of-core" FFT algorithms for very large n [389], but these algorithms appear to have been largely superseded in practice by both the gigabytes of memory now common on personal computers and, for extremely large n, by algorithms for distributed-memory parallel computers.

[5] This includes the registers: on current "x86" processors, the user-visible instruction set (with a small number of floating-point registers) is internally translated at runtime to RISC-like "μ-ops" with a much larger number of physical **rename registers** that are allocated automatically.

[6] More generally, one can assume that a **cache line** of L consecutive data items are loaded into the cache at once, in order to exploit spatial locality. The ideal-cache model in this case requires that the cache be **tall**: $Z = \Omega\left(L^2\right)$ [137].

other datum, and the ideal-cache model assumes that the optimal replacement strategy is used [20]: the new datum replaces the datum that will not be needed for the longest time in the future; in practice, this can be simulated to within a factor of two by replacing the least-recently used datum [137], but ideal replacement is much simpler to analyze. Armed with this ideal-cache model, we can now understand some basic features of FFT implementations that remain essentially true even on real cache architectures. In particular, we want to know the **cache complexity**, the number $Q(n; Z)$ of cache misses for an FFT of size n with an ideal cache of size Z, and what algorithm choices reduce this complexity.

First, consider a textbook radix-2 algorithm, which divides n by 2 at each stage and operates breadth-first as in Figure 11.2(left), performing all butterflies of a given size at a time. If $n > Z$, then each pass over the array incurs $\Theta(n)$ cache misses to reload the data, and there are $log_2 n$ passes, for $\Theta(nlog_2 n)$ cache misses in total—no temporal locality at all is exploited!

One traditional solution to this problem is **blocking**: the computation is divided into maximal blocks that fit into the cache, and the computations for each block are completed before moving on to the next block. Here, a block of Z numbers can fit into the cache[7] (not including storage for twiddle factors and so on), and thus the natural unit of computation is a sub-FFT of size Z. Since each of these blocks involves $\Theta(ZlogZ)$ arithmetic operations, and there are $\Theta(nlogn)$ operations overall, there must be $\Theta\left(\frac{n}{Z}log_Z n\right)$ such blocks. More explicitly, one could use a radix-Z Cooley-Tukey algorithm, breaking n down by factors of Z [or $\Theta(Z)$] until a size Z is reached: each stage requires n/Z blocks, and there are $log_Z n$ stages, again giving $\Theta\left(\frac{n}{Z}log_Z n\right)$ blocks overall. Since each block requires Z cache misses to load it into cache, the cache complexity Q_b of such a blocked algorithm is

$$Q_b(n; Z) = \Theta(nlog_Z n). \qquad (11.3)$$

In fact, this complexity is rigorously **optimal** for Cooley-Tukey FFT algorithms [184], and immediately points us towards **large radices** (not radix 2!) to exploit caches effectively in FFTs.

However, there is one shortcoming of any blocked FFT algorithm: it is **cache aware**, meaning that the implementation depends explicitly on the cache size Z. The implementation must be modified (e.g. changing the radix) to adapt to different machines as the cache size changes. Worse, as mentioned above, actual machines have multiple levels of cache, and to exploit these one must perform multiple levels of blocking, each parameterized by the corresponding cache size. In the above example, if there were a smaller and faster cache of size $z < Z$, the size-Z sub-FFTs should themselves be performed via radix-z Cooley-Tukey using blocks of size z. And so on. There are two paths out of these difficulties: one is self-optimization, where the implementation automatically adapts itself to the hardware

[7] Of course, $O(n)$ additional storage may be required for twiddle factors, the output data (if the FFT is not in-place), and so on, but these only affect the n that fits into cache by a constant factor and hence do not impact cache-complexity analysis. We won't worry about such constant factors in this section.

(implicitly including any cache sizes), as described in "Adaptive Composition of FFT Algorithms" (Section 11.5: Adaptive Composition of FFT Algorithms); the other is to exploit **cache-oblivious** algorithms. FFTW employs both of these techniques.

The goal of cache-obliviousness is to structure the algorithm so that it exploits the cache without having the cache size as a parameter: the same code achieves the same asymptotic cache complexity regardless of the cache size Z. An **optimal cache-oblivious** algorithm achieves the **optimal** cache complexity (that is, in an asymptotic sense, ignoring constant factors). Remarkably, optimal cache-oblivious algorithms exist for many problems, such as matrix multiplication, sorting, transposition, and FFTs [137]. Not all cache-oblivious algorithms are optimal, of course—for example, the textbook radix-2 algorithm discussed above is "pessimal" cache-oblivious (its cache complexity is independent of Z because it always achieves the worst case!).

For instance, Figure 11.2(right) and the algorithm of p. **??** shows a way to obliviously exploit the cache with a radix-2 Cooley-Tukey algorithm, by ordering the computation depth-first rather than breadth-first. That is, the DFT of size n is divided into two DFTs of size $n/2$, and one DFT of size $n/2$ is **completely finished** before doing **any** computations for the second DFT of size $n/2$. The two subtransforms are then combined using $n/2$ radix-2 butterflies, which requires a pass over the array and (hence n cache misses if $n > Z$). This process is repeated recursively until a base-case (e.g. size 2) is reached. The cache complexity $Q_2(n;Z)$ of this algorithm satisfies the recurrence

$$Q_2(n;Z) = \{ \begin{array}{cc} n & n \leq Z \\ 2Q_2(n/2;Z) + \Theta(n) & \text{otherwise} \end{array} . \tag{11.4}$$

The key property is this: once the recursion reaches a size $n \leq Z$, the subtransform fits into the cache and no further misses are incurred. The algorithm does not "know" this and continues subdividing the problem, of course, but all of those further subdivisions are in-cache because they are performed in the same depth-first branch of the tree. The solution of (11.4) is

$$Q_2(n;Z) = \Theta(nlog[n/Z]) . \tag{11.5}$$

This is worse than the theoretical optimum $Q_b(n;Z)$ from (11.3), but it is cache-oblivious (Z never entered the algorithm) and exploits at least **some** temporal locality.[8] On the other hand, when it is combined with FFTW's self-optimization and larger radices in "Adaptive Composition of FFT Algorithms" (Section 11.5: Adaptive Composition of FFT Algorithms), this algorithm actually performs very well until n becomes extremely large. By itself, however, the algorithm of p. **??** must be modified to attain adequate performance for reasons that have nothing to do with the cache. These practical issues are discussed further in "Cache-obliviousness in practice" (Section 11.4.2: Cache-obliviousness in practice).

[8]This advantage of depth-first recursive implementation of the radix-2 FFT was pointed out many years ago by Singleton (where the "cache" was core memory) [341].

There exists a different recursive FFT that is **optimal** cache-oblivious, however, and that is the radix-$\sqrt{n}$ "four-step" Cooley-Tukey algorithm (again executed recursively, depth-first) [137]. The cache complexity Q_o of this algorithm satisfies the recurrence:

$$Q_o(n;Z) = \{ \begin{array}{cc} n & n \leq Z \\ 2\sqrt{n} Q_o(\sqrt{n};Z) + \Theta(n) & \text{otherwise} \end{array} . \quad (11.6)$$

That is, at each stage one performs $\sqrt{n}$ DFTs of size $\sqrt{n}$ (recursively), then multiplies by the $\Theta(n)$ twiddle factors (and does a matrix transposition to obtain in-order output), then finally performs another $\sqrt{n}$ DFTs of size $\sqrt{n}$. The solution of (11.6) is $Q_o(n;Z) = \Theta(n log_Z n)$, the same as the optimal cache complexity (11.3)!

These algorithms illustrate the basic features of most optimal cache-oblivious algorithms: they employ a recursive divide-and-conquer strategy to subdivide the problem until it fits into cache, at which point the subdivision continues but no further cache misses are required. Moreover, a cache-oblivious algorithm exploits all levels of the cache in the same way, so an optimal cache-oblivious algorithm exploits a multi-level cache optimally as well as a two-level cache [137]: the multi-level "blocking" is implicit in the recursion.

11.4.2 Cache-obliviousness in practice

Even though the radix-$\sqrt{n}$ algorithm is optimal cache-oblivious, it does not follow that FFT implementation is a solved problem. The optimality is only in an asymptotic sense, ignoring constant factors, $O(n)$ terms, etcetera, all of which can matter a great deal in practice. For small or moderate n, quite different algorithms may be superior, as discussed in "Memory strategies in FFTW" (Section 11.4.3: Memory strategies in FFTW). Moreover, real caches are inferior to an ideal cache in several ways. The unsurprising consequence of all this is that cache-obliviousness, like any complexity-based algorithm property, does not absolve one from the ordinary process of software optimization. At best, it reduces the amount of memory/cache tuning that one needs to perform, structuring the implementation to make further optimization easier and more portable.

Perhaps most importantly, one needs to perform an optimization that has almost nothing to do with the caches: the recursion must be "coarsened" to amortize the function-call overhead and to enable compiler optimization. For example, the simple pedagogical code of the algorithm in p. **??** recurses all the way down to $n = 1$, and hence there are $\approx 2n$ function calls in total, so that every data point incurs a two-function-call overhead on average. Moreover, the compiler cannot fully exploit the large register sets and instruction-level parallelism of modern processors with an $n = 1$ function body.[9] These problems can be effectively erased, however, simply by making the base cases larger, e.g. the recursion could stop when

[9]In principle, it might be possible for a compiler to automatically coarsen the recursion, similar to how compilers can partially unroll loops. We are currently unaware of any general-purpose compiler that performs this optimization, however.

$n = 32$ is reached, at which point a highly optimized hard-coded FFT of that size would be executed. In FFTW, we produced this sort of large base-case using a specialized code-generation program described in "Generating Small FFT Kernels" (Section 11.6: Generating Small FFT Kernels).

One might get the impression that there is a strict dichotomy that divides cache-aware and cache-oblivious algorithms, but the two are not mutually exclusive in practice. Given an implementation of a cache-oblivious strategy, one can further optimize it for the cache characteristics of a particular machine in order to improve the constant factors. For example, one can tune the radices used, the transition point between the radix-$\sqrt{n}$ algorithm and the bounded-radix algorithm, or other algorithmic choices as described in "Memory strategies in FFTW" (Section 11.4.3: Memory strategies in FFTW). The advantage of starting cache-aware tuning with a cache-oblivious approach is that the starting point already exploits all levels of the cache to some extent, and one has reason to hope that good performance on one machine will be more portable to other architectures than for a purely cache-aware "blocking" approach. In practice, we have found this combination to be very successful with FFTW.

11.4.3 Memory strategies in FFTW

The recursive cache-oblivious strategies described above form a useful starting point, but FFTW supplements them with a number of additional tricks, and also exploits cache-obliviousness in less-obvious forms.

We currently find that the general radix-$\sqrt{n}$ algorithm is beneficial only when n becomes very large, on the order of $2^{20} \approx 10^6$. In practice, this means that we use at most a single step of radix-$\sqrt{n}$ (two steps would only be used for $n \gtrsim 2^{40}$). The reason for this is that the implementation of radix $\sqrt{n}$ is less efficient than for a bounded radix: the latter has the advantage that an entire radix butterfly can be performed in hard-coded loop-free code within local variables/registers, including the necessary permutations and twiddle factors.

Thus, for more moderate n, FFTW uses depth-first recursion with a bounded radix, similar in spirit to the algorithm of p. **??** but with much larger radices (radix 32 is common) and base cases (size 32 or 64 is common) as produced by the code generator of "Generating Small FFT Kernels" (Section 11.6: Generating Small FFT Kernels). The self-optimization described in "Adaptive Composition of FFT Algorithms" (Section 11.5: Adaptive Composition of FFT Algorithms) allows the choice of radix and the transition to the radix-$\sqrt{n}$ algorithm to be tuned in a cache-aware (but entirely automatic) fashion.

For small n (including the radix butterflies and the base cases of the recursion), hard-coded FFTs (FFTW's **codelets**) are employed. However, this gives rise to an interesting problem: a codelet for (e.g.) $n = 64$ is ~ 2000 lines long, with hundreds of variables and over 1000 arithmetic operations that can be executed in many orders, so what order should be chosen? The key problem here is the efficient use of the CPU registers, which essentially form a nearly ideal, fully associative cache. Normally, one relies on the compiler for all code scheduling and register allocation, but but the compiler needs help with such long blocks of code (indeed, the general register-allocation problem is NP-complete). In particular, FFTW's generator knows more about the code than the compiler—the generator knows it is an FFT, and therefore it can use an optimal cache-oblivious schedule (analogous to the radix-$\sqrt{n}$ algorithm) to order the code independent of the number of registers [128]. The compiler is then used only for local "cache-aware" tuning (both for register allocation and the CPU pipeline).[10] As a practical matter, one consequence of this scheduler is that FFTW's machine-independent codelets are no slower than machine-specific codelets generated by an automated search and optimization over many possible codelet implementations, as performed by the SPIRAL project [420].

(When implementing hard-coded base cases, there is another choice because a loop of small transforms is always required. Is it better to implement a hard-coded FFT of size 64, for example, or an unrolled loop of four size-16 FFTs, both of which operate on the same amount of data? The former should be more efficient because it performs more computations with the same amount of data, thanks to the $logn$ factor in the FFT's $nlogn$ complexity.)

In addition, there are many other techniques that FFTW employs to supplement the basic recursive strategy, mainly to address the fact that cache implementations strongly favor accessing consecutive data—thanks to cache lines, limited associativity, and direct mapping using low-order address bits (accessing data at power-of-two intervals in memory, which is distressingly common in FFTs, is thus especially prone to cache-line conflicts). Unfortunately, the known FFT algorithms inherently involve some non-consecutive access (whether mixed with the computation or in separate bit-reversal/transposition stages). There are many optimizations in FFTW to address this. For example, the data for several butterflies at a time can be copied to a small buffer before computing and then copied back, where the copies and computations involve more consecutive access than doing the computation directly in-place. Or, the input data for the subtransform can be copied from (discontiguous) input to (contiguous) output before performing the subtransform in-place (see "Indirect plans" (Section 11.5.2.4: Indirect plans)), rather than performing the subtransform directly out-of-place (as in algorithm 1 (p. ??)). Or, the order of loops can be interchanged in order to push the outermost loop from the first radix step [the ℓ_2 loop in (11.2)] down to the leaves, in order to make the input access more consecutive (see "Discussion" (Section 11.5.2.6: Discussion)). Or, the twiddle factors can be computed using a smaller look-up table (fewer memory loads) at the cost of more arithmetic (see "Numerical Accuracy in FFTs" (Section 11.7: Numerical Accuracy in FFTs)). The choice of whether to use any of these techniques, which come into play mainly for moderate n ($2^{13} < n < 2^{20}$), is made by the self-optimizing planner as described in the next section.

[10]One practical difficulty is that some "optimizing" compilers will tend to greatly re-order the code, destroying FFTW's optimal schedule. With GNU gcc, we circumvent this problem by using compiler flags that explicitly disable certain stages of the optimizer.

11.5 Adaptive Composition of FFT Algorithms

As alluded to several times already, FFTW implements a wide variety of FFT algorithms (mostly rearrangements of Cooley-Tukey) and selects the "best" algorithm for a given n automatically. In this section, we describe how such self-optimization is implemented, and especially how FFTW's algorithms are structured as a composition of algorithmic fragments. These techniques in FFTW are described in greater detail elsewhere [134], so here we will focus only on the essential ideas and the motivations behind them.

An FFT algorithm in FFTW is a composition of algorithmic steps called a **plan**. The algorithmic steps each solve a certain class of **problems** (either solving the problem directly or recursively breaking it into sub-problems of the same type). The choice of plan for a given problem is determined by a **planner** that selects a composition of steps, either by runtime measurements to pick the fastest algorithm, or by heuristics, or by loading a pre-computed plan. These three pieces: problems, algorithmic steps, and the planner, are discussed in the following subsections.

11.5.1 The problem to be solved

In early versions of FFTW, the only choice made by the planner was the sequence of radices [131], and so each step of the plan took a DFT of a given size n, possibly with discontiguous input/output, and reduced it (via a radix r) to DFTs of size n/r, which were solved recursively. That is, each step solved the following problem: given a size n, an **input pointer I**, an **input stride** ι, an **output pointer O**, and an **output stride** o, it computed the DFT of $\mathbf{I}[\ell\iota]$ for $0 \leq \ell < n$ and stored the result in $\mathbf{O}[ko]$ for $0 \leq k < n$. However, we soon found that we could not easily express many interesting algorithms within this framework; for example, **in-place** ($\mathbf{I} = \mathbf{O}$) FFTs that do not require a separate bit-reversal stage [195], [375], [297], [166]. It became clear that the key issue was not the choice of algorithms, as we had first supposed, but the definition of the problem to be solved. Because only problems that can be expressed can be solved, the representation of a problem determines an outer bound to the space of plans that the planner can explore, and therefore it ultimately constrains FFTW's performance.

The difficulty with our initial $(n, \mathbf{I}, \iota, \mathbf{O}, o)$ problem definition was that it forced each algorithmic step to address only a single DFT. In fact, FFTs break down DFTs into **multiple** smaller DFTs, and it is the **combination** of these smaller transforms that is best addressed by many algorithmic choices, especially to rearrange the order of memory accesses between the subtransforms. Therefore, we redefined our notion of a problem in FFTW to be not a single DFT, but rather a **loop** of DFTs, and in fact **multiple nested loops** of DFTs. The following sections describe some of the new algorithmic steps that such a problem definition enables, but first we will define the problem more precisely.

DFT problems in FFTW are expressed in terms of structures called I/O tensors,[11] which in turn are described in terms of ancillary structures called I/O dimensions. An **I/O dimension** d is a triple $d = (n, \iota, o)$, where n is a non-negative integer called the **length**, ι is an integer called the **input stride**, and o is an integer called the **output stride**. An **I/O tensor** $t = \{d_1, d_2, ..., d_\rho\}$ is a set of I/O dimensions. The non-negative integer $\rho = |t|$ is called the **rank** of the I/O tensor. A **DFT problem**, denoted by $dft\,(\mathbf{N}, \mathbf{V}, \mathbf{I}, \mathbf{O})$, consists of two I/O tensors $\mathbf{N}$ and $\mathbf{V}$, and of two **pointers** $\mathbf{I}$ and $\mathbf{O}$. Informally, this describes $|\mathbf{V}|$ nested loops of $|\mathbf{N}|$-dimensional DFTs with input data starting at memory location $\mathbf{I}$ and output data starting at $\mathbf{O}$.

For simplicity, let us consider only one-dimensional DFTs, so that $\mathbf{N} = \{(n, \iota, o)\}$ implies a *DFT* of length n on input data with stride ι and output data with stride o, much like in the original FFTW as described above. The main new feature is then the addition of zero or more "loops" $\mathbf{V}$. More formally, $dft\,(\mathbf{N}, \{(n, \iota, o)\} \cup \mathbf{V}, \mathbf{I}, \mathbf{O})$ is recursively defined as a "loop" of n problems: for all $0 \leq k < n$, do all computations in $dft\,(\mathbf{N}, \mathbf{V}, \mathbf{I} + k \cdot \iota, \mathbf{O} + k \cdot o)$. The case of multi-dimensional DFTs is defined more precisely elsewhere [134], but essentially each I/O dimension in $\mathbf{N}$ gives one dimension of the transform.

We call $\mathbf{N}$ the **size** of the problem. The **rank** of a problem is defined to be the rank of its size (i.e., the dimensionality of the DFT). Similarly, we call $\mathbf{V}$ the **vector size** of the problem, and the **vector rank** of a problem is correspondingly defined to be the rank of its vector size. Intuitively, the vector size can be interpreted as a set of "loops" wrapped around a single DFT, and we therefore refer to a single I/O dimension of $\mathbf{V}$ as a **vector loop**. (Alternatively, one can view the problem as describing a DFT over a $|\mathbf{V}|$-dimensional vector space.) The problem does not specify the order of execution of these loops, however, and therefore FFTW is free to choose the fastest or most convenient order.

11.5.1.1 DFT problem examples

A more detailed discussion of the space of problems in FFTW can be found in [134] , but a simple understanding can be gained by examining a few examples demonstrating that the I/O tensor representation is sufficiently general to cover many situations that arise in practice, including some that are not usually considered to be instances of the DFT.

A single one-dimensional DFT of length n, with stride-1 input $\mathbf{X}$ and output $\mathbf{Y}$, as in (11.1), is denoted by the problem $dft\,(\{(n, 1, 1)\}, \{\}, \mathbf{X}, \mathbf{Y})$ (no loops: vector-rank zero).

As a more complicated example, suppose we have an $n_1 \times n_2$ matrix $\mathbf{X}$ stored as n_1 consecutive blocks of contiguous length-n_2 rows (this is called **row-major** format). The in-place DFT of all the **rows** of this matrix would be denoted by the problem $dft\,(\{(n_2, 1, 1)\}, \{(n_1, n_2, n_2)\}, \mathbf{X}, \mathbf{X})$: a length-$n_1$ loop of size-n_2 contiguous DFTs,

[11] I/O tensors are unrelated to the tensor-product notation used by some other authors to describe FFT algorithms [389], [296].

where each iteration of the loop offsets its input/output data by a stride n_2. Conversely, the in-place DFT of all the **columns** of this matrix would be denoted by $dft(\{(n_1, n_2, n_2)\}, \{(n_2, 1, 1)\}, \mathbf{X}, \mathbf{X})$: compared to the previous example, **N** and **V** are swapped. In the latter case, each DFT operates on discontiguous data, and FFTW might well choose to interchange the loops: instead of performing a loop of DFTs computed individually, the subtransforms themselves could act on n_2-component vectors, as described in "The space of plans in FFTW" (Section 11.5.2: The space of plans in FFTW).

A size-1 DFT is simply a copy $Y[0] = X[0]$, and here this can also be denoted by $\mathbf{N} = \{\}$ (rank zero, a "zero-dimensional" DFT). This allows FFTW's problems to represent many kinds of copies and permutations of the data within the same problem framework, which is convenient because these sorts of operations arise frequently in FFT algorithms. For example, to copy n consecutive numbers from **I** to **O**, one would use the rank-zero problem $dft(\{\}, \{(n, 1, 1)\}, \mathbf{I}, \mathbf{O})$. More interestingly, the in-place **transpose** of an $n_1 \times n_2$ matrix **X** stored in row-major format, as described above, is denoted by $dft(\{\}, \{(n_1, n_2, 1), (n_2, 1, n_1)\}, \mathbf{X}, \mathbf{X})$ (rank zero, vector-rank two).

11.5.2 The space of plans in FFTW

Here, we describe a subset of the possible plans considered by FFTW; while not exhaustive [134], this subset is enough to illustrate the basic structure of FFTW and the necessity of including the vector loop(s) in the problem definition to enable several interesting algorithms. The plans that we now describe usually perform some simple "atomic" operation, and it may not be apparent how these operations fit together to actually compute DFTs, or why certain operations are useful at all. We shall discuss those matters in "Discussion" (Section 11.5.2.6: Discussion).

Roughly speaking, to solve a general DFT problem, one must perform three tasks. First, one must reduce a problem of arbitrary vector rank to a set of loops nested around a problem of vector rank 0, i.e., a single (possibly multi-dimensional) DFT. Second, one must reduce the multi-dimensional DFT to a sequence of of rank-1 problems, i.e., one-dimensional DFTs; for simplicity, however, we do not consider multi-dimensional DFTs below. Third, one must solve the rank-1, vector rank-0 problem by means of some DFT algorithm such as Cooley-Tukey. These three steps need not be executed in the stated order, however, and in fact, almost every permutation and interleaving of these three steps leads to a correct DFT plan. The choice of the set of plans explored by the planner is critical for the usability of the FFTW system: the set must be large enough to contain the fastest possible plans, but it must be small enough to keep the planning time acceptable.

11.5.2.1 Rank-0 plans

The rank-0 problem $dft(\{\}, \mathbf{V}, \mathbf{I}, \mathbf{O})$ denotes a permutation of the input array into the output array. FFTW does not solve arbitrary rank-0 problems, only the following two special cases that arise in practice.

- When $|\mathbf{V}| = 1$ and $\mathbf{I} \neq \mathbf{O}$, FFTW produces a plan that copies the input array into the output array. Depending on the strides, the plan consists of a loop or, possibly, of a call to the ANSI C function `memcpy`, which is specialized to copy contiguous regions of memory.
- When $|\mathbf{V}| = 2$, $\mathbf{I} = \mathbf{O}$, and the strides denote a matrix-transposition problem, FFTW creates a plan that transposes the array in-place. FFTW implements the square transposition $dft(\{\}, \{(n, \iota, o), (n, o, \iota)\}, \mathbf{I}, \mathbf{O})$ by means of the cache-oblivious algorithm from [137], which is fast and, in theory, uses the cache optimally regardless of the cache size (using principles similar to those described in the section "FFTs and the Memory Hierarchy" (Section 11.4: FFTs and the Memory Hierarchy)). A generalization of this idea is employed for non-square transpositions with a large common factor or a small difference between the dimensions, adapting algorithms from [100].

11.5.2.2 Rank-1 plans

Rank-1 DFT problems denote ordinary one-dimensional Fourier transforms. FFTW deals with most rank-1 problems as follows.

11.5.2.2.1 Direct plans

When the DFT rank-1 problem is "small enough" (usually, $n \leq 64$), FFTW produces a **direct plan** that solves the problem directly. These plans operate by calling a fragment of C code (a **codelet**) specialized to solve problems of one particular size, whose generation is described in "Generating Small FFT Kernels" (Section 11.6: Generating Small FFT Kernels). More precisely, the codelets compute a loop ($|\mathbf{V}| \leq 1$) of small DFTs.

11.5.2.2.2 Cooley-Tukey plans

For problems of the form $dft(\{(n, \iota, o)\}, \mathbf{V}, \mathbf{I}, \mathbf{O})$ where $n = rm$, FFTW generates a plan that implements a radix-r Cooley-Tukey algorithm "Review of the Cooley-Tukey FFT" (Section 11.2: Review of the Cooley-Tukey FFT). Both decimation-in-time and decimation-in-frequency plans are supported, with both small fixed radices (usually, $r \leq 64$) produced by the codelet generator "Generating Small FFT Kernels" (Section 11.6: Generating Small FFT Kernels) and also arbitrary radices (e.g. radix-$\sqrt{n}$).

The most common case is a **decimation in time** (**DIT**) plan, corresponding to a **radix** $r = n_2$ (and thus $m = n_1$) in the notation of "Review of the Cooley-Tukey FFT" (Section 11.2: Review of the Cooley-Tukey FFT): it first solves $dft(\{(m, r \cdot \iota, o)\}, \mathbf{V} \cup \{(r, \iota, m \cdot o)\}, \mathbf{I}, \mathbf{O})$, then multiplies the output array $\mathbf{O}$ by the twiddle factors, and finally solves $dft(\{(r, m \cdot o, m \cdot o)\}, \mathbf{V} \cup \{(m, o, o)\}, \mathbf{O}, \mathbf{O})$. For performance, the last two steps are not planned independently, but are fused together in a single "twiddle" codelet—a fragment of C code that multiplies its input by the twiddle factors and performs a DFT of size r, operating in-place on $\mathbf{O}$.

11.5.2.3 Plans for higher vector ranks

These plans extract a vector loop to reduce a DFT problem to a problem of lower vector rank, which is then solved recursively. Any of the vector loops of $\mathbf{V}$ could be extracted in this way, leading to a number of possible plans corresponding to different loop orderings.

Formally, to solve $dft(\mathbf{N}, \mathbf{V}, \mathbf{I}, \mathbf{O})$, where $\mathbf{V} = \{(n, \iota, o)\} \cup \mathbf{V}_1$, FFTW generates a loop that, for all k such that $0 \leq k < n$, invokes a plan for $dft(\mathbf{N}, \mathbf{V}_1, \mathbf{I} + k \cdot \iota, \mathbf{O} + k \cdot o)$.

11.5.2.4 Indirect plans

Indirect plans transform a DFT problem that requires some data shuffling (or discontiguous operation) into a problem that requires no shuffling plus a rank-0 problem that performs the shuffling.

Formally, to solve $dft(\mathbf{N}, \mathbf{V}, \mathbf{I}, \mathbf{O})$ where $|\mathbf{N}| > 0$, FFTW generates a plan that first solves $dft(\{\}, \mathbf{N} \cup \mathbf{V}, \mathbf{I}, \mathbf{O})$, and then solves $dft(copy - o(\mathbf{N}), copy - o(\mathbf{V}), \mathbf{O}, \mathbf{O})$. Here we define $copy - o(t)$ to be the I/O tensor $\{(n, o, o) \mid (n, \iota, o) \in t\}$: that is, it replaces the input strides with the output strides. Thus, an indirect plan first rearranges/copies the data to the output, then solves the problem in place.

11.5.2.5 Plans for prime sizes

As discussed in "Goals and Background of the FFTW Project" (Section 11.3: Goals and Background of the FFTW Project), it turns out to be surprisingly useful to be able to handle large prime n (or large prime factors). **Rader plans** implement the algorithm from [309] to compute one-dimensional DFTs of prime size in $\Theta(nlogn)$ time. **Bluestein plans** implement Bluestein's "chirp-z" algorithm, which can also handle prime n in $\Theta(nlogn)$ time [35], [305], [278]. **Generic plans** implement a naive $\Theta(n^2)$ algorithm (useful for $n \lesssim 100$).

11.5.2.6 Discussion

Although it may not be immediately apparent, the combination of the recursive rules in "The space of plans in FFTW" (Section 11.5.2: The space of plans in FFTW) can produce a number of useful algorithms. To illustrate these compositions, we discuss three particular issues: depth- vs. breadth-first, loop reordering, and in-place transforms.

---

size-30 DFT, depth-first:
{ loop 3
{ { size-5 direct codelet, vector size 2
{ { size-2 twiddle codelet, vector size 5
size-3 twiddle codelet, vector size 10

size-30 DFT, breadth-first:
{ { loop 3
{ { size-5 direct codelet, vector size 2
{ { loop 3
{ { size-2 twiddle codelet, vector size 5
size-3 twiddle codelet, vector size 10

Figure 11.3: Two possible decompositions for a size-30 DFT, both for the arbitrary choice of DIT radices 3 then 2 then 5, and prime-size codelets. Items grouped by a "{" result from the plan for a single sub-problem. In the depth-first case, the vector rank was reduced to zero as per "Plans for higher vector ranks" (Section 11.5.2.3: Plans for higher vector ranks) before decomposing sub-problems, and vice-versa in the breadth-first case.

As discussed previously in sections "Review of the Cooley-Tukey FFT" (Section 11.2: Review of the Cooley-Tukey FFT) and "Understanding FFTs with an ideal cache" (Section 11.4.1: Understanding FFTs with an ideal cache), the same Cooley-Tukey decomposition can be executed in either traditional breadth-first order or in recursive depth-first order, where the latter has some theoretical cache advantages. FFTW is explicitly recursive, and thus it can naturally employ a depth-first order. Because its sub-problems contain a vector loop that can be executed in a variety of orders, however, FFTW can also employ breadth-first traversal. In particular, a 1d algorithm resembling the traditional breadth-first Cooley-Tukey would result from applying "Cooley-Tukey plans" (Section 11.5.2.2.2: Cooley-Tukey plans) to completely factorize the problem size before applying the loop rule "Plans for higher vector ranks" (Section 11.5.2.3: Plans for higher vector ranks) to reduce the vector ranks, whereas depth-first traversal would result from applying the loop rule before factorizing each subtransform. These two possibilities are illustrated by an example in Figure 11.3.

Another example of the effect of loop reordering is a style of plan that we sometimes call **vector recursion** (unrelated to "vector-radix" FFTs [114]). The basic idea is that, if one has a loop (vector-rank 1) of transforms, where the vector stride is smaller than the transform size, it is advantageous to push the loop towards the leaves of the transform decomposition, while otherwise maintaining recursive depth-first ordering, rather than looping "outside" the transform; i.e., apply the usual FFT to "vectors" rather than numbers. Limited forms of this idea have appeared for computing multiple FFTs on vector processors (where the loop in question maps directly to a hardware vector) [372]. For example, Cooley-Tukey produces a unit **input**-stride vector loop at the top-level DIT decomposition, but with a large **output** stride; this difference in strides makes it non-obvious whether vector recursion is advantageous for the sub-problem, but for large transforms we often observe the planner to choose this possibility.

In-place 1d transforms (with no separate bit reversal pass) can be obtained as follows by a combination DIT and DIF plans "Cooley-Tukey plans" (Section 11.5.2.2.2: Cooley-Tukey plans) with transposes "Rank-0 plans" (Section 11.5.2.1: Rank-0 plans). First, the transform is decomposed via a radix-p DIT plan into a vector of p transforms of size qm, then these are decomposed in turn by a radix-q DIF plan into a vector (rank 2) of $p \times q$ transforms of size m. These transforms of size m have input and output at different places/strides in the original array, and so cannot be solved independently. Instead, an indirect plan "Indirect plans" (Section 11.5.2.4: Indirect plans) is used to express the sub-problem as pq in-place transforms of size m, followed or preceded by an $m \times p \times q$ rank-0 transform. The latter sub-problem is easily seen to be m in-place $p \times q$ transposes (ideally square, i.e. $p = q$). Related strategies for in-place transforms based on small transposes were described in [195], [375], [297], [166]; alternating DIT/DIF, without concern for in-place operation, was also considered in [255], [322].

11.5.3 The FFTW planner

Given a problem and a set of possible plans, the basic principle behind the FFTW planner is straightforward: construct a plan for each applicable algorithmic step, time the execution of these plans, and select the fastest one. Each algorithmic step may break the problem into subproblems, and the fastest plan for each subproblem is constructed in the same way. These timing measurements can either be performed at runtime, or alternatively the plans for a given set of sizes can be precomputed and loaded at a later time.

A direct implementation of this approach, however, faces an exponential explosion of the number of possible plans, and hence of the planning time, as n increases. In order to reduce the planning time to a manageable level, we employ several heuristics to reduce the space of possible plans that must be compared. The most important of these heuristics is **dynamic programming** [96]: it optimizes each sub-problem locally, independently of the larger context (so that the "best" plan for a given sub-problem is re-used whenever that sub-problem is encountered). Dynamic programming is not guaranteed to find the fastest plan, because the performance of plans is context-dependent on real machines (e.g., the contents of the cache depend on the preceding computations); however, this approximation works reasonably well in practice and greatly reduces the planning time. Other approximations, such as restrictions on the types of loop-reorderings that are considered "Plans for higher vector ranks" (Section 11.5.2.3: Plans for higher vector ranks), are described in [134].

Alternatively, there is an **estimate mode** that performs no timing measurements whatsoever, but instead minimizes a heuristic cost function. This can reduce the planner time by several orders of magnitude, but with a significant penalty observed in plan efficiency; e.g., a penalty of 20% is typical for moderate $n \lesssim 2^{13}$, whereas a factor of 2–3 can be suffered for large $n \gtrsim 2^{16}$ [134]. Coming up with a better heuristic plan is an interesting open research question; one difficulty is that, because FFT algorithms depend on factorization, knowing a good plan for n does not immediately help one find a good plan for nearby n.

11.6 Generating Small FFT Kernels

The base cases of FFTW's recursive plans are its **codelets**, and these form a critical component of FFTW's performance. They consist of long blocks of highly optimized, straight-line code, implementing many special cases of the DFT that give the planner a large space of plans in which to optimize. Not only was it impractical to write numerous codelets by hand, but we also needed to rewrite them many times in order to explore different algorithms and optimizations. Thus, we designed a special-purpose "FFT compiler" called **genfft** that produces the codelets automatically from an abstract description. genfft is summarized in this section and described in more detail by [128].

A typical codelet in FFTW computes a DFT of a small, fixed size n (usually, $n \leq 64$), possibly with the input or output multiplied by twiddle factors "Cooley-Tukey plans" (Section 11.5.2.2.2: Cooley-Tukey plans). Several other kinds of codelets can be produced by genfft , but we will focus here on this common case.

In principle, all codelets implement some combination of the Cooley-Tukey algorithm from (11.2) and/or some other DFT algorithm expressed by a similarly compact formula. However, a high-performance implementation of the DFT must address many more concerns than (11.2) alone suggests. For example, (11.2) contains multiplications by 1 that are more efficient to omit. (11.2) entails a run-time factorization of n, which can be precomputed if n is known in advance. (11.2) operates on complex numbers, but breaking the complex-number abstraction into real and imaginary components turns out to expose certain non-obvious optimizations. Additionally, to exploit the long pipelines in current processors, the recursion implicit in (11.2) should be unrolled and re-ordered to a significant degree. Many further optimizations are possible if the complex input is known in advance to be purely real (or imaginary). Our design goal for genfft was to keep the expression of the DFT algorithm independent of such concerns. This separation allowed us to experiment with various DFT algorithms and implementation strategies independently and without (much) tedious rewriting.

genfft is structured as a compiler whose input consists of the kind and size of the desired codelet, and whose output is C code. genfft operates in four phases: creation, simplification, scheduling, and unparsing.

In the **creation** phase, genfft produces a representation of the codelet in the form of a directed acyclic graph (**dag**). The dag is produced according to well-known DFT algorithms: Cooley-Tukey (11.2), prime-factor [278], split-radix [422], [107], [391], [230], [114], and Rader

[309]. Each algorithm is expressed in a straightforward math-like notation, using complex numbers, with no attempt at optimization. Unlike a normal FFT implementation, however, the algorithms here are evaluated symbolically and the resulting symbolic expression is represented as a dag, and in particular it can be viewed as a **linear network** [98] (in which the edges represent multiplication by constants and the vertices represent additions of the incoming edges).

In the **simplification** phase, genfft applies local rewriting rules to each node of the dag in order to simplify it. This phase performs algebraic transformations (such as eliminating multiplications by 1) and common-subexpression elimination. Although such transformations can be performed by a conventional compiler to some degree, they can be carried out here to a greater extent because genfft can exploit the specific problem domain. For example, two equivalent subexpressions can always be detected, even if the subexpressions are written in algebraically different forms, because all subexpressions compute linear functions. Also, genfft can exploit the property that **network transposition** (reversing the direction of every edge) computes the transposed linear operation [98], in order to transpose the network, simplify, and then transpose back—this turns out to expose additional common subexpressions [128]. In total, these simplifications are sufficiently powerful to derive DFT algorithms specialized for real and/or symmetric data automatically from the complex algorithms. For example, it is known that when the input of a DFT is real (and the output is hence conjugate-symmetric), one can save a little over a factor of two in arithmetic cost by specializing FFT algorithms for this case—with genfft , this specialization can be done entirely automatically, pruning the redundant operations from the dag, to match the lowest known operation count for a real-input FFT starting only from the complex-data algorithm [128], [202]. We take advantage of this property to help us implement real-data DFTs [128], [134], to exploit machine-specific "SIMD" instructions "SIMD instructions" (Section 11.6.1: SIMD instructions) [134], and to generate codelets for the discrete cosine (DCT) and sine (DST) transforms [128], [202]. Furthermore, by experimentation we have discovered additional simplifications that improve the speed of the generated code. One interesting example is the elimination of negative constants [128]: multiplicative constants in FFT algorithms often come in positive/negative pairs, but every C compiler we are aware of will generate separate load instructions for positive and negative versions of the same constants.[12] We thus obtained a 10–15% speedup by making all constants positive, which involves propagating minus signs to change additions into subtractions or vice versa elsewhere in the dag (a daunting task if it had to be done manually for tens of thousands of lines of code).

In the **scheduling** phase, genfft produces a topological sort of the dag (a **schedule**). The goal of this phase is to find a schedule such that a C compiler can subsequently perform a good register allocation. The scheduling algorithm used by genfft offers certain theoretical guarantees because it has its foundations in the theory of cache-oblivious algorithms [137] (here, the registers are viewed as a form of cache), as described in "Memory strategies in

[12]Floating-point constants must be stored explicitly in memory; they cannot be embedded directly into the CPU instructions like integer "immediate" constants.

FFTW" (Section 11.4.3: Memory strategies in FFTW). As a practical matter, one consequence of this scheduler is that FFTW's machine-independent codelets are no slower than machine-specific codelets generated by SPIRAL [420].

In the stock genfft implementation, the schedule is finally unparsed to C. A variation from [127] implements the rest of a compiler back end and outputs assembly code.

11.6.1 SIMD instructions

Unfortunately, it is impossible to attain nearly peak performance on current popular processors while using only portable C code. Instead, a significant portion of the available computing power can only be accessed by using specialized SIMD (single-instruction multiple data) instructions, which perform the same operation in parallel on a data vector. For example, all modern "x86" processors can execute arithmetic instructions on "vectors" of four single-precision values (SSE instructions) or two double-precision values (SSE2 instructions) at a time, assuming that the operands are arranged consecutively in memory and satisfy a 16-byte alignment constraint. Fortunately, because nearly all of FFTW's low-level code is produced by genfft , machine-specific instructions could be exploited by modifying the generator—the improvements are then automatically propagated to all of FFTW's codelets, and in particular are not limited to a small set of sizes such as powers of two.

SIMD instructions are superficially similar to "vector processors", which are designed to perform the same operation in parallel on an all elements of a data array (a "vector"). The performance of "traditional" vector processors was best for long vectors that are stored in contiguous memory locations, and special algorithms were developed to implement the DFT efficiently on this kind of hardware [372], [166]. Unlike in vector processors, however, the SIMD vector length is small and fixed (usually 2 or 4). Because microprocessors depend on caches for performance, one cannot naively use SIMD instructions to simulate a long-vector algorithm: while on vector machines long vectors generally yield better performance, the performance of a microprocessor drops as soon as the data vectors exceed the capacity of the cache. Consequently, SIMD instructions are better seen as a restricted form of instruction-level parallelism than as a degenerate flavor of vector parallelism, and different DFT algorithms are required.

The technique used to exploit SIMD instructions in genfft is most easily understood for vectors of length two (e.g., SSE2). In this case, we view a **complex** DFT as a pair of **real** DFTs:

$$\mathrm{DFT}\,(A + i \cdot B) = \mathrm{DFT}\,(A) + i \cdot \mathrm{DFT}\,(B)\ , \tag{11.7}$$

where A and B are two real arrays. Our algorithm computes the two real DFTs in parallel using SIMD instructions, and then it combines the two outputs according to (11.7). This SIMD algorithm has two important properties. First, if the data is stored as an array of complex numbers, as opposed to two separate real and imaginary arrays, the SIMD loads and stores always operate on correctly-aligned contiguous locations, even if the the complex numbers themselves have a non-unit stride. Second, because the algorithm finds two-way parallelism in the real and imaginary parts of a single DFT (as opposed to performing two DFTs in parallel), we can completely parallelize DFTs of any size, not just even sizes or powers of 2.

11.7 Numerical Accuracy in FFTs

An important consideration in the implementation of any practical numerical algorithm is numerical accuracy: how quickly do floating-point roundoff errors accumulate in the course of the computation? Fortunately, FFT algorithms for the most part have remarkably good accuracy characteristics. In particular, for a DFT of length n computed by a Cooley-Tukey algorithm with finite-precision floating-point arithmetic, the **worst-case** error growth is $O(logn)$ [139], [373] and the mean error growth for random inputs is only $O\left(\sqrt{logn}\right)$ [326], [373]. This is so good that, in practical applications, a properly implemented FFT will rarely be a significant contributor to the numerical error.

The amazingly small roundoff errors of FFT algorithms are sometimes explained incorrectly as simply a consequence of the reduced number of operations: since there are fewer operations compared to a naive $O(n^2)$ algorithm, the argument goes, there is less accumulation of roundoff error. The real reason, however, is more subtle than that, and has to do with the **ordering** of the operations rather than their number. For example, consider the computation of only the output $Y[0]$ in the radix-2 algorithm of p. ??, ignoring all of the other outputs of the FFT. $Y[0]$ is the sum of all of the inputs, requiring $n-1$ additions. The FFT does not change this requirement, it merely changes the order of the additions so as to re-use some of them for other outputs. In particular, this radix-2 DIT FFT computes $Y[0]$ as follows: it first sums the even-indexed inputs, then sums the odd-indexed inputs, then adds the two sums; the even- and odd-indexed inputs are summed recursively by the same procedure. This process is sometimes called **cascade summation**, and even though it still requires $n-1$ total additions to compute $Y[0]$ by itself, its roundoff error grows much more slowly than simply adding $X[0]$, $X[1]$, $X[2]$ and so on in sequence. Specifically, the roundoff error when adding up n floating-point numbers in sequence grows as $O(n)$ in the worst case, or as $O(\sqrt{n})$ on average for random inputs (where the errors grow according to a random walk), but simply reordering these n-1 additions into a cascade summation yields $O(\log n)$ worst-case and $O\left(\sqrt{\log n}\right)$ average-case error growth [182].

However, these encouraging error-growth rates **only** apply if the trigonometric "twiddle" factors in the FFT algorithm are computed very accurately. Many FFT implementations, including FFTW and common manufacturer-optimized libraries, therefore use precomputed tables of twiddle factors calculated by means of standard library functions (which compute trigonometric constants to roughly machine precision). The other common method to compute twiddle factors is to use a trigonometric recurrence formula—this saves memory (and cache), but almost all recurrences have errors that grow as $O(\sqrt{n})$, $O(n)$, or even $O(n^2)$

[374], which lead to corresponding errors in the FFT. For example, one simple recurrence is $e^{i(k+1)\theta} = e^{ik\theta}e^{i\theta}$, multiplying repeatedly by $e^{i\theta}$ to obtain a sequence of equally spaced angles, but the errors when using this process grow as $O(n)$ [374]. A common improved recurrence is $e^{i(k+1)\theta} = e^{ik\theta} + e^{ik\theta}\left(e^{i\theta} - 1\right)$, where the small quantity[13] $e^{i\theta} - 1 = cos(\theta) - 1 + isin(\theta)$ is computed using $cos(\theta) - 1 = -2sin^2(\theta/2)$ [341]; unfortunately, the error using this method still grows as $O(\sqrt{n})$ [374], far worse than logarithmic.

There are, in fact, trigonometric recurrences with the same logarithmic error growth as the FFT, but these seem more difficult to implement efficiently; they require that a table of $\Theta(logn)$ values be stored and updated as the recurrence progresses [42], [374]. Instead, in order to gain at least some of the benefits of a trigonometric recurrence (reduced memory pressure at the expense of more arithmetic), FFTW includes several ways to compute a much smaller twiddle table, from which the desired entries can be computed accurately on the fly using a bounded number (usually < 3) of complex multiplications. For example, instead of a twiddle table with n entries ω_n^k, FFTW can use two tables with $\Theta(\sqrt{n})$ entries each, so that ω_n^k is computed by multiplying an entry in one table (indexed with the low-order bits of k) by an entry in the other table (indexed with the high-order bits of k).

There are a few non-Cooley-Tukey algorithms that are known to have worse error characteristics, such as the "real-factor" algorithm [313], [114], but these are rarely used in practice (and are not used at all in FFTW). On the other hand, some commonly used algorithms for type-I and type-IV discrete cosine transforms [372], [290], [73] have errors that we observed to grow as $\sqrt{n}$ even for accurate trigonometric constants (although we are not aware of any theoretical error analysis of these algorithms), and thus we were forced to use alternative algorithms [134].

To measure the accuracy of FFTW, we compare against a slow FFT implemented in arbitrary-precision arithmetic, while to verify the correctness we have found the $O(nlogn)$ self-test algorithm of [122] very useful.

11.8 Concluding Remarks

It is unlikely that many readers of this chapter will ever have to implement their own fast Fourier transform software, except as a learning exercise. The computation of the DFT, much like basic linear algebra or integration of ordinary differential equations, is so central to numerical computing and so well-established that robust, flexible, highly optimized libraries are widely available, for the most part as free/open-source software. And yet there are many other problems for which the algorithms are not so finalized, or for which algorithms are published but the implementations are unavailable or of poor quality. Whatever new problems one comes across, there is a good chance that the chasm between theory and efficient implementation will be just as large as it is for FFTs, unless computers become

[13]In an FFT, the twiddle factors are powers of ω_n, so θ is a small angle proportional to $1/n$ and $e^{i\theta}$ is close to 1.

much simpler in the future. For readers who encounter such a problem, we hope that these lessons from FFTW will be useful:

- Generality and portability should almost always come first.
- The number of operations, up to a constant factor, is less important than the order of operations.
- Recursive algorithms with large base cases make optimization easier.
- Optimization, like any tedious task, is best automated.
- Code generation reconciles high-level programming with low-level performance.

We should also mention one final lesson that we haven't discussed in this chapter: you can't optimize in a vacuum, or you end up congratulating yourself for making a slow program slightly faster. We started the FFTW project after downloading a dozen FFT implementations, benchmarking them on a few machines, and noting how the winners varied between machines and between transform sizes. Throughout FFTW's development, we continued to benefit from repeated benchmarks against the dozens of high-quality FFT programs available online, without which we would have thought FFTW was "complete" long ago.

11.9 Acknowledgements

SGJ was supported in part by the Materials Research Science and Engineering Center program of the National Science Foundation under award DMR-9400334; MF was supported in part by the Defense Advanced Research Projects Agency (DARPA) under contract No. NBCH30390004. We are also grateful to Sidney Burrus for the opportunity to contribute this chapter, and for his continual encouragement—dating back to his first kind words in 1997 for the initial FFT efforts of two graduate students venturing outside their fields.

Chapter 12

Algorithms for Data with Restrictions[1]

12.1 Algorithms for Real Data

Many applications involve processing real data. It is inefficient to simply use a complex FFT on real data because arithmetic would be performed on the zero imaginary parts of the input, and, because of symmetries, output values would be calculated that are redundant. There are several approaches to developing special algorithms or to modifying complex algorithms for real data.

There are two methods which use a complex FFT in a special way to increase efficiency [39], [359]. The first method uses a length-N complex FFT to compute two length-N real FFTs by putting the two real data sequences into the real and the imaginary parts of the input to a complex FFT. Because transforms of real data have even real parts and odd imaginary parts, it is possible to separate the transforms of the two inputs with 2N-4 extra additions. This method requires, however, that two inputs be available at the same time.

The second method [359] uses the fact that the last stage of a decimation-in-time radix-2 FFT combines two independent transforms of length N/2 to compute a length-N transform. If the data are real, the two half length transforms are calculated by the method described above and the last stage is carried out to calculate the total length-N FFT of the real data. It should be noted that the half-length FFT does not have to be calculated by a radix-2 FFT. In fact, it should be calculated by the most efficient complex-data algorithm possible, such as the SRFFT or the PFA. The separation of the two half-length transforms and the computation of the last stage requires $N-6$ real multiplications and $(5/2)\,N-6$ real additions [359].

It is possible to derive more efficient real-data algorithms directly rather than using a complex FFT. The basic idea is from Bergland [21], [22] and Sande [325] which, at each stage, uses the symmetries of a constant radix Cooley-Tukey FFT to minimize arithmetic and storage. In the usual derivation [275] of the radix-2 FFT, the length-N transform is written as the

[1]This content is available online at <http://cnx.org/content/m16338/1.7/>.

combination of the length-N/2 DFT of the even indexed data and the length-N/2 DFT of the odd indexed data. If the input to each half-length DFT is real, the output will have Hermitian symmetry. Hence the output of each stage can be arranged so that the results of that stage stores the complex DFT with the real part located where half of the DFT would have gone, and the imaginary part located where the conjugate would have gone. This removes most of the redundant calculations and storage but slightly complicates the addressing. The resulting butterfly structure for this algorithm [359] resembles that for the fast Hartley transform [353]. The complete algorithm has one half the number of multiplications and N-2 fewer than half the additions of the basic complex FFT. Applying this approach to the split-radix FFT gives a particularly interesting algorithm [103], [359], [111].

Special versions of both the PFA and WFTA can also be developed for real data. Because the operations in the stages of the PFA can be commuted, it is possible to move the combination of the transform of the real part of the input and imaginary part to the last stage. Because the imaginary part of the input is zero, half of the algorithm is simply omitted. This results in the number of multiplications required for the real transform being exactly half of that required for complex data and the number of additions being about N less than half that required for the complex case because adding a pure real number to a pure imaginary number does not require an actual addition. Unfortunately, the indexing and data transfer becomes somewhat more complicated [179], [359]. A similar approach can be taken with the WFTA [179], [359], [284].

12.2 Special Algorithms for input Data that is mostly Zero, for Calculating only a few Outputs, or where the Sampling is not Uniform

In some cases, most of the data to be transformed are zero. It is clearly wasteful to do arithmetic on that zero data. Another special case is when only a few DFT values are needed. It is likewise wasteful to calculate outputs that are not needed. We use a process called "pruning" to remove the unneeded operations.

In other cases, the data are non-uniform sampling of a continuous time signal [13].

12.3 Algorithms for Approximate DFTs

There are applications where approximations to the DFT are all that is needed.[161], [163]

Chapter 13

Convolution Algorithms[1]

13.1 Fast Convolution by the FFT

One of the main applications of the FFT is to do convolution more efficiently than the direct calculation from the definition which is:

$$y(n) = \sum h(m)\ x(n-m) \tag{13.1}$$

which, with a change of variables, can also be written as:

$$y(n) = \sum x(m)\ h(n-m) \tag{13.2}$$

This is often used to filter a signal $x(n)$ with a filter whose impulse response is $h(n)$. Each output value $y(n)$ requires N multiplications and $N-1$ additions if $y(n)$ and $h(n)$ have N terms. So, for N output values, on the order of N^2 arithmetic operations are required.

Because the DFT converts convolution to multiplication:

$$DFT\{y(n)\} \ = \ DFT\{h(n)\}\ DFT\{x(n)\} \tag{13.3}$$

can be calculated with the FFT and bring the order of arithmetic operations down to $Nlog(N)$ which can be significant for large N.

This approach, which is called "fast convolutions", is a form of block processing since a whole block or segment of $x(n)$ must be available to calculate even one output value, $y(n)$. So, a time delay of one block length is always required. Another problem is the filtering use of convolution is usually non-cyclic and the convolution implemented with the DFT is cyclic. This is dealt with by appending zeros to $x(n)$ and $h(n)$ such that the output of the cyclic convolution gives one block of the output of the desired non-cyclic convolution.

For filtering and some other applications, one wants "on going" convolution where the filter response $h(n)$ may be finite in length or duration, but the input $x(n)$ is of arbitrary length.

[1] This content is available online at <http://cnx.org/content/m16339/1.10/>.

Two methods have traditionally used to break the input into blocks and use the FFT to convolve the block so that the output that would have been calculated by directly implementing (13.1) or (13.2) can be constructed efficiently. These are called "overlap-add" and "over-lap save".

13.1.1 Fast Convolution by Overlap-Add

In order to use the FFT to convolve (or filter) a long input sequence $x(n)$ with a finite length-M impulse response, $h(n)$, we partition the input sequence in segments or blocks of length L. Because convolution (or filtering) is linear, the output is a linear sum of the result of convolving the first block with $h(n)$ plus the result of convolving the second block with $h(n)$, plus the rest. Each of these block convolutions can be calculated by using the FFT. The output is the inverse FFT of the product of the FFT of $x(n)$ and the FFT of $h(n)$. Since the number of arithmetic operation to calculate the convolution directly is on the order of M^2 and, if done with the FFT, is on the order of $Mlog(M)$, there can be a great savings by using the FFT for large M.

The reason this procedure is not totally straightforward, is the length of the output of convolving a length-L block with a length-M filter is of length $L+M-1$. This means the output blocks cannot simply be concatenated but must be overlapped and added, hence the name for this algorithm is "Overlap-Add".

The second issue that must be taken into account is the fact that the overlap-add steps need non-cyclic convolution and convolution by the FFT is cyclic. This is easily handled by appending $L-1$ zeros to the impulse response and $M-1$ zeros to each input block so that all FFTs are of length $M+L-1$. This means there is no aliasing and the implemented cyclic convolution gives the same output as the desired non-cyclic convolution.

The savings in arithmetic can be considerable when implementing convolution or performing FIR digital filtering. However, there are two penalties. The use of blocks introduces a delay of one block length. None of the first block of output can be calculated until all of the first block of input is available. This is not a problem for "off line" or "batch" processing but can be serious for real-time processing. The second penalty is the memory required to store and process the blocks. The continuing reduction of memory cost often removes this problem.

The efficiency in terms of number of arithmetic operations per output point increases for large blocks because of the $Mlog(M)$ requirements of the FFT. However, the blocks become very large ($L>>M$), much of the input block will be the appended zeros and efficiency is lost. For any particular application, taking the particular filter and FFT algorithm being used and the particular hardware being used, a plot of efficiency vs. block length, L should be made and L chosen to maximize efficiency given any other constraints that are applicable.

Usually, the block convolutions are done by the FFT, but they could be done by any efficient, finite length method. One could use "rectangular transforms" or "number-theoretic transforms". A generalization of this method is presented later in the notes.

13.1.2 Fast Convolution by Overlap-Save

An alternative approach to the Overlap-Add can be developed by starting with segmenting the output rather than the input. If one considers the calculation of a block of output, it is seen that not only the corresponding input block is needed, but part of the preceding input block also needed. Indeed, one can show that a length $M + L - 1$ segment of the input is needed for each output block. So, one saves the last part of the preceding block and concatenates it with the current input block, then convolves that with $h(n)$ to calculate the current output

13.2 Block Processing, a Generalization of Overlap Methods

Convolution is intimately related to the DFT. It was shown in The DFT as Convolution or Filtering (Chapter 5) that a prime length DFT could be converted to cyclic convolution. It has been long known [276] that convolution can be calculated by multiplying the DFTs of signals.

An important question is what is the fastest method for calculating digital convolution. There are several methods that each have some advantage. The earliest method for fast convolution was the use of sectioning with overlap-add or overlap-save and the FFT [276], [300], [66]. In most cases the convolution is of real data and, therefore, real-data FFTs should be used. That approach is still probably the fastest method for longer convolution on a general purpose computer or microprocessor. The shorter convolutions should simply be calculated directly.

13.3 Introduction

The partitioning of long or infinite strings of data into shorter sections or blocks has been used to allow application of the FFT to realize on-going or continuous convolution [368], [181]. This section develops the idea of block processing and shows that it is a generalization of the overlap-add and overlap-save methods [368], [147]. They further generalize the idea to a multidimensional formulation of convolution [3], [47]. Moving in the opposite direction, it is shown that, rather than partitioning a string of scalars into blocks and then into blocks of blocks, one can partition a scalar number into blocks of bits and then include the operation of multiplication in the signal processing formulation. This is called distributed arithmetic [45] and, since it describes operations at the bit level, is completely general. These notes try to present a coherent development of these ideas.

13.4 Block Signal Processing

In this section the usual convolution and recursion that implements FIR and IIR discrete-time filters are reformulated in terms of vectors and matrices. Because the same data is partitioned and grouped in a variety of ways, it is important to have a consistent notation in order to be clear. The nth element of a data sequence is expressed $h(n)$ or, in some cases to simplify, h_n. A block or finite length column vector is denoted $\underline{h}_n$ with n indicating the nth block or section of a longer vector. A matrix, square or rectangular, is indicated by an upper case letter such as H with a subscript if appropriate.

13.4.1 Block Convolution

The operation of a finite impulse response (FIR) filter is described by a finite convolution as

$$y(n) = \sum_{k=0}^{L-1} h(k)\, x(n-k) \tag{13.4}$$

where $x(n)$ is causal, $h(n)$ is causal and of length L, and the time index n goes from zero to infinity or some large value. With a change of index variables this becomes

$$y(n) = \sum h(n-k)\, x(k) \tag{13.5}$$

which can be expressed as a matrix operation by

$$\begin{bmatrix} y_0 \\ y_1 \\ y_2 \\ \vdots \end{bmatrix} = \begin{bmatrix} h_0 & 0 & 0 & \cdots & 0 \\ h_1 & h_0 & 0 & & \\ h_2 & h_1 & h_0 & & \\ \vdots & & & & \vdots \end{bmatrix} \begin{bmatrix} x_0 \\ x_1 \\ x_2 \\ \vdots \end{bmatrix}. \tag{13.6}$$

The H matrix of impulse response values is partitioned into N by N square sub matrices and the X and Y vectors are partitioned into length-N blocks or sections. This is illustrated for $N = 3$ by

$$H_0 = \begin{bmatrix} h_0 & 0 & 0 \\ h_1 & h_0 & 0 \\ h_2 & h_1 & h_0 \end{bmatrix} \qquad H_1 = \begin{bmatrix} h_3 & h_2 & h_1 \\ h_4 & h_3 & h_2 \\ h_5 & h_4 & h_3 \end{bmatrix} \qquad \text{etc.} \tag{13.7}$$

$$\underline{x}_0 = \begin{bmatrix} x_0 \\ x_1 \\ x_2 \end{bmatrix} \qquad \underline{x}_1 = \begin{bmatrix} x_3 \\ x_4 \\ x_5 \end{bmatrix} \qquad \underline{y}_0 = \begin{bmatrix} y_0 \\ y_1 \\ y_2 \end{bmatrix} \qquad \text{etc.} \tag{13.8}$$

Substituting these definitions into (13.6) gives

$$\begin{bmatrix} \underline{y}_0 \\ \underline{y}_1 \\ \underline{y}_2 \\ \vdots \end{bmatrix} = \begin{bmatrix} H_0 & 0 & 0 & \cdots & 0 \\ H_1 & H_0 & 0 & & \\ H_2 & H_1 & H_0 & & \\ \vdots & & & & \vdots \end{bmatrix} \begin{bmatrix} \underline{x}_0 \\ \underline{x}_1 \\ \underline{x}_2 \\ \vdots \end{bmatrix} \tag{13.9}$$

The general expression for the nth output block is

$$\underline{y}_n = \sum_{k=0}^{n} H_{n-k}\, \underline{x}_k \tag{13.10}$$

which is a vector or block convolution. Since the matrix-vector multiplication within the block convolution is itself a convolution, (13.10) is a sort of convolution of convolutions and the finite length matrix-vector multiplication can be carried out using the FFT or other fast convolution methods.

The equation for one output block can be written as the product

$$\underline{y}_2 = [H_2 H_1 H_0] \begin{bmatrix} \underline{x}_0 \\ \underline{x}_1 \\ \underline{x}_2 \end{bmatrix} \tag{13.11}$$

and the effects of one input block can be written

$$\begin{bmatrix} H_0 \\ H_1 \\ H_2 \end{bmatrix} \underline{x}_1 = \begin{bmatrix} \underline{y}_0 \\ \underline{y}_1 \\ \underline{y}_2 \end{bmatrix}. \tag{13.12}$$

These are generalize statements of overlap save and overlap add [368], [147]. The block length can be longer, shorter, or equal to the filter length.

13.4.2 Block Recursion

Although less well-known, IIR filters can also be implemented with block processing [145], [74], [396], [43], [44]. The block form of an IIR filter is developed in much the same way as for the block convolution implementation of the FIR filter. The general constant coefficient difference equation which describes an IIR filter with recursive coefficients a_l, convolution coefficients b_k, input signal $x(n)$, and output signal $y(n)$ is given by

$$y(n) = \sum_{l=1}^{N-1} a_l\, y_{n-l} + \sum_{k=0}^{M-1} b_k\, x_{n-k} \tag{13.13}$$

using both functional notation and subscripts, depending on which is easier and clearer. The impulse response $h(n)$ is

$$h(n) = \sum_{l=1}^{N-1} a_l \, h(n-l) + \sum_{k=0}^{M-1} b_k \, \delta(n-k) \tag{13.14}$$

which can be written in matrix operator form

$$\begin{bmatrix} 1 & 0 & 0 & \cdots & 0 \\ a_1 & 1 & 0 & & \\ a_2 & a_1 & 1 & & \\ a_3 & a_2 & a_1 & & \\ 0 & a_3 & a_2 & & \\ \vdots & & & & \vdots \end{bmatrix} \begin{bmatrix} h_0 \\ h_1 \\ h_2 \\ h_3 \\ h_4 \\ \vdots \end{bmatrix} = \begin{bmatrix} b_0 \\ b_1 \\ b_2 \\ b_3 \\ 0 \\ \vdots \end{bmatrix} \tag{13.15}$$

In terms of N by N submatrices and length-N blocks, this becomes

$$\begin{bmatrix} A_0 & 0 & 0 & \cdots & 0 \\ A_1 & A_0 & 0 & & \\ 0 & A_1 & A_0 & & \\ \vdots & & & & \vdots \end{bmatrix} \begin{bmatrix} \underline{h}_0 \\ \underline{h}_1 \\ \underline{h}_2 \\ \vdots \end{bmatrix} = \begin{bmatrix} \underline{b}_0 \\ \underline{b}_1 \\ 0 \\ \vdots \end{bmatrix} \tag{13.16}$$

From this formulation, a block recursive equation can be written that will generate the impulse response block by block.

$$A_0 \, \underline{h}_n + A_1 \, \underline{h}_{n-1} = 0 \text{ for } n \geq 2 \tag{13.17}$$

$$\underline{h}_n = -A_0^{-1} A_1 \, \underline{h}_{n-1} = K \, \underline{h}_{n-1} \text{ for } n \geq 2 \tag{13.18}$$

with initial conditions given by

$$\underline{h}_1 = -A_0^{-1} A_1 A_0^{-1} \, \underline{b}_0 + A_0^{-1} \, \underline{b}_1 \tag{13.19}$$

This can also be written to generate the square partitions of the impulse response matrix by

$$H_n = K H_{n-1} \text{ for } n \geq 2 \tag{13.20}$$

with initial conditions given by

$$H_1 = K A_0^{-1} \, B_0 + A_0^{-1} \, B_1 \tag{13.21}$$

ane $K = -A_0^{-1} A_1$. This recursively generates square submatrices of H similar to those defined in (13.7) and (13.9) and shows the basic structure of the dynamic system.

Next, we develop the recursive formulation for a general input as described by the scalar difference equation (13.14) and in matrix operator form by

$$\begin{bmatrix} 1 & 0 & 0 & \cdots & 0 \\ a_1 & 1 & 0 & & \\ a_2 & a_1 & 1 & & \\ a_3 & a_2 & a_1 & & \\ 0 & a_3 & a_2 & & \\ \vdots & & & & \vdots \end{bmatrix} \begin{bmatrix} y_0 \\ y_1 \\ y_2 \\ y_3 \\ y_4 \\ \vdots \end{bmatrix} = \begin{bmatrix} b_0 & 0 & 0 & \cdots & 0 \\ b_1 & b_0 & 0 & & \\ b_2 & b_1 & b_0 & & \\ 0 & b_2 & b_1 & & \\ 0 & 0 & b_2 & & \\ \vdots & & & & \vdots \end{bmatrix} \begin{bmatrix} x_0 \\ x_1 \\ x_2 \\ x_3 \\ x_4 \\ \vdots \end{bmatrix} \tag{13.22}$$

which, after substituting the definitions of the sub matrices and assuming the block length is larger than the order of the numerator or denominator, becomes

$$\begin{bmatrix} A_0 & 0 & 0 & \cdots & 0 \\ A_1 & A_0 & 0 & & \\ 0 & A_1 & A_0 & & \\ \vdots & & & & \vdots \end{bmatrix} \begin{bmatrix} \underline{y}_0 \\ \underline{y}_1 \\ \underline{y}_2 \\ \vdots \end{bmatrix} = \begin{bmatrix} B_0 & 0 & 0 & \cdots & 0 \\ B_1 & B_0 & 0 & & \\ 0 & B_1 & B_0 & & \\ \vdots & & & & \vdots \end{bmatrix} \begin{bmatrix} \underline{x}_0 \\ \underline{x}_1 \\ \underline{x}_2 \\ \vdots \end{bmatrix}. \tag{13.23}$$

From the partitioned rows of (13.24), one can write the block recursive relation

$$A_0 \, \underline{y}_{n+1} + A_1 \, \underline{y}_n = B_0 \, \underline{x}_{n+1} + B_1 \, \underline{x}_n \tag{13.24}$$

Solving for $\underline{y}_{n+1}$ gives

$$\underline{y}_{n+1} = -A_0^{-1} A_1 \, \underline{y}_n + A_0^{-1} B_0 \, \underline{x}_{n+1} + A_0^{-1} B_1 \, \underline{x}_n \tag{13.25}$$

$$\underline{y}_{n+1} = K \, \underline{y}_n + H_0 \, \underline{x}_{n+1} + \tilde{H}_1 \, \underline{x}_n \tag{13.26}$$

which is a first order vector difference equation [43], [44]. This is the fundamental block recursive algorithm that implements the original scalar difference equation in (13.14). It has several important characteristics.

- The block recursive formulation is similar to a state variable equation but the states are blocks or sections of the output [44], [220], [427], [428].
- The eigenvalues of K are the poles of the original scalar problem raised to the N power plus others that are zero. The longer the block length, the "more stable" the filter is, i.e. the further the poles are from the unit circle [43], [44], [427], [15], [16].

- If the block length were shorter than the denominator, the vector difference equation would be higher than first order. There would be a non zero A_2. If the block length were shorter than the numerator, there would be a non zero B_2 and a higher order block convolution operation. If the block length were one, the order of the vector equation would be the same as the scalar equation. They would be the same equation.

- The actual arithmetic that goes into the calculation of the output is partly recursive and partly convolution. The longer the block, the more the output is calculated by convolution and, the more arithmetic is required.
- It is possible to remove the zero eigenvalues in K by making K rectangular or square and N by N This results in a form even more similar to a state variable formulation [240], [44]. This is briefly discussed below in section 2.3.
- There are several ways of using the FFT in the calculation of the various matrix products in (13.25) and in (13.27) and (13.28). Each has some arithmetic advantage for various forms and orders of the original equation. It is also possible to implement some of the operations using rectangular transforms, number theoretic transforms, distributed arithmetic, or other efficient convolution algorithms [44], [427], [54], [48], [426], [286].
- By choosing the block length equal to the period, a periodically time varying filter can be made block time invariant. In other words, all the time varying characteristics are moved to the finite matrix multiplies which leave the time invariant properties at the block level. This allows use of z-transform and other time-invariant methods to be used for stability analysis and frequency response analysis [244], [245]. It also turns out to be related to filter banks and multi-rate filters [222], [221], [97].

13.4.3 Block State Formulation

It is possible to reduce the size of the matrix operators in the block recursive description (13.26) to give a form even more like a state variable equation [240], [44], [428]. If K in (13.26) has several zero eigenvalues, it should be possible to reduce the size of K until it has full rank. That was done in [44] and the result is

$$\underline{z}_n = K_1 \, \underline{z}_{n-1} + K_2 \, \underline{x}_n \tag{13.27}$$

$$\underline{y}_n = H_1 \, \underline{z}_{n-1} + H_0 \, \underline{x}_n \tag{13.28}$$

where H_0 is the same N by N convolution matrix, N_1 is a rectangular L by N partition of the convolution matrix H, K_1 is a square N by N matrix of full rank, and K_2 is a rectangular N by L matrix.

This is now a minimal state equation whose input and output are blocks of the original input and output. Some of the matrix multiplications can be carried out using the FFT or other techniques.

13.4.4 Block Implementations of Digital Filters

The advantage of the block convolution and recursion implementations is a possible improvement in arithmetic efficiency by using the FFT or other fast convolution methods for some of the multiplications in (13.10) or (13.25) [246], [247]. There is the reduction of quantization effects due to an effective decrease in the magnitude of the eigenvalues and the possibility of easier parallel implementation for IIR filters. The disadvantages are a delay of at least one block length and an increased memory requirement.

These methods could also be used in the various filtering methods for evaluating the DFT. This the chirp z-transform, Rader's method, and Goertzel's algorithm.

13.4.5 Multidimensional Formulation

This process of partitioning the data vectors and the operator matrices can be continued by partitioning (13.10) and (13.24) and creating blocks of blocks to give a higher dimensional structure. One should use index mapping ideas rather than partitioned matrices for this approach [3], [47].

13.4.6 Periodically Time-Varying Discrete-Time Systems

Most time-varying systems are periodically time-varying and this allows special results to be obtained. If the block length is set equal to the period of the time variations, the resulting block equations are time invariant and all to the time varying characteristics are contained in the matrix multiplications. This allows some of the tools of time invariant systems to be used on periodically time-varying systems.

The PTV system is analyzed in [425], [97], [81], [244], the filter analysis and design problem, which includes the decimation–interpolation structure, is addressed in [126], [245], [222], and the bandwidth compression problem in [221]. These structures can take the form of filter banks [387].

13.4.7 Multirate Filters, Filter Banks, and Wavelets

Another area that is related to periodically time varying systems and to block processing is filter banks [387], [152]. Recently the area of perfect reconstruction filter banks has been further developed and shown to be closely related to wavelet based signal analysis [97], [99], [151], [387]. The filter bank structure has several forms with the polyphase and lattice being particularly interesting.

An idea that has some elements of multirate filters, perfect reconstruction, and distributed arithmetic is given in [142], [140], [141]. Parks has noted that design of multirate filters has some elements in common with complex approximation and of 2-D filter design [337], [338] and is looking at using Tang's method for these designs.

13.4.8 Distributed Arithmetic

Rather than grouping the individual scalar data values in a discrete-time signal into blocks, the scalar values can be partitioned into groups of bits. Because multiplication of integers, multiplication of polynomials, and discrete-time convolution are the same operations, the bit-level description of multiplication can be mixed with the convolution of the signal processing. The resulting structure is called distributed arithmetic [45], [402]. It can be used to create an efficient table look-up scheme to implement an FIR or IIR filter using no multiplications by fetching previously calculated partial products which are stored in a table. Distributed arithmetic, block processing, and multi-dimensional formulations can be combined into an integrated powerful description to implement digital filters and processors. There may be a new form of distributed arithmetic using the ideas in [140], [141].

13.5 Direct Fast Convolution and Rectangular Transforms

A relatively new approach uses index mapping directly to convert a one dimensional convolution into a multidimensional convolution [47], [8]. This can be done by either a type-1 or type-2 map. The short convolutions along each dimension are then done by Winograd's optimal algorithms. Unlike for the case of the DFT, there is no savings of arithmetic from the index mapping alone. All the savings comes from efficient short algorithms. In the case of index mapping with convolution, the multiplications must be nested together in the center of the algorithm in the same way as for the WFTA. There is no equivalent to the PFA structure for convolution. The multidimensional convolution can not be calculated by row and column convolutions as the DFT was by row and column DFTs.

It would first seem that applying the index mapping and optimal short algorithms directly to convolution would be more efficient than using DFTs and converting them to convolution to be calculated by the same optimal algorithms. In practical algorithms, however, the DFT method seems to be more efficient [286].

A method that is attractive for special purpose hardware uses distributed arithmetic [45]. This approach uses a table look up of precomputed partial products to produce a system that does convolution without requiring multiplications [79].

Another method that requires special hardware uses number theoretic transforms [31], [237], [265] to calculate convolution. These transforms are defined over finite fields or rings with arithmetic performed modulo special numbers. These transforms have rather limited flexibility, but when they can be used, they are very efficient.

13.6 Number Theoretic Transforms for Convolution

13.6.1 Results from Number Theory

A basic review of the number theory useful for signal processing algorithms will be given here with specific emphasis on the congruence theory for number theoretic transforms [279], [165], [260], [237], [328].

13.6.2 Number Theoretic Transforms

Here we look at the conditions placed on a general linear transform in order for it to support cyclic convolution. The form of a linear transformation of a length-N sequence of number is given by

$$X(k) = \sum_{n=0}^{N-1} t(n,k)\, x(n) \tag{13.29}$$

for $k = 0, 1, \cdots, (N-1)$. The definition of cyclic convolution of two sequences is given by

$$y(n) = \sum_{m=0}^{N-1} x(m)\, h(n-m) \tag{13.30}$$

for $n = 0, 1, \cdots, (N-1)$ and all indices evaluated modulo N. We would like to find the properties of the transformation such that it will support the cyclic convolution. This means that if $X(k)$, $H(k)$, and $Y(k)$ are the transforms of $x(n)$, $h(n)$, and $y(n)$ respectively,

$$Y(k) = X(k)\, H(k). \tag{13.31}$$

The conditions are derived by taking the transform defined in (13.4) of both sides of equation (13.5) which gives

$$Y(k) = \sum_{n=0}^{N-1} t(n,k) \sum_{m=0}^{N-1} x(m)\, h(n-m) \tag{13.32}$$

$$= \sum_{m=0}^{N-1}\sum_{n=0}^{N-1} x(m)\, h(n-m)\, t(n,k). \tag{13.33}$$

Making the change of index variables, $l = n - m$, gives

$$= \sum_{m=0}^{N-1}\sum_{l=0}^{N-1} x(m)\, h(l)\, t(l+m,k). \tag{13.34}$$

But from (13.6), this must be

$$Y(k) = \sum_{n=0}^{N-1} x(n)\, t(n,k) \sum_{m=0}^{N-1} x(m)\, t(m,k) \tag{13.35}$$

$$= \sum_{m=0}^{N-1}\sum_{l=0}^{N-1} x(m)\, h(l)\, t(n,k)\, t(l,k). \tag{13.36}$$

This must be true for all $x(n)$, $h(n)$, and k, therefore from (13.9) and (13.11) we have

$$t(m+l,k) = t(m,k)\, t(l,k) \tag{13.37}$$

For $l=0$ we have

$$t(m,k) = t(m,k)\, t(0,k) \tag{13.38}$$

and, therefore, $t(0,k)=1$. For $l=m$ we have

$$t(2m,k) = t(m,k)\, t(m,k) = t^2(m,k) \tag{13.39}$$

For $l=pm$ we likewise have

$$t(pm,k) = t^p(m,k) \tag{13.40}$$

and, therefore,

$$t^N(m,k) = t(Nm,k) = t(0,k) = 1. \tag{13.41}$$

But

$$t(m,k) = t^m(1,k) = t^k(m,1), \tag{13.42}$$

therefore,

$$t(m,k) = t^{mk}(1,1). \tag{13.43}$$

Defining $t(1,1)=\alpha$ gives the form for our general linear transform (13.4) as

$$X(k) = \sum_{n=0}^{N-1} \alpha^{nk}\, x(n) \tag{13.44}$$

where α is a root of order N , which means that N is the smallest integer such that $\alpha^N=1$.

Theorem 1 The transform (13.13) supports cyclic convolution if and only if α is a root of order N and N^{-1} is defined.

This is discussed in [2], [4].

Theorem 2 The transform (13.13) supports cyclic convolution if and only if

$$N|O(M) \tag{13.45}$$

where

$$O(M) = gcd\{p_1 - 1, p_2 - 1, \cdots, p_l - 1\} \tag{13.46}$$

and

$$M = p_1^{r_1} p_2^{r_2} \cdots p_l^{r_l}. \tag{13.47}$$

This theorem is a more useful form of Theorem 1. Notice that $N_{max} = O(M)$.

One needs to find appropriate N, M, and α such that

- N should be appropriate for a fast algorithm and handle the desired sequence lengths.
- M should allow the desired dynamic range of the signals and should allow simple modular arithmetic.
- α should allow a simple multiplication for $\alpha^{nk} x(n)$.

We see that if M is even, it has a factor of 2 and, therefore, $O(M) = N_{max} = 1$ which implies M should be odd. If M is prime the $O(M) = M - 1$ which is as large as could be expected in a field of M integers. For $M = 2^k - 1$, let k be a composite $k = pq$ where p is prime. Then $2^p - 1$ divides $2^{pq} - 1$ and the maximum possible length of the transform will be governed by the length possible for $2^p - 1$. Therefore, only the prime k need be considered interesting. Numbers of this form are know as Mersenne numbers and have been used by Rader [311]. For Mersenne number transforms, it can be shown that transforms of length at least $2p$ exist and the corresponding $\alpha = -2$. Mersenne number transforms are not of as much interest because $2p$ is not highly composite and, therefore, we do not have FFT-type algorithms.

For $M = 2^k + 1$ and k odd, 3 divides $2^k + 1$ and the maximum possible transform length is 2. Thus we consider only even k. Let $k = s2^t$, where s is an odd integer. Then 2^{2^t} divides $2^{s2^t} + 1$ and the length of the possible transform will be governed by the length possible for $2^{2^t} + 1$. Therefore, integers of the form $M = 2^{2^t} + 1$ are of interest. These numbers are known as Fermat numbers [311]. Fermat numbers are prime for $0 \leq t \leq 4$ and are composite for all $t \geq 5$.

Since Fermat numbers up to F_4 are prime, $O(F_t) = 2^b$ where $b = 2^t$ and we can have a Fermat number transform for any length $N = 2^m$ where $m \leq b$. For these Fermat primes the integer $\alpha = 3$ is of order $N = 2^b$ allowing the largest possible transform length. The integer $\alpha = 2$ is of order $N = 2b = 2^{t+1}$. This is particularly attractive since α to a power is multiplied times the data values in (13.4).

The following table gives possible parameters for various Fermat number moduli.

t	b	$M = F_t$	N_2	$N_{\sqrt{2}}$	N_{max}	α for N_{max}
3	8	2^8+1	16	32	256	3
4	16	$2^{16}+1$	32	64	65536	3
5	32	$2^{32}+1$	64	128	128	$\sqrt{2}$
6	64	$2^{64}+1$	128	256	256	$\sqrt{2}$

Table 13.1

This table gives values of N for the two most important values of α which are 2 and $\sqrt{2}$. The second column give the approximate number of bits in the number representation. The third column gives the Fermat number modulus, the fourth is the maximum convolution length for $\alpha = 2$, the fifth is the maximum length for $\alpha = \sqrt{2}$, the sixth is the maximum length for any α, and the seventh is the α for that maximum length. Remember that the first two rows have a Fermat number modulus which is prime and second two rows have a composite Fermat number as modulus. Note the differences.

The books, articles, and presentations that discuss NTT and related topics are [209], [237], [265], [31], [253], [257], [288], [312], [311], [1], [55], [2], [4]. A recent book discusses NT in a signal processing context [215].

Chapter 14

Comments: Fast Fourier Transforms[1]

14.1 Other work and Results

This section comes from a note describing results on efficient algorithms to calculate the discrete Fourier transform (DFT) that were collected over years. Perhaps the most interesting is the discovery that the Cooley-Tukey FFT was described by Gauss in 1805 [175]. That gives some indication of the age of research on the topic, and the fact that a 1995 compiled bibliography [363] on efficient algorithms contains over 3400 entries indicates its volume. Three IEEE Press reprint books contain papers on the FFT [303], [84], [85]. An excellent general purpose FFT program has been described in [132], [129] and is used in Matlab and available over the internet.

In addition to this book there are several others [238], [266], [25], [170], [383], [254], [33], [37], [345] that give a good modern theoretical background for the FFT, one book [67] that gives the basic theory plus both FORTRAN and TMS 320 assembly language programs, and other books [219], [348], [70] that contain chapters on advanced FFT topics. A good up-to-date, on-line reference with both theory and programming techniques is in [11]. The history of the FFT is outlined in [87], [175] and excellent survey articles can be found in [115], [93]. The foundation of much of the modern work on efficient algorithms was done by S. Winograd. These results can be found in [412], [415], [418]. An outline and discussion of his theorems can be found in [219] as well as [238], [266], [25], [170].

Efficient FFT algorithms for length-2^M were described by Gauss and discovered in modern times by Cooley and Tukey [91]. These have been highly developed and good examples of FORTRAN programs can be found in [67]. Several new algorithms have been published that require the least known amount of total arithmetic [423], [108], [104], [229], [394], [71]. Of these, the split-radix FFT [108], [104], [392], [366] seems to have the best structure for programming, and an efficient program has been written [351] to implement it. A mixture of decimation-in-time and decimation-in-frequency with very good efficiency is given in [323], [324] and one called the Sine-Cosine FT [71]. Recently a modification to the split-radix algo-

[1]This content is available online at <http://cnx.org/content/m16434/1.8/>.

rithm has been described [203] that has a slightly better total arithmetic count. Theoretical bounds on the number of multiplications required for the FFT based on Winograd's theories are given in [170], [172]. Schemes for calculating an in-place, in-order radix-2 FFT are given in [17], [19], [196], [379]. Discussion of various forms of unscramblers is given in [51], [321], [186], [123], [318], [400], [424], [370], [315]. A discussion of the relation of the computer architecture, algorithm and compiler can be found in [251], [242]. A modification to allow lengths of $N = q\,2^m$ for q odd is given in [24].

The "other" FFT is the prime factor algorithm (PFA) which uses an index map originally developed by Thomas and by Good. The theory of the PFA was derived in [214] and further developed and an efficient in-order and in-place program given in [58], [67]. More results on the PFA are given in [377], [378], [379], [380], [364]. A method has been developed to use dynamic programming to design optimal FFT programs that minimize the number of additions and data transfers as well as multiplications [191]. This new approach designs custom algorithms for a particular computer architecture. An efficient and practical development of Winograd's ideas has given a design method that does not require the rather difficult Chinese remainder theorem [219], [199] for short prime length FFT's. These ideas have been used to design modules of length 11, 13, 17, 19, and 25 [189]. Other methods for designing short DFT's can be found in [376], [223]. A use of these ideas with distributed arithmetic and table look-up rather than multiplication is given in [80]. A program that implements the nested Winograd Fourier transform algorithm (WFTA) is given in [238] but it has not proven as fast or as versatile as the PFA [58]. An interesting use of the PFA was announced [75] in searching for large prime numbers.

These efficient algorithms can not only be used on DFT's but on other transforms with a similar structure. They have been applied to the discrete Hartley transform [354], [36] and the discrete cosine transform [394], [401], [314].

The fast Hartley transform has been proposed as a superior method for real data analysis but that has been shown not to be the case. A well-designed real-data FFT [360] is always as good as or better than a well-designed Hartley transform [354], [113], [289], [386], [371]. The Bruun algorithm [41], [369] also looks promising for real data applications as does the Rader-Brenner algorithm [310], [76], [386]. A novel approach to calculating the inverse DFT is given in [109].

General length algorithms include [340], [143], [125]. For lengths that are not highly composite or prime, the chirp z-transform in a good candidate [67], [307] for longer lengths and an efficient order-N^2 algorithm called the QFT [343], [157], [160] for shorter lengths. A method which automatically generates near-optimal prime length Winograd based programs has been given in [199], [330], [332], [334], [336]. This gives the same efficiency for shorter lengths (i.e. $N \leq 19$) and new algorithms for much longer lengths and with well-structured algorithms. Another approach is given in [285]. Special methods are available for very long lengths [183], [365]. A very interesting general length FFT system called the FFTW has been developed by Frigo and Johnson at MIT. It uses a library of efficient "codelets" which are composed for a very efficient calculation of the DFT on a wide variety of computers [132], [129], [136]. For most lengths and on most computers, this is the fastest FFT today. Surprisingly, it uses a recursive program structure. The FFTW won the 1999 Wilkinson Prize for Numerical Software.

The use of the FFT to calculate discrete convolution was one of its earliest uses. Although the more direct rectangular transform [9] would seem to be more efficient, use of the FFT or PFA is still probably the fastest method on a general purpose computer or DSP chip [287], [360], [113], [241]. On special purpose hardware or special architectures, the use of distributed arithmetic [80] or number theoretic transforms [5] may be even faster. Special algorithms for use with the short-time Fourier transform [346] and for the calculation of a few DFT values [349], [316], [347] and for recursive implementation [399], [129] have also been developed. An excellent analysis of efficient programming the FFT on DSP microprocessors is given in [243], [242]. Formulations of the DFT in terms of tensor or Kronecker products look promising for developing algorithms for parallel and vector computer architectures [361], [383], [200], [390], [385], [154], [153].

Various approaches to calculating approximate DFTs have been based on cordic methods, short word lengths, or some form of pruning. A new method that uses the characteristics of the signals being transformed has combined the discrete wavelet transform (DWT) combined with the DFT to give an approximate FFT with $O(N)$ multiplications [162], [164], [69] for certain signal classes. A similar approach has been developed using filter banks [339], [185].

The study of efficient algorithms not only has a long history and large bibliography, it is still an exciting research field where new results are used in practical applications.

More information can be found on the Rice DSP Group's web page[2]

[2]http://www-dsp.rice.edu

Chapter 15

Conclusions: Fast Fourier Transforms[1]

This book has developed a class of efficient algorithms based on index mapping and polynomial algebra. This provides a framework from which the Cooley-Tukey FFT, the split-radix FFT, the PFA, and WFTA can be derived. Even the programs implementing these algorithms can have a similar structure. Winograd's theorems were presented and shown to be very powerful in both deriving algorithms and in evaluating them. The simple radix-2 FFT provides a compact, elegant means for efficiently calculating the DFT. If some elaboration is allowed, significant improvement can be had from the split-radix FFT, the radix-4 FFT or the PFA. If multiplications are expensive, the WFTA requires the least of all.

Several method for transforming real data were described that are more efficient than directly using a complex FFT. A complex FFT can be used for real data by artificially creating a complex input from two sections of real input. An alternative and slightly more efficient method is to construct a special FFT that utilizes the symmetries at each stage.

As computers move to multiprocessors and multicore, writing and maintaining efficient programs becomes more and more difficult. The highly structured form of FFTs allows automatic generation of very efficient programs that are tailored specifically to a particular DSP or computer architecture.

For high-speed convolution, the traditional use of the FFT or PFA with blocking is probably the fastest method although rectangular transforms, distributed arithmetic, or number theoretic transforms may have a future with special VLSI hardware.

The ideas presented in these notes can also be applied to the calculation of the discrete Hartley transform [355], [112], the discrete cosine transform [119], [395], and to number theoretic transforms [32], [239], [267].

There are many areas for future research. The relationship of hardware to algorithms, the proper use of multiple processors, the proper design and use of array processors and vector processors are all open. There are still many unanswered questions in multi-dimensional

[1]This content is available online at <http://cnx.org/content/m16340/1.7/>.

algorithms where a simple extension of one-dimensional methods will not suffice.

Appendix 1: FFT Flowgraphs[1]

16.1 Signal Flow Graphs of Cooley-Tukey FFTs

The following four figures are flow graphs for Radix-2 Cooley-Tukey FFTs. The first is a length-16, decimation-in-frequency Radix-2 FFT with the input data in order and output data scrambled. The first stage has 8 length-2 "butterflies" (which overlap in the figure) followed by 8 multiplications by powers of W which are called "twiddle factors". The second stage has 2 length-8 FFTs which are each calculated by 4 butterflies followed by 4 multiplies. The third stage has 4 length-4 FFTs, each calculated by 2 butterflies followed by 2 multiplies and the last stage is simply 8 butterflies followed by trivial multiplies by one. This flow graph should be compared with the index map in Polynomial Description of Signals (Chapter 4), the polynomial decomposition in The DFT as Convolution or Filtering (Chapter 5), and the program in Appendix 3. In the program, the butterflies and twiddle factor multiplications are done together in the inner most loop. The outer most loop indexes through the stages. If the length of the FFT is a power of two, the number of stages is that power (log N).

The second figure below is a length-16, decimation-in-time FFT with the input data scrambled and output data in order. The first stage has 8 length-2 "butterflies" followed by 8 twiddle factors multiplications. The second stage has 4 length-4 FFTs which are each calculated by 2 butterflies followed by 2 multiplies. The third stage has 2 length-8 FFTs, each calculated by 4 butterflies followed by 8 multiplies and the last stage is simply 8 length-2 butterflies. This flow graph should be compared with the index map in Polynomial Description of Signals (Chapter 4), the polynomial decomposition in The DFT as Convolution or Filtering (Chapter 5), and the program in Appendix 3 (Chapter 18). Here, the FFT must be preceded by a scrambler.

The third and fourth figures below are a length-16 decimation-in-frequency and a decimation-in-time but, in contrast to the figures above, the DIF has the output in order which requires a scrambled input and the DIT has the input in order which requires the output be unscrambled. Compare with the first two figures. Note the order of the twiddle factors. The number of additions and multiplications in all four flow graphs is the same and the structure of the three-loop program which executes the flow graph is the same.

[1] This content is available online at <http://cnx.org/content/m16352/1.11/>.

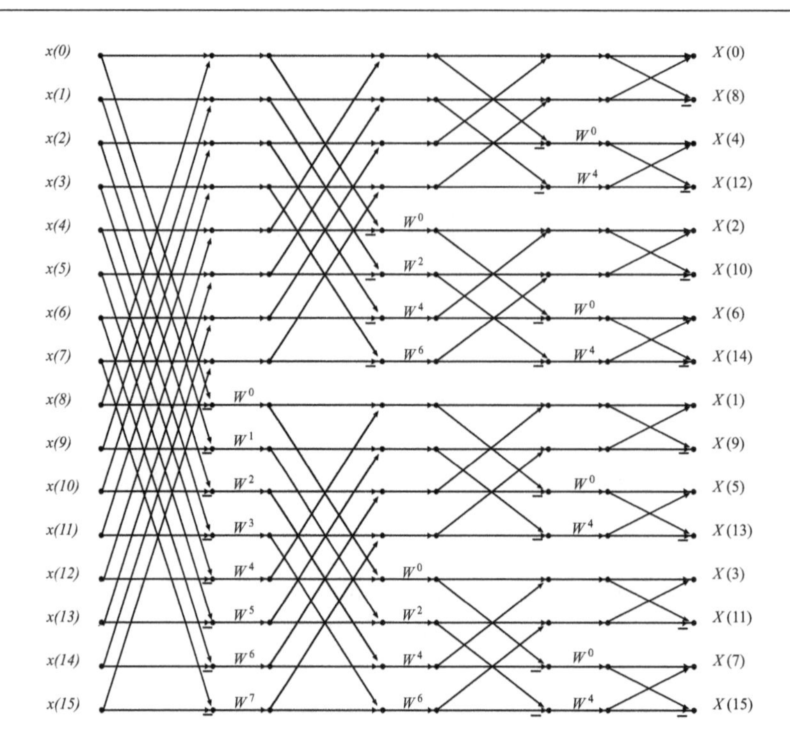

Figure 16.1: Length-16, Decimation-in-Frequency, In-order input, Radix-2 FFT

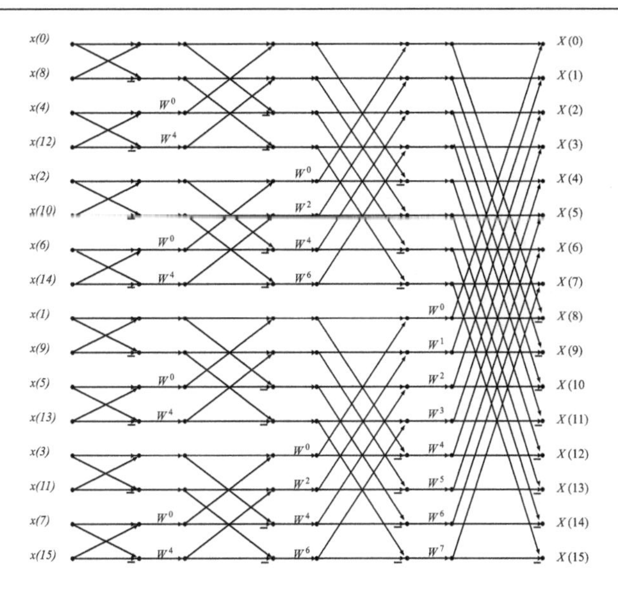

Figure 16.2: Length-16, Decimation-in-Time, In-order output, Radix-2 FFT

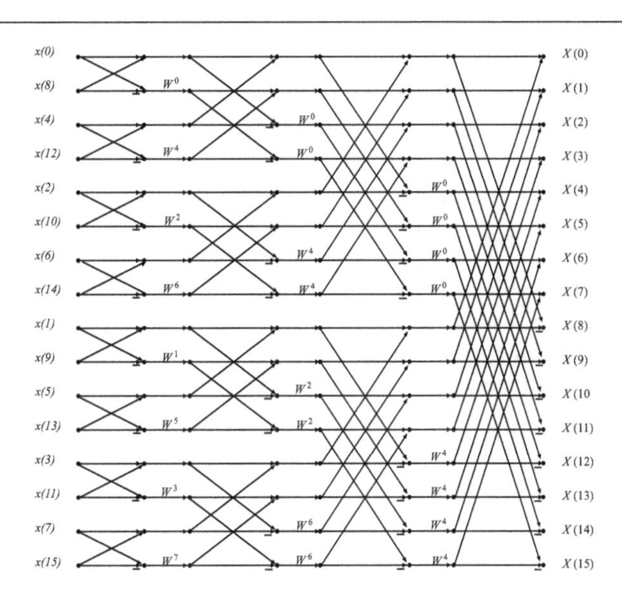

Figure 16.3: Length-16, alternate Decimation-in-Frequency, In-order output, Radix-2 FFT

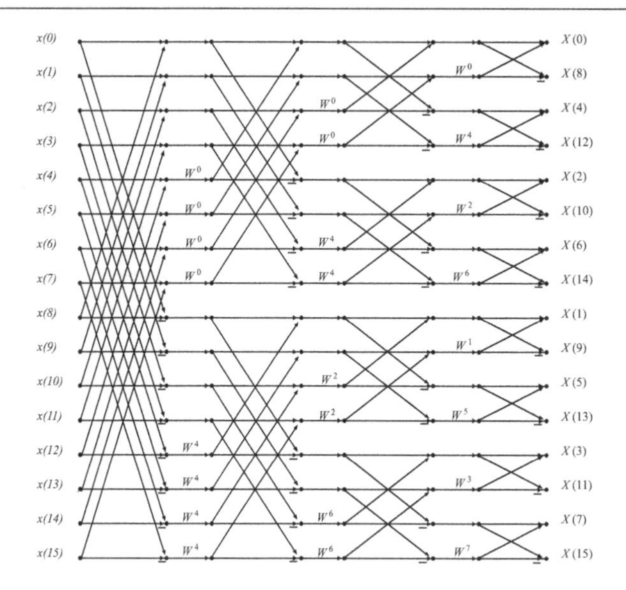

Figure 16.4: Length-16, alternate Decimation-in-Time, In-order input, Radix-2 FFT

The following is a length-16, decimation-in-frequency Radix-4 FFT with the input data in order and output data scrambled. There are two stages with the first stage having 4 length-4 "butterflies" followed by 12 multiplications by powers of W which are called "twiddle factors. The second stage has 4 length-4 FFTs which are each calculated by 4 butterflies followed by 4 multiplies. Note, each stage here looks like two stages but it is one and there is only one place where twiddle factor multiplications appear. This flow graph should be compared with the index map in Polynomial Description of Signals (Chapter 4), the polynomial decomposition in The DFT as Convolution or Filtering (Chapter 5), and the program in Appendix 3 (Chapter 18). Log to the base 4 of 16 is 2. The total number of twiddle factor multiplication here is 12 compared to 24 for the radix-2. The unscrambler is a base-four reverse order counter rather than a bit reverse counter, however, a modification of the radix four butterflies will allow a bit reverse counter to be used with the radix-4 FFT as with the radix-2.

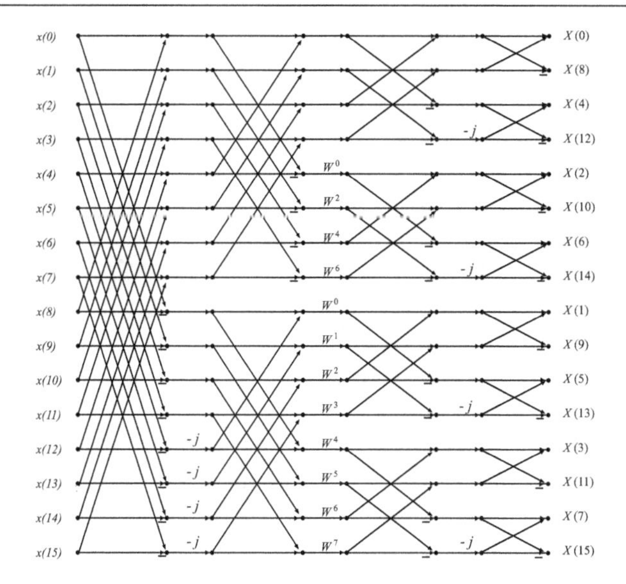

Figure 16.5: Length-16, Decimation-in-Frequency, In-order input, Radix-4 FFT

The following two flowgraphs are length-16, decimation-in-frequency Split Radix FFTs with the input data in order and output data scrambled. Because the "butterflies" are L shaped, the stages do not progress uniformly like the Radix-2 or 4. These two figures are the same with the first drawn in a way to compare with the Radix-2 and 4, and the second to illustrate the L shaped butterflies. These flow graphs should be compared with the index map in Polynomial Description of Signals (Chapter 4) and the program in Appendix 3 (Chapter 18). Because of the non-uniform stages, the program indexing is more complicated. Although the number of twiddle factor multiplications is 12 as was the radix-4 case, for longer lengths, the split-radix has slightly fewer multiplications than the radix-4.

Because the structures of the radix-2, radix-4, and split-radix FFTs are the same, the number of data additions is same for all of them. However, each complex twiddle factor multiplication requires two real additions (and four real multiplications) the number of additions will be fewer for the structures with fewer multiplications.

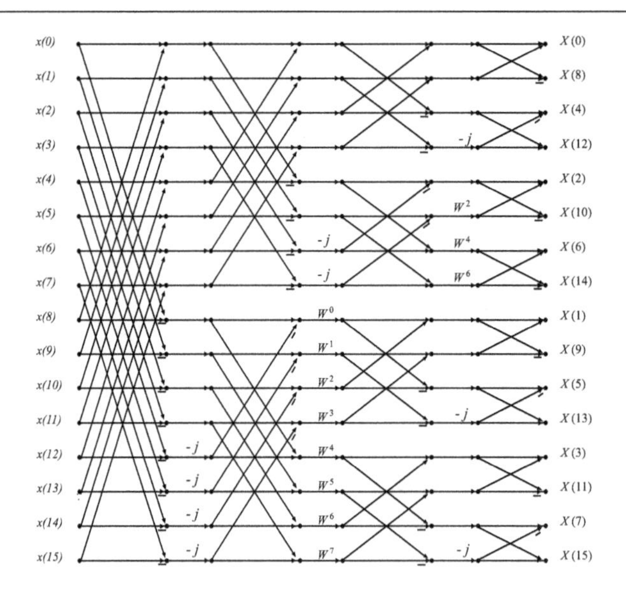

Figure 16.6: Length-16, Decimation-in-Frequency, In-order input, Split-Radix FFT

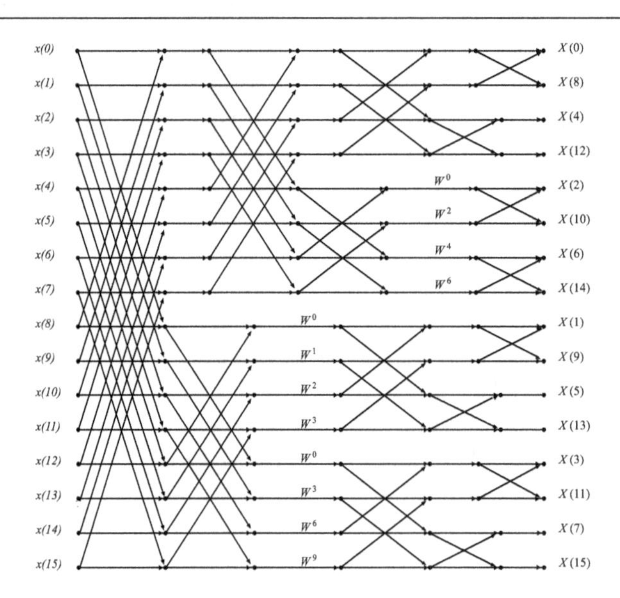

Figure 16.7: Length-16, Decimation-in-Frequency, Split-Radix with special BFs FFT

Appendix 2: Operation Counts for General Length FFT[1]

17.1 Figures

The Glassman-Ferguson FFT is a compact implementation of a mixed-radix Cooley-Tukey FFT with the short DFTs for each factor being calculated by a Goertzel-like algorithm. This means there are twiddle factor multiplications even when the factors are relatively prime, however, the indexing is simple and compact. It will calculate the DFT of a sequence of any length but is efficient only if the length is highly composite. The figures contain plots of the number of floating point multiplications plus additions vs. the length of the FFT. The numbers on the vertical axis have relative meaning but no absolute meaning.

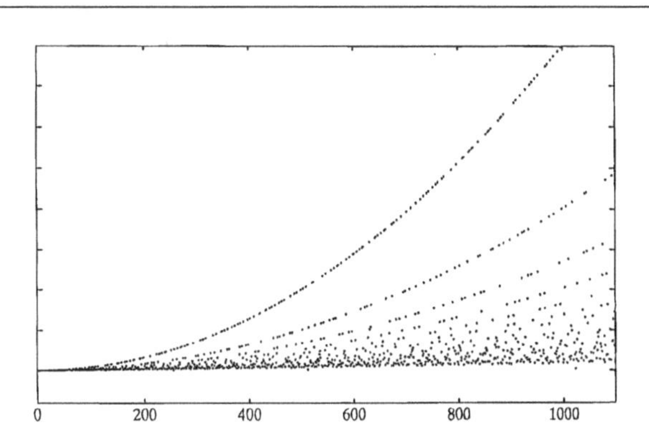

Figure 17.1: Flop-Count vs Length for the Glassman-Ferguson FFT

Note the parabolic shape of the curve for certain values. The upper curve is for prime lengths, the next one is for lengths that are two times a prime, and the next one is for lengths that are for three times a prime, etc. The shape of the lower boundary is roughly N log N. The

[1] This content is available online at <http://cnx.org/content/m16353/1.8/>.

program that generated these two figures used a Cooley-Tukey FFT if the length is two to a power which accounts for the points that are below the major lower boundary.

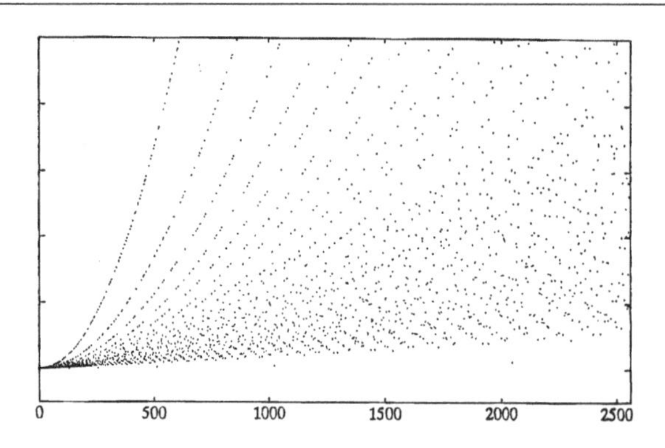

Figure 17.2: Flop-Count vs Length for the Glassman-Ferguson FFT

Appendix 3: FFT Computer Programs[1]

18.1 Goertzel Algorithm

A FORTRAN implementation of the first-order Goertzel algorithm with in-order input as given in () and [68] is given below.

[1]This content is available online at <http://cnx.org/content/m17397/1.5/>.

```
      C-----------------------------------------------
C   GOERTZEL'S  DFT  ALGORITHM
C   First order, input inorder
C   C. S. BURRUS,   SEPT 1983
C---------------------------------------------
    SUBROUTINE DFT(X,Y,A,B,N)
    REAL X(260), Y(260), A(260), B(260)
    Q = 6.283185307179586/N
    DO 20 J=1, N
       C  = COS(Q*(J-1))
       S  = SIN(Q*(J-1))
       AT = X(1)
       BT = Y(1)
       DO 30 I = 2, N
          T  = C*AT - S*BT + X(I)
          BT = C*BT + S*AT + Y(I)
          AT = T
30     CONTINUE
       A(J) = C*AT - S*BT
       B(J) = C*BT + S*AT
20  CONTINUE
    RETURN
    END
```

Listing 18.1: First Order Goertzel Algorithm

18.2 Second Order Goertzel Algorithm

Below is the program for a second order Goertzel algorithm.

```
      C--------------------------------------------
C   GOERTZEL'S  DFT  ALGORITHM
C   Second order, input inorder
C   C. S. BURRUS,   SEPT 1983
C---------------------------------------------
    SUBROUTINE DFT(X,Y,A,B,N)
    REAL X(260), Y(260), A(260), B(260)
C
    Q = 6.283185307179586/N
    DO 20 J = 1, N
       C  = COS(Q*(J-1))
       S  = SIN(Q*(J-1))
       CC = 2*C
       A2 = 0
       B2 = 0
       A1 = X(1)
       B1 = Y(1)
       DO 30 I = 2, N
          T  = A1
          A1 = CC*A1 - A2 + X(I)
          A2 = T
          T  = B1
          B1 = CC*B1 - B2 + Y(I)
          B2 = T
30     CONTINUE
       A(J) = C*A1 - A2 - S*B1
       B(J) = C*B1 - B2 + S*A1
20  CONTINUE
C
    RETURN
    END
```

Listing 18.2: Second Order Goertzel Algorithm

18.3 Second Order Goertzel Algorithm 2

Second order Goertzel algorithm that calculates two outputs at a time.

```
    C-----------------------------------------------------
C GOERTZEL'S  DFT  ALGORITHM,  Second order
C Input inorder, output by twos;  C.S. Burrus, SEPT 1991
C-----------------------------------------------------
    SUBROUTINE DFT(X,Y,A,B,N)
    REAL X(260), Y(260), A(260), B(260)
    Q = 6.283185307179586/N
    DO 20 J = 1, N/2 + 1
       C  = COS(Q*(J-1))
       S  = SIN(Q*(J-1))
       CC = 2*C
       A2 = 0
       B2 = 0
       A1 = X(1)
       B1 = Y(1)
       DO 30 I = 2, N
          T  = A1
          A1 = CC*A1 - A2 + X(I)
          A2 = T
          T  = B1
          B1 = CC*B1 - B2 + Y(I)
          B2 = T
30     CONTINUE
       A2  = C*A1 - A2
       T   = S*B1
       A(J)     = A2 - T
       A(N-J+2) = A2 + T
       B2  = C*B1 - B2
       T   = S*A1
       B(J)     = B2 + T
       B(N-J+2) = B2 - T
20  CONTINUE
    RETURN
    END
```

Figure. Second Order Goertzel Calculating Two Outputs at a Time

18.4 Basic QFT Algorithm

A FORTRAN implementation of the basic QFT algorithm is given below to show how the theory is implemented. The program is written for clarity, not to minimize the number of floating point operations.

```
     C
   SUBROUTINE QDFT(X,Y,XX,YY,NN)
   REAL X(0:260),Y(0:260),XX(0:260),YY(0:260)
   C
   N1 = NN - 1
   N2 = N1/2
   N21 = NN/2
   Q   = 6.283185308/NN
   DO 2 K = 0, N21
      SSX = X(0)
      SSY = Y(0)
      SDX = 0
      SDY = 0
      IF (MOD(NN,2).EQ.0) THEN
         SSX = SSX + COS(3.1426*K)*X(N21)
         SSY = SSY + COS(3.1426*K)*Y(N21)
      ENDIF
      DO 3 N = 1, N2
         SSX = SSX + (X(N) + X(NN-N))*COS(Q*N*K)
         SSY = SSY + (Y(N) + Y(NN-N))*COS(Q*N*K)
         SDX = SDX + (X(N) - X(NN-N))*SIN(Q*N*K)
         SDY = SDY + (Y(N) - Y(NN-N))*SIN(Q*N*K)
3     CONTINUE
      XX(K) = SSX + SDY
      YY(K) = SSY - SDX
      XX(NN-K) = SSX - SDY
      YY(NN-K) = SSY + SDX
2  CONTINUE
   RETURN
   END
```

Listing 18.3: Simple QFT Fortran Program

18.5 Basic Radix-2 FFT Algorithm

Below is the Fortran code for a simple Decimation-in-Frequency, Radix-2, one butterfly Cooley-Tukey FFT followed by a bit-reversing unscrambler.

```
    C
C   A COOLEY-TUKEY RADIX-2, DIF  FFT PROGRAM
C   COMPLEX INPUT DATA IN ARRAYS X AND Y
C      C. S. BURRUS, RICE UNIVERSITY, SEPT 1983
C---------------------------------------------------------
    SUBROUTINE FFT (X,Y,N,M)
    REAL X(1), Y(1)
C--------------MAIN FFT LOOPS-----------------------------
C
    N2 = N
    DO 10 K = 1, M
        N1 = N2
        N2 = N2/2
        E  = 6.283185307179586/N1
        A  = 0
        DO 20 J = 1, N2
        C = COS (A)
        S = SIN (A)
        A = J*E
        DO 30 I = J, N, N1
                    L = I + N2
                    XT   = X(I) - X(L)
                    X(I) = X(I) + X(L)
                    YT   = Y(I) - Y(L)
                    Y(I) = Y(I) + Y(L)
                    X(L) = C*XT + S*YT
                    Y(L) = C*YT - S*XT
   30           CONTINUE
   20       CONTINUE
   10   CONTINUE
C
C------------DIGIT REVERSE COUNTER-----------------
  100   J = 1
    N1 = N - 1
    DO 104 I=1, N1
        IF (I.GE.J) GOXTO 101
        XT = X(J)
        X(J) = X(I)
```

```
        X(I) = XT
        XT   = Y(J)
        Y(J) = Y(I)
        Y(I) = XT
  101       K = N/2
  102       IF (K.GE.J) GOTO 103
        J = J - K
        K = K/2
        GOTO 102
  103       J = J + K
  104   CONTINUE
    RETURN
    END
```

Figure: Radix-2, DIF, One Butterfly Cooley-Tukey FFT

18.6 Basic DIT Radix-2 FFT Algorithm

Below is the Fortran code for a simple Decimation-in-Time, Radix-2, one butterfly Cooley-Tukey FFT preceeded by a bit-reversing scrambler.

```
    C
C   A COOLEY-TUKEY RADIX-2, DIT  FFT PROGRAM
C   COMPLEX INPUT DATA IN ARRAYS X AND Y
C      C. S. BURRUS, RICE UNIVERSITY, SEPT 1985
C
C---------------------------------------------------------
    SUBROUTINE FFT (X,Y,N,M)
    REAL X(1), Y(1)
C------------DIGIT REVERSE COUNTER-----------------
C
  100   J = 1
    N1 = N - 1
    DO 104 I=1, N1
        IF (I.GE.J) GOTO 101
            XT = X(J)
            X(J) = X(I)
            X(I) = XT
            XT   = Y(J)
            Y(J) = Y(I)
            Y(I) = XT
```

```
  101        K = N/2
  102        IF (K.GE.J) GOTO 103
        J = J - K
        K = K/2
        GOTO 102
  103        J = J + K
  104   CONTINUE
C--------------MAIN FFT LOOPS-----------------------------
C
    N2 = 1
    DO 10 K = 1, M
        E  = 6.283185307179586/(2*N2)
        A  = 0
        DO 20 J = 1, N2
        C = COS (A)
        S = SIN (A)
        A = J*E
        DO 30 I = J, N, 2*N2
                    L = I + N2
                    XT = C*X(L) + S*Y(L)
                    YT = C*Y(L) - S*X(L)
                    X(L) = X(I) - XT
                    X(I) = X(I) + XT
                    Y(L) = Y(I) - YT
                    Y(I) = Y(I) + YT
   30            CONTINUE
   20        CONTINUE
        N2 = N2+N2
   10   CONTINUE
C
    RETURN
    END
```

18.7 DIF Radix-2 FFT Algorithm

Below is the Fortran code for a Decimation-in-Frequency, Radix-2, three butterfly Cooley-Tukey FFT followed by a bit-reversing unscrambler.

```
    C   A COOLEY-TUKEY RADIX 2, DIF  FFT PROGRAM
C   THREE-BF, MULT BY  1  AND  J  ARE REMOVED
C   COMPLEX INPUT DATA IN ARRAYS X AND Y
C   TABLE LOOK-UP OF W VALUES
```

```
C       C. S. BURRUS, RICE UNIVERSITY, SEPT 1983
C---------------------------------------------------------
      SUBROUTINE FFT (X,Y,N,M,WR,WI)
      REAL X(1), Y(1), WR(1), WI(1)
C--------------MAIN FFT LOOPS-----------------------------
C
      N2 = N
      DO 10 K = 1, M
          N1 = N2
          N2 = N2/2
          JT = N2/2 + 1
          DO 1 I = 1, N, N1
          L = I + N2
          T    = X(I) - X(L)
          X(I) = X(I) + X(L)
          X(L) = T
          T    = Y(I) - Y(L)
          Y(I) = Y(I) + Y(L)
          Y(L) = T
   1          CONTINUE
          IF (K.EQ.M) GOTO 10
          IE  = N/N1
          IA  = 1
          DO 20 J = 2, N2
          IA = IA + IE
          IF (J.EQ.JT) GOTO 50
          C = WR(IA)
          S = WI(IA)
          DO 30 I = J, N, N1
              L = I + N2
              T    = X(I) - X(L)
              X(I) = X(I) + X(L)
              TY   = Y(I) - Y(L)
              Y(I) = Y(I) + Y(L)
              X(L) = C*T + S*TY
              Y(L) = C*TY - S*T
   30         CONTINUE
          GOTO 25
   50         DO 40 I = J, N, N1
              L = I + N2
              T    = X(I) - X(L)
              X(I) = X(I) + X(L)
```

```
            TY   = Y(I) - Y(L)
            Y(I) = Y(I) + Y(L)
            X(L) = TY
            Y(L) =-T
   40       CONTINUE
   25       A = J*E
   20       CONTINUE
   10   CONTINUE
C------------DIGIT REVERSE COUNTER Goes here----------
    RETURN
    END
```

18.8 Basic DIF Radix-4 FFT Algorithm

Below is the Fortran code for a simple Decimation-in-Frequency, Radix-4, one butterfly Cooley-Tukey FFT to be followed by an unscrambler.

```
    C   A COOLEY-TUKEY RADIX-4 DIF  FFT PROGRAM
C   COMPLEX INPUT DATA IN ARRAYS X AND Y
C   LENGTH IS  N = 4 ** M
C     C. S. BURRUS, RICE UNIVERSITY, SEPT 1983
C---------------------------------------------------------
    SUBROUTINE  FFT4 (X,Y,N,M)
    REAL X(1), Y(1)
C--------------MAIN FFT LOOPS-----------------------------
    N2 = N
    DO 10 K = 1, M
        N1 = N2
        N2 = N2/4
        E = 6.283185307179586/N1
        A = 0
C--------------------MAIN BUTTERFLIES-------------------
        DO 20 J=1, N2
            B    = A + A
            C    = A + B
            CO1  = COS(A)
            CO2  = COS(B)
            CO3  = COS(C)
            SI1  = SIN(A)
            SI2  = SIN(B)
            SI3  = SIN(C)
```

```
            A     = J*E
C---------------BUTTERFLIES WITH SAME W--------------
            DO 30 I=J, N, N1
            I1 = I  + N2
            I2 = I1 + N2
            I3 = I2 + N2
            R1 = X(I ) + X(I2)
            R3 = X(I ) - X(I2)
            S1 = Y(I ) + Y(I2)
            S3 = Y(I ) - Y(I2)
            R2 = X(I1) + X(I3)
            R4 = X(I1) - X(I3)
            S2 = Y(I1) + Y(I3)
            S4 = Y(I1) - Y(I3)
            X(I) = R1 + R2
            R2   = R1 - R2
            R1   = R3 - S4
            R3   = R3 + S4
            Y(I) = S1 + S2
            S2   = S1 - S2
            S1   = S3 + R4
            S3   = S3 - R4
            X(I1) = CO1*R3 + SI1*S3
            Y(I1) = CO1*S3 - SI1*R3
            X(I2) = CO2*R2 + SI2*S2
            Y(I2) = CO2*S2 - SI2*R2
            X(I3) = CO3*R1 + SI3*S1
            Y(I3) = CO3*S1 - SI3*R1
 30             CONTINUE
 20         CONTINUE
 10     CONTINUE
C-----------DIGIT REVERSE COUNTER goes here-----
    RETURN
    END
```

18.9 Basic DIF Radix-4 FFT Algorithm

Below is the Fortran code for a Decimation-in-Frequency, Radix-4, three butterfly Cooley-Tukey FFT followed by a bit-reversing unscrambler. Twiddle factors are precalculated and stored in arrays WR and WI.

```
    C
```

```
C   A COOLEY-TUKEY RADIX-4 DIF  FFT PROGRAM
C   THREE BF, MULTIPLICATIONS BY  1, J, ETC. ARE REMOVED
C   COMPLEX INPUT DATA IN ARRAYS X AND Y
C   LENGTH IS  N = 4 ** M
C   TABLE LOOKUP OF W VALUES
C
C     C. S. BURRUS, RICE UNIVERSITY,  SEPT 1983
C
C---------------------------------------------------------
C
    SUBROUTINE  FFT4 (X,Y,N,M,WR,WI)
    REAL X(1), Y(1), WR(1), WI(1)
    DATA C21 / 0.707106778 /
C
C--------------MAIN FFT LOOPS-----------------------------
C
    N2 = N
    DO 10 K = 1, M
        N1 = N2
        N2 = N2/4
        JT = N2/2 + 1
C---------------SPECIAL BUTTERFLY FOR W = 1---------------
        DO 1 I = 1, N, N1
            I1 = I  + N2
            I2 = I1 + N2
            I3 = I2 + N2
            R1 = X(I ) + X(I2)
            R3 = X(I ) - X(I2)
            S1 = Y(I ) + Y(I2)
            S3 = Y(I ) - Y(I2)
            R2 = X(I1) + X(I3)
            R4 = X(I1) - X(I3)
            S2 = Y(I1) + Y(I3)
            S4 = Y(I1) - Y(I3)
C
            X(I) = R1 + R2
            X(I2)= R1 - R2
            X(I3)= R3 - S4
            X(I1)= R3 + S4
C
            Y(I) = S1 + S2
            Y(I2)= S1 - S2
```

```
          Y(I3)= S3 + R4
          Y(I1)= S3 - R4
C
  1       CONTINUE
      IF (K.EQ.M) GOTO 10
      IE = N/N1
      IA1 = 1

C--------------GENERAL BUTTERFLY-----------------
      DO 20 J = 2, N2
      IA1  = IA1 + IE
      IF (J.EQ.JT) GOTO 50
      IA2  = IA1 + IA1 - 1
          IA3  = IA2 + IA1 - 1
          CO1  = WR(IA1)
          CO2  = WR(IA2)
          CO3  = WR(IA3)
          SI1  = WI(IA1)
          SI2  = WI(IA2)
          SI3  = WI(IA3)
C----------------BUTTERFLIES WITH SAME W---------------
          DO 30 I = J, N, N1
          I1 = I  + N2
          I2 = I1 + N2
          I3 = I2 + N2
          R1 = X(I ) + X(I2)
          R3 = X(I ) - X(I2)
          S1 = Y(I ) + Y(I2)
          S3 = Y(I ) - Y(I2)
          R2 = X(I1) + X(I3)
          R4 = X(I1) - X(I3)
          S2 = Y(I1) + Y(I3)
          S4 = Y(I1) - Y(I3)
C
          X(I) = R1 + R2
          R2   = R1 - R2
          R1   = R3 - S4
          R3   = R3 + S4
C
          Y(I) = S1 + S2
          S2   = S1 - S2
          S1   = S3 + R4
```

```
            S3   = S3 - R4
C
            X(I1) = CO1*R3 + SI1*S3
            Y(I1) = CO1*S3 - SI1*R3
            X(I2) = CO2*R2 + SI2*S2
            Y(I2) = CO2*S2 - SI2*R2
            X(I3) = CO3*R1 + SI3*S1
            Y(I3) = CO3*S1 - SI3*R1
  30            CONTINUE
        GOTO 20
C------------------SPECIAL BUTTERFLY FOR  W = J-----------
  50            DO 40 I = J, N, N1
            I1 = I  + N2
            I2 = I1 + N2
            I3 = I2 + N2
            R1 = X(I ) + X(I2)
            R3 = X(I ) - X(I2)
            S1 = Y(I ) + Y(I2)
            S3 = Y(I ) - Y(I2)
            R2 = X(I1) + X(I3)
            R4 = X(I1) - X(I3)
            S2 = Y(I1) + Y(I3)
            S4 = Y(I1) - Y(I3)
C
            X(I) = R1 + R2
            Y(I2)=-R1 + R2
            R1   = R3 - S4
            R3   = R3 + S4
C
            Y(I) = S1 + S2
            X(I2)= S1 - S2
            S1   = S3 + R4
            S3   = S3 - R4
C
            X(I1) = (S3 + R3)*C21
            Y(I1) = (S3 - R3)*C21
            X(I3) = (S1 - R1)*C21
            Y(I3) =-(S1 + R1)*C21
  40            CONTINUE
  20        CONTINUE
  10    CONTINUE
C-----------DIGIT REVERSE COUNTER----------
```

```
  100    J = 1
    N1 = N - 1
    DO 104 I = 1, N1
        IF (I.GE.J) GOTO 101
        R1   = X(J)
        X(J) = X(I)
        X(I) = R1
        R1   = Y(J)
        Y(J) = Y(I)
        Y(I) = R1
 101        K = N/4
 102        IF (K*3.GE.J) GOTO 103
                J = J - K*3
                K = K/4
                GOTO 102
 103        J = J + K
 104    CONTINUE
    RETURN
    END
```

18.10 Basic DIF Split Radix FFT Algorithm

Below is the Fortran code for a simple Decimation-in-Frequency, Split-Radix, one butterfly FFT to be followed by a bit-reversing unscrambler.

```
    C   A DUHAMEL-HOLLMANN SPLIT RADIX  FFT PROGRAM
C   FROM: ELECTRONICS LETTERS, JAN. 5, 1984
C   COMPLEX INPUT DATA IN ARRAYS X AND Y
C   LENGTH IS  N = 2 ** M
C     C. S. BURRUS, RICE UNIVERSITY, MARCH 1984
C
C---------------------------------------------------------
    SUBROUTINE  FFT (X,Y,N,M)
    REAL X(1), Y(1)
C--------------MAIN FFT LOOPS-----------------------------
C
    N1 = N
    N2 = N/2
    IP = 0
    IS = 1
    A  = 6.283185307179586/N
```

```
    DO 10 K = 1, M-1
        JD = N1 + N2
        N1 = N2
        N2 = N2/2
        J0 = N1*IP + 1
        IP = 1 - IP
        DO 20 J = J0, N, JD
            JS = 0
        JT = J + N2 - 1
            DO 30 I = J, JT
            JSS= JS*IS
            JS = JS + 1
                C1 = COS(A*JSS)
                C3 = COS(3*A*JSS)
                S1 = -SIN(A*JSS)
                S3 = -SIN(3*A*JSS)
            I1 = I  + N2
            I2 = I1 + N2
            I3 = I2 + N2
            R1     = X(I ) + X(I2)
            R2     = X(I ) - X(I2)
            R3     = X(I1) - X(I3)
            X(I2) = X(I1) + X(I3)
            X(I1) = R1
C
            R1     = Y(I ) + Y(I2)
            R4     = Y(I ) - Y(I2)
            R5     = Y(I1) - Y(I3)
            Y(I2) = Y(I1) + Y(I3)
            Y(I1) = R1
C
            R1     = R2 - R5
            R2     = R2 + R5
            R5     = R4 + R3
            R4     = R4 - R3
C
            X(I)   = C1*R1 + S1*R5
            Y(I)   = C1*R5 - S1*R1
            X(I3) = C3*R2 + S3*R4
            Y(I3) = C3*R4 - S3*R2
  30            CONTINUE
  20        CONTINUE
```

```
        IS = IS + IS
 10     CONTINUE
    IP = 1 - IP
    JO = 2 - IP
    DO 5 I = JO, N-1, 3
       I1 = I + 1
       R1     = X(I) + X(I1)
       X(I1) = X(I) - X(I1)
       X(I)   = R1
       R1     = Y(I) + Y(I1)
       Y(I1) = Y(I) - Y(I1)
       Y(I)   = R1
  5     CONTINUE
    RETURN
    END
```

18.11 DIF Split Radix FFT Algorithm

Below is the Fortran code for a simple Decimation-in-Frequency, Split-Radix, two butterfly FFT to be followed by a bit-reversing unscrambler. Twiddle factors are precalculated and stored in arrays WR and WI.

```
    C-------------------------------------------------------------C
C       A DUHAMEL-HOLLMAN SPLIT RADIX FFT                         C
C       REF: ELECTRONICS LETTERS, JAN. 5, 1984                    C
C       COMPLEX INPUT AND OUTPUT DATA IN ARRAYS X AND Y           C
C       LENGTH IS N = 2 ** M,  OUTPUT IN BIT-REVERSED ORDER       C
C   TWO BUTTERFLIES TO REMOVE MULTS BY UNITY                  C
C       SPECIAL LAST TWO STAGES                                   C
C   TABLE LOOK-UP OF SINE AND COSINE VALUES               C
C       C.S. BURRUS,       RICE UNIV.       APRIL 1985            C
C-------------------------------------------------------------C
C
        SUBROUTINE FFT(X,Y,N,M,WR,WI)
        REAL X(1),Y(1),WR(1),WI(1)
    C81= 0.707106778
        N2 = 2*N
        DO  10 K = 1, M-3
        IS  = 1
        ID  = N2
        N2 = N2/2
        N4 = N2/4
```

```
 2      DO 1 I0 = IS, N-1, ID
        I1 = I0 + N4
        I2 = I1 + N4
            I3 = I2 + N4
        R1     = X(I0) - X(I2)
        X(I0) = X(I0) + X(I2)
        R2     = Y(I1) - Y(I3)
        Y(I1) = Y(I1) + Y(I3)
        X(I2) = R1 + R2
        R2     = R1 - R2
        R1     = X(I1) - X(I3)
        X(I1) = X(I1) + X(I3)
            X(I3) = R2
            R2     = Y(I0) - Y(I2)
                Y(I0) = Y(I0) + Y(I2)
            Y(I2) =-R1 + R2
            Y(I3) = R1 + R2
  1     CONTINUE
        IS = 2*ID - N2 + 1
        ID = 4*ID
            IF (IS.LT.N) GOTO 2
        IE  = N/N2
            IA1 = 1
            DO  20 J = 2, N4
                IA1 = IA1 + IE
                IA3 = 3*IA1 - 2
                CC1 = WR(IA1)
                SS1 = WI(IA1)
                CC3 = WR(IA3)
                SS3 = WI(IA3)
                IS  = J
                ID  = 2*N2
 40             DO 30 I0 = IS, N-1, ID
                    I1 = I0 + N4
                    I2 = I1 + N4
                    I3 = I2 + N4
C
                    R1     = X(I0) - X(I2)
                    X(I0) = X(I0) + X(I2)
                    R2     = X(I1) - X(I3)
                    X(I1) = X(I1) + X(I3)
                    S1     = Y(I0) - Y(I2)
```

```
                    Y(I0) = Y(I0) + Y(I2)
                    S2    = Y(I1) - Y(I3)
                    Y(I1) = Y(I1) + Y(I3)
C
                    S3    = R1 - S2
                    R1    - R1 + S2
                    S2    = R2 - S1
                    R2    = R2 + S1
                    X(I2) = R1*CC1 - S2*SS1
                    Y(I2) =-S2*CC1 - R1*SS1
                    X(I3) = S3*CC3 + R2*SS3
                    Y(I3) = R2*CC3 - S3*SS3
 30             CONTINUE
                IS = 2*ID - N2 + J
                ID = 4*ID
                IF (IS.LT.N) GOTO 40
  20        CONTINUE
  10    CONTINUE
C
        IS =  1
        ID = 32
  50    DO 60 I = IS, N, ID
            I0    = I + 8
            DO 15 J = 1, 2
               R1 = X(I0)   + X(I0+2)
               R3 = X(I0)   - X(I0+2)
               R2 = X(I0+1) + X(I0+3)
               R4 = X(I0+1) - X(I0+3)
               X(I0)   = R1 + R2
               X(I0+1) = R1 - R2
               R1 = Y(I0)   + Y(I0+2)
               S3 = Y(I0)   - Y(I0+2)
               R2 = Y(I0+1) + Y(I0+3)
               S4 = Y(I0+1) - Y(I0+3)
               Y(I0)   = R1 + R2
               Y(I0+1) = R1 - R2
               Y(I0+2) = S3 - R4
               Y(I0+3) = S3 + R4
               X(I0+2) = R3 + S4
               X(I0+3) = R3 - S4
               I0 = I0 + 4
  15        CONTINUE
```

```
  60    CONTINUE
            IS = 2*ID - 15
            ID = 4*ID
        IF (IS.LT.N) GOTO 50
C
        IS =  1
        ID = 16
  55    DO 65 I0 = IS, N, ID
                R1 = X(I0)   + X(I0+4)
            R5 = X(I0)   - X(I0+4)
            R2 = X(I0+1) + X(I0+5)
            R6 = X(I0+1) - X(I0+5)
            R3 = X(I0+2) + X(I0+6)
            R7 = X(I0+2) - X(I0+6)
            R4 = X(I0+3) + X(I0+7)
            R8 = X(I0+3) - X(I0+7)
            T1 = R1 - R3
            R1 = R1 + R3
            R3 = R2 - R4
            R2 = R2 + R4
            X(I0)   = R1 + R2
            X(I0+1) = R1 - R2
C
            R1 = Y(I0)   + Y(I0+4)
            S5 = Y(I0)   - Y(I0+4)
            R2 = Y(I0+1) + Y(I0+5)
            S6 = Y(I0+1) - Y(I0+5)
            S3 = Y(I0+2) + Y(I0+6)
            S7 = Y(I0+2) - Y(I0+6)
            R4 = Y(I0+3) + Y(I0+7)
            S8 = Y(I0+3) - Y(I0+7)
            T2 = R1 - S3
            R1 = R1 + S3
            S3 = R2 - R4
            R2 = R2 + R4
            Y(I0)   = R1 + R2
            Y(I0+1) = R1 - R2
            X(I0+2) = T1 + S3
            X(I0+3) = T1 - S3
            Y(I0+2) = T2 - R3
            Y(I0+3) = T2 + R3
C
```

```
            R1 = (R6 - R8)*C81
            R6 = (R6 + R8)*C81
            R2 = (S6 - S8)*C81
            S6 = (S6 + S8)*C81
C
            T1 = R5 - R1
            R5 = R5 + R1
            R8 = R7 - R6
            R7 = R7 + R6
            T2 = S5 - R2
            S5 = S5 + R2
            S8 = S7 - S6
            S7 = S7 + S6
            X(I0+4) = R5 + S7
            X(I0+7) = R5 - S7
            X(I0+5) = T1 + S8
            X(I0+6) = T1 - S8
            Y(I0+4) = S5 - R7
            Y(I0+7) = S5 + R7
            Y(I0+5) = T2 - R8
            Y(I0+6) = T2 + R8
  65    CONTINUE
            IS = 2*ID - 7
            ID = 4*ID
        IF (IS.LT.N) GOTO 55
C
C------------BIT REVERSE COUNTER-----------------
C
  100   J = 1
        N1 = N - 1
        DO 104 I=1, N1
            IF (I.GE.J) GOTO 101
            XT = X(J)
            X(J) = X(I)
            X(I) = XT
            XT   = Y(J)
            Y(J) = Y(I)
            Y(I) = XT
  101       K = N/2
  102       IF (K.GE.J) GOTO 103
                J = J - K
                K = K/2
```

```
              GOTO 102
 103        J = J + K
 104    CONTINUE
        RETURN
        END
```

18.12 Prime Factor FFT Algorithm

Below is the Fortran code for a Prime-Factor Algorithm (PFA) FFT allowing factors of the length of 2, 3, 4, 5, and 7. It is followed by an unscrambler.

```
    C---------------------------------------------------
C
C   A PRIME FACTOR FFT PROGRAM WITH GENERAL MODULES
C   COMPLEX INPUT DATA IN ARRAYS  X AND Y
C   COMPLEX OUTPUT IN  A AND B
C   LENGTH  N  WITH  M  FACTORS IN ARRAY  NI
C     N = NI(1)*NI(2)* ... *NI(M)
C   UNSCRAMBLING CONSTANT  UNSC
C     UNSC = N/NI(1) + N/NI(2) +...+ N/NI(M), MOD N
C      C. S. BURRUS, RICE UNIVERSITY, JAN 1987
C
C--------------------------------------------------
C
    SUBROUTINE PFA(X,Y,N,M,NI,A,B,UNSC)
C
    INTEGER  NI(4), I(16), UNSC
        REAL X(1), Y(1), A(1), B(1)
C
    DATA  C31, C32  / -0.86602540,-1.50000000 /
    DATA  C51, C52  /  0.95105652,-1.53884180 /
    DATA  C53, C54  / -0.36327126, 0.55901699 /
    DATA  C55       / -1.25  /
    DATA  C71, C72  / -1.16666667,-0.79015647 /
    DATA  C73, C74  /  0.055854267, 0.7343022 /
    DATA  C75, C76  /  0.44095855,-0.34087293 /
    DATA  C77, C78  /  0.53396936, 0.87484229 /
C
C-----------------NESTED LOOPS----------------------
C
    DO 10 K=1, M
          N1 = NI(K)
```

```
      N2 = N/N1
      DO 15 J=1, N, N1
             IT   = J
             DO 30 L=1, N1
                I(L) = IT
            A(L) = X(IT)
            B(L) = Y(IT)
                IT = IT + N2
                IF (IT.GT.N)  IT = IT - N
   30        CONTINUE
             GOTO (20,102,103,104,105,20,107), N1
C
C----------------WFTA N=2--------------------------------
C
  102   R1    = A(1)
    A(1)  = R1 + A(2)
    A(2)  = R1 - A(2)
C
    R1   = B(1)
    B(1) = R1 + B(2)
    B(2) = R1 - B(2)
C
    GOTO 20
C----------------WFTA N=3--------------------------------
C
  103   R2 = (A(2) - A(3)) * C31
    R1 =  A(2) + A(3)
    A(1)= A(1) + R1
    R1  = A(1) + R1 * C32
C
    S2 = (B(2) - B(3)) * C31
    S1 =  B(2) + B(3)
    B(1)= B(1) + S1
    S1  = B(1) + S1 * C32
C
    A(2) = R1 - S2
    A(3) = R1 + S2
    B(2) = S1 + R2
    B(3) = S1 - R2
C
    GOTO 20
C
```

```
C----------------WFTA N=4---------------------------------
C
  104   R1 = A(1) + A(3)
    T1 = A(1) - A(3)
    R2 = A(2) + A(4)
    A(1) = R1 + R2
    A(3) = R1 - R2
C
    R1 = B(1) + B(3)
    T2 = B(1) - B(3)
    R2 = B(2) + B(4)
    B(1) = R1 + R2
    B(3) = R1 - R2
C
    R1 = A(2) - A(4)
    R2 = B(2) - B(4)
C
    A(2) = T1 + R2
    A(4) = T1 - R2
    B(2) = T2 - R1
    B(4) = T2 + R1
C
    GOTO 20
C
C----------------WFTA N=5--------------------------------
C
  105   R1 = A(2) + A(5)
    R4 = A(2) - A(5)
    R3 = A(3) + A(4)
    R2 = A(3) - A(4)
C
    T = (R1 - R3) * C54
    R1 = R1 + R3
    A(1) = A(1) + R1
    R1   = A(1) + R1 * C55
C
    R3 = R1 - T
    R1 = R1 + T
C
    T = (R4 + R2) * C51
    R4 =  T + R4 * C52
    R2 =  T + R2 * C53
```

```
C
      S1 = B(2) + B(5)
      S4 = B(2) - B(5)
      S3 = B(3) + B(4)
      S2 = B(3) - B(4)
C
      T = (S1 - S3) * C54
      S1 = S1 + S3
      B(1) = B(1) + S1
      S1   = B(1) + S1 * C55
C
      S3 =  S1 - T
      S1 =  S1 + T
C
      T = (S4 + S2) * C51
      S4 =  T + S4 * C52
      S2 =  T + S2 * C53
C
      A(2) = R1 + S2
      A(5) = R1 - S2
      A(3) = R3 - S4
      A(4) = R3 + S4
C
      B(2) = S1 - R2
      B(5) = S1 + R2
      B(3) = S3 + R4
      B(4) = S3 - R4
C
      GOTO 20
C-----------------WFTA N=7--------------------------
C
  107   R1 = A(2) + A(7)
      R6 = A(2) - A(7)
      S1 = B(2) + B(7)
      S6 = B(2) - B(7)
      R2 = A(3) + A(6)
      R5 = A(3) - A(6)
      S2 = B(3) + B(6)
      S5 = B(3) - B(6)
      R3 = A(4) + A(5)
      R4 = A(4) - A(5)
      S3 = B(4) + B(5)
```

```
      S4 = B(4) - B(5)
C
      T3 = (R1 - R2) * C74
      T  = (R1 - R3) * C72
      R1 = R1 + R2 + R3
      A(1) = A(1) + R1
      R1   = A(1) + R1 * C71
      R2 =(R3 - R2) * C73
      R3 = R1 - T + R2
      R2 = R1 - R2 - T3
      R1 = R1 + T + T3
      T = (R6 - R5) * C78
      T3 =(R6 + R4) * C76
      R6 =(R6 + R5 - R4) * C75
      R5 =(R5 + R4) * C77
      R4 = R6 - T3 + R5
      R5 = R6 - R5 - T
      R6 = R6 + T3 + T
C
      T3 = (S1 - S2) * C74
      T  = (S1 - S3) * C72
      S1 =  S1 + S2 + S3
      B(1) = B(1) + S1
      S1   = B(1) + S1 * C71
      S2 =(S3 - S2) * C73
      S3 = S1 - T  + S2
      S2 = S1 - S2 - T3
      S1 = S1 + T  + T3
      T  = (S6 - S5) * C78
      T3 = (S6 + S4) * C76
      S6 = (S6 + S5 - S4) * C75
      S5 = (S5 + S4) * C77
      S4 = S6 - T3 + S5
      S5 = S6 - S5 - T
      S6 = S6 + T3 + T
C
      A(2) = R3 + S4
      A(7) = R3 - S4
      A(3) = R1 + S6
      A(6) = R1 - S6
      A(4) = R2 - S5
      A(5) = R2 + S5
```

```
      B(4) = S2 + R5
      B(5) = S2 - R5
      B(2) = S3 - R4
      B(7) = S3 + R4
      B(3) = S1 - R6
      B(6) = S1 + R6
C
  20          IT   = J
              DO 31 L=1, N1
                 I(L) = IT
             X(IT) = A(L)
             Y(IT) = B(L)
                 IT = IT + N2
                 IF (IT.GT.N)  IT = IT - N
  31          CONTINUE
  15        CONTINUE
  10      CONTINUE
C
C-----------------UNSCRAMBLING----------------------
C
      L = 1
      DO 2 K=1, N
         A(K) = X(L)
         B(K) = Y(L)
            L = L + UNSC
            IF (L.GT.N)  L = L - N
   2  CONTINUE
      RETURN
      END
C
```

18.13 In Place, In Order Prime Factor FFT Algorithm

Below is the Fortran code for a Prime-Factor Algorithm (PFA) FFT allowing factors of the length of 2, 3, 4, 5, 7, 8, 9, and 16. It is both in-place and in-order, so requires no unscrambler.

```
      C
C     A PRIME FACTOR FFT PROGRAM
C     IN-PLACE AND IN-ORDER
C     COMPLEX INPUT DATA IN ARRAYS  X AND Y
C     LENGTH  N  WITH  M  FACTORS IN ARRAY  NI
```

```
C       N = NI(1)*NI(2)*...*NI(M)
C   REDUCED TEMP STORAGE IN SHORT WFTA MODULES
C   Has modules 2,3,4,5,7,8,9,16
C   PROGRAM BY  C. S. BURRUS,  RICE UNIVERSITY
C                  SEPT 1983
C---------------------------------------------------
C
    SUBROUTINE PFA(X,Y,N,M,NI)
    INTEGER  NI(4), I(16), IP(16), LP(16)
        REAL X(1), Y(1)
    DATA  C31, C32  / -0.86602540,-1.50000000 /
    DATA  C51, C52  /  0.95105652,-1.53884180 /
    DATA  C53, C54  / -0.36327126, 0.55901699 /
    DATA  C55       / -1.25  /
    DATA  C71, C72  / -1.16666667,-0.79015647 /
    DATA  C73, C74  /  0.055854267, 0.7343022 /
    DATA  C75, C76  /  0.44095855,-0.34087293 /
    DATA  C77, C78  /  0.53396936, 0.87484229 /
    DATA  C81       /  0.70710678 /
    DATA  C95       / -0.50000000 /
    DATA  C92, C93  /  0.93969262, -0.17364818 /
    DATA  C94, C96  /  0.76604444, -0.34202014 /
    DATA  C97, C98  / -0.98480775, -0.64278761 /
    DATA  C162,C163 /  0.38268343,  1.30656297 /
    DATA  C164,C165 /  0.54119610,  0.92387953 /
C
C-----------------NESTED LOOPS----------------------------------
C
    DO 10 K=1, M
          N1 = NI(K)
       N2 = N/N1
       L  = 1
       N3 = N2 - N1*(N2/N1)
       DO 15 J = 1, N1
          LP(J) = L
              L = L + N3
          IF (L.GT.N1) L = L - N1
   15      CONTINUE
C
       DO 20 J=1, N, N1
          IT   = J
          DO 30 L=1, N1
```

```
            I(L) = IT
            IP(LP(L)) = IT
            IT = IT + N2
            IF (IT.GT.N)  IT = IT - N
  30         CONTINUE
        GOTO (20,102,103,104,105,20,107,108,109,
     +            20,20,20,20,20,20,116),N1

C-----------------WFTA N=2--------------------------------
C
 102   R1        = X(I(1))
    X(I(1))  = R1 + X(I(2))
    X(I(2))  = R1 - X(I(2))
C
    R1       = Y(I(1))
    Y(IP(1)) = R1 + Y(I(2))
    Y(IP(2)) = R1 - Y(I(2))
C
    GOTO 20
C
C-----------------WFTA N=3--------------------------------
C
 103   R2 = (X(I(2)) - X(I(3))) * C31
    R1 =  X(I(2)) + X(I(3))
    X(I(1))= X(I(1)) + R1
    R1     = X(I(1)) + R1 * C32
C
    S2 = (Y(I(2)) - Y(I(3))) * C31
    S1 =  Y(I(2)) + Y(I(3))
    Y(I(1))= Y(I(1)) + S1
    S1     = Y(I(1)) + S1 * C32
C
    X(IP(2)) = R1 - S2
    X(IP(3)) = R1 + S2
    Y(IP(2)) = S1 + R2
    Y(IP(3)) = S1 - R2
C
    GOTO 20
C
C-----------------WFTA N=4---------------------------------
C
 104   R1 = X(I(1)) + X(I(3))
```

```
      T1 = X(I(1)) - X(I(3))
      R2 = X(I(2)) + X(I(4))
      X(IP(1)) = R1 + R2
      X(IP(3)) = R1 - R2
C
      R1 = Y(I(1)) + Y(I(3))
      T2 = Y(I(1)) - Y(I(3))
      R2 = Y(I(2)) + Y(I(4))
      Y(IP(1)) = R1 + R2
      Y(IP(3)) = R1 - R2
C
      R1 = X(I(2)) - X(I(4))
      R2 = Y(I(2)) - Y(I(4))
C
      X(IP(2)) = T1 + R2
      X(IP(4)) = T1 - R2
      Y(IP(2)) = T2 - R1
      Y(IP(4)) = T2 + R1
C
      GOTO 20

C----------------WFTA N=5--------------------------------
C
  105   R1 = X(I(2)) + X(I(5))
      R4 = X(I(2)) - X(I(5))
      R3 = X(I(3)) + X(I(4))
      R2 = X(I(3)) - X(I(4))
C
      T = (R1 - R3) * C54
      R1 = R1 + R3
      X(I(1)) = X(I(1)) + R1
      R1      = X(I(1)) + R1 * C55
C
      R3 = R1 - T
      R1 = R1 + T
C
      T = (R4 + R2) * C51
      R4 =  T + R4 * C52
      R2 =  T + R2 * C53
C
      S1 = Y(I(2)) + Y(I(5))
      S4 = Y(I(2)) - Y(I(5))
```

```
      S3 = Y(I(3)) + Y(I(4))
      S2 = Y(I(3)) - Y(I(4))
C
      T = (S1 - S3) * C54
      S1 = S1 + S3
      Y(I(1)) = Y(I(1)) + S1
      S1      = Y(I(1)) + S1 * C55
C
      S3 =  S1 - T
      S1 =  S1 + T
C
      T = (S4 + S2) * C51
      S4 =  T + S4 * C52
      S2 =  T + S2 * C53
C
      X(IP(2)) = R1 + S2
      X(IP(5)) = R1 - S2
      X(IP(3)) = R3 - S4
      X(IP(4)) = R3 + S4
C
      Y(IP(2)) = S1 - R2
      Y(IP(5)) = S1 + R2
      Y(IP(3)) = S3 + R4
      Y(IP(4)) = S3 - R4
C
      GOTO 20

C-----------------WFTA N=7-------------------------
C
  107   R1 = X(I(2)) + X(I(7))
      R6 = X(I(2)) - X(I(7))
      S1 = Y(I(2)) + Y(I(7))
      S6 = Y(I(2)) - Y(I(7))
      R2 = X(I(3)) + X(I(6))
      R5 = X(I(3)) - X(I(6))
      S2 = Y(I(3)) + Y(I(6))
      S5 = Y(I(3)) - Y(I(6))
      R3 = X(I(4)) + X(I(5))
      R4 = X(I(4)) - X(I(5))
      S3 = Y(I(4)) + Y(I(5))
      S4 = Y(I(4)) - Y(I(5))
C
```

```
      T3 = (R1 - R2) * C74
      T  = (R1 - R3) * C72
      R1 = R1 + R2 + R3
      X(I(1)) = X(I(1)) + R1
      R1      = X(I(1)) + R1 * C71
      R2 =(R3 - R2) * C73
      R3 = R1 - T + R2
      R2 = R1 - R2 - T3
      R1 = R1 + T + T3
      T = (R6 - R5) * C78
      T3 =(R6 + R4) * C76
      R6 =(R6 + R5 - R4) * C75
      R5 =(R5 + R4) * C77
      R4 = R6 - T3 + R5
      R5 = R6 - R5 - T
      R6 = R6 + T3 + T
C
      T3 = (S1 - S2) * C74
      T  = (S1 - S3) * C72
      S1 =  S1 + S2 + S3
      Y(I(1)) = Y(I(1)) + S1
      S1      = Y(I(1)) + S1 * C71
      S2 =(S3 - S2) * C73
      S3 = S1 - T  + S2
      S2 = S1 - S2 - T3
      S1 = S1 + T  + T3
      T  = (S6 - S5) * C78
      T3 = (S6 + S4) * C76
      S6 = (S6 + S5 - S4) * C75
      S5 = (S5 + S4) * C77
      S4 = S6 - T3 + S5
      S5 = S6 - S5 - T
      S6 = S6 + T3 + T
C
      X(IP(2)) = R3 + S4
      X(IP(7)) = R3 - S4
      X(IP(3)) = R1 + S6
      X(IP(6)) = R1 - S6
      X(IP(4)) = R2 - S5
      X(IP(5)) = R2 + S5
      Y(IP(4)) = S2 + R5
      Y(IP(5)) = S2 - R5
```

```
      Y(IP(2)) = S3 - R4
      Y(IP(7)) = S3 + R4
      Y(IP(3)) = S1 - R6
      Y(IP(6)) = S1 + R6
C
      GOTO 20

C-----------------WFTA N=8-------------------------
C
  108   R1 = X(I(1)) + X(I(5))
      R2 = X(I(1)) - X(I(5))
      R3 = X(I(2)) + X(I(8))
      R4 = X(I(2)) - X(I(8))
      R5 = X(I(3)) + X(I(7))
      R6 = X(I(3)) - X(I(7))
      R7 = X(I(4)) + X(I(6))
      R8 = X(I(4)) - X(I(6))
      T1 = R1 + R5
      T2 = R1 - R5
      T3 = R3 + R7
      R3 =(R3 - R7) * C81
      X(IP(1)) = T1 + T3
      X(IP(5)) = T1 - T3
      T1 = R2 + R3
      T3 = R2 - R3
      S1 = R4 - R8
      R4 =(R4 + R8) * C81
      S2 = R4 + R6
      S3 = R4 - R6
      R1 = Y(I(1)) + Y(I(5))
      R2 = Y(I(1)) - Y(I(5))
      R3 = Y(I(2)) + Y(I(8))
      R4 = Y(I(2)) - Y(I(8))
      R5 = Y(I(3)) + Y(I(7))
      R6 = Y(I(3)) - Y(I(7))
      R7 = Y(I(4)) + Y(I(6))
      R8 = Y(I(4)) - Y(I(6))
      T4 = R1 + R5
      R1 = R1 - R5
      R5 = R3 + R7
      R3 =(R3 - R7) * C81
      Y(IP(1)) = T4 + R5
```

```
      Y(IP(5)) = T4 - R5
      R5 = R2 + R3
      R2 = R2 - R3
      R3 = R4 - R8
      R4 =(R4 + R8) * C81
      R7 = R4 + R6
      R4 = R4 - R6
      X(IP(2)) = T1 + R7
      X(IP(8)) = T1 - R7
      X(IP(3)) = T2 + R3
      X(IP(7)) = T2 - R3
      X(IP(4)) = T3 + R4
      X(IP(6)) = T3 - R4
      Y(IP(2)) = R5 - S2
      Y(IP(8)) = R5 + S2
      Y(IP(3)) = R1 - S1
      Y(IP(7)) = R1 + S1
      Y(IP(4)) = R2 - S3
      Y(IP(6)) = R2 + S3
C
      GOTO 20

C-----------------WFTA N=9----------------------
C
  109   R1 = X(I(2)) + X(I(9))
      R2 = X(I(2)) - X(I(9))
      R3 = X(I(3)) + X(I(8))
      R4 = X(I(3)) - X(I(8))
      R5 = X(I(4)) + X(I(7))
      T8 =(X(I(4)) - X(I(7))) * C31
      R7 = X(I(5)) + X(I(6))
      R8 = X(I(5)) - X(I(6))
      T0 = X(I(1)) + R5
      T7 = X(I(1)) + R5 * C95
      R5 = R1 + R3 + R7
      X(I(1)) = T0 + R5
      T5 = T0 + R5 * C95
      T3 = (R3 - R7) * C92
      R7 = (R1 - R7) * C93
      R3 = (R1 - R3) * C94
      T1 = T7 + T3 + R3
      T3 = T7 - T3 - R7
```

```
      T7 = T7 + R7 - R3
      T6 = (R2 - R4 + R8) * C31
      T4 = (R4 + R8) * C96
      R8 = (R2 - R8) * C97
      R2 = (R2 + R4) * C98
      T2 = T8 + T4 + R2
      T4 = T8 - T4 - R8
      T8 = T8 + R8 - R2
C
      R1 = Y(I(2)) + Y(I(9))
      R2 = Y(I(2)) - Y(I(9))
      R3 = Y(I(3)) + Y(I(8))
      R4 = Y(I(3)) - Y(I(8))
      R5 = Y(I(4)) + Y(I(7))
      R6 =(Y(I(4)) - Y(I(7))) * C31
      R7 = Y(I(5)) + Y(I(6))
      R8 = Y(I(5)) - Y(I(6))
      T0 = Y(I(1)) + R5
      T9 = Y(I(1)) + R5 * C95
      R5 = R1 + R3 + R7
      Y(I(1)) = T0 + R5
      R5 = T0 + R5 * C95
      T0 = (R3 - R7) * C92
      R7 = (R1 - R7) * C93
      R3 = (R1 - R3) * C94
      R1 = T9 + T0 + R3
      T0 = T9 - T0 - R7
      R7 = T9 + R7 - R3
      R9 = (R2 - R4 + R8) * C31
      R3 = (R4 + R8) * C96
      R8 = (R2 - R8) * C97
      R4 = (R2 + R4) * C98
      R2 = R6 + R3 + R4
      R3 = R6 - R8 - R3
      R8 = R6 + R8 - R4
C
      X(IP(2)) = T1 - R2
      X(IP(9)) = T1 + R2
      Y(IP(2)) = R1 + T2
      Y(IP(9)) = R1 - T2
      X(IP(3)) = T3 + R3
      X(IP(8)) = T3 - R3
```

```
      Y(IP(3)) = T0 - T4
      Y(IP(8)) = T0 + T4
      X(IP(4)) = T5 - R9
      X(IP(7)) = T5 + R9
      Y(IP(4)) = R5 + T6
      Y(IP(7)) = R5 - T6
      X(IP(5)) = T7 - R8
      X(IP(6)) = T7 + R8
      Y(IP(5)) = R7 + T8
      Y(IP(6)) = R7 - T8
C
      GOTO 20

C------------------WFTA N=16------------------------
C
  116   R1 = X(I(1)) + X(I(9))
      R2 = X(I(1)) - X(I(9))
      R3 = X(I(2)) + X(I(10))
      R4 = X(I(2)) - X(I(10))
      R5 = X(I(3)) + X(I(11))
      R6 = X(I(3)) - X(I(11))
      R7 = X(I(4)) + X(I(12))
      R8 = X(I(4)) - X(I(12))
      R9 = X(I(5)) + X(I(13))
      R10= X(I(5)) - X(I(13))
      R11 = X(I(6)) + X(I(14))
      R12 = X(I(6)) - X(I(14))
      R13 = X(I(7)) + X(I(15))
      R14 = X(I(7)) - X(I(15))
      R15 = X(I(8)) + X(I(16))
      R16 = X(I(8)) - X(I(16))
      T1 = R1 + R9
      T2 = R1 - R9
      T3 = R3 + R11
      T4 = R3 - R11
      T5 = R5 + R13
      T6 = R5 - R13
      T7 = R7 + R15
      T8 = R7 - R15
      R1 = T1 + T5
      R3 = T1 - T5
      R5 = T3 + T7
```

```
R7 = T3 - T7
X(IP( 1)) = R1 + R5
X(IP( 9)) = R1 - R5
T1 = C81 * (T4 + T8)
T5 = C81 * (T4 - T8)
R9 = T2 + T5
R11= T2 - T5
R13 = T6 + T1
R15 = T6 - T1
T1 = R4 + R16
T2 = R4 - R16
T3 = C81 * (R6 + R14)
T4 = C81 * (R6 - R14)
T5 = R8 + R12
T6 = R8 - R12
T7 = C162 * (T2 - T6)
T2 = C163 * T2 - T7
T6 = C164 * T6 - T7
T7 = R2 + T4
T8 = R2 - T4
R2 = T7 + T2
R4 = T7 - T2
R6 = T8 + T6
R8 = T8 - T6
T7 = C165 * (T1 + T5)
T2 = T7 - C164 * T1
T4 = T7 - C163 * T5
T6 = R10 + T3
T8 = R10 - T3
R10 = T6 + T2
R12 = T6 - T2
R14 = T8 + T4
R16 = T8 - T4
R1 = Y(I(1)) + Y(I(9))
S2 = Y(I(1)) - Y(I(9))
S3 = Y(I(2)) + Y(I(10))
S4 = Y(I(2)) - Y(I(10))
R5 = Y(I(3)) + Y(I(11))
S6 = Y(I(3)) - Y(I(11))
S7 = Y(I(4)) + Y(I(12))
S8 = Y(I(4)) - Y(I(12))
S9 = Y(I(5)) + Y(I(13))
```

```
S10= Y(I(5)) - Y(I(13))
S11 = Y(I(6)) + Y(I(14))
S12 = Y(I(6)) - Y(I(14))
S13 = Y(I(7)) + Y(I(15))
S14 = Y(I(7)) - Y(I(15))
S15 = Y(I(8)) + Y(I(16))
S16 = Y(I(8)) - Y(I(16))
T1 = R1 + S9
T2 = R1 - S9
T3 = S3 + S11
T4 = S3 - S11
T5 = R5 + S13
T6 = R5 - S13
T7 = S7 + S15
T8 = S7 - S15
R1 = T1 + T5
S3 = T1 - T5
R5 = T3 + T7
S7 = T3 - T7
Y(IP( 1)) = R1 + R5
Y(IP( 9)) = R1 - R5
X(IP( 5)) = R3 + S7
X(IP(13)) = R3 - S7
Y(IP( 5)) = S3 - R7
Y(IP(13)) = S3 + R7
T1 = C81 * (T4 + T8)
T5 = C81 * (T4 - T8)
S9 = T2 + T5
S11= T2 - T5
S13 = T6 + T1
S15 = T6 - T1
T1 = S4 + S16
T2 = S4 - S16
T3 = C81 * (S6 + S14)
T4 = C81 * (S6 - S14)
T5 = S8 + S12
T6 = S8 - S12
T7 = C162 * (T2 - T6)
T2 = C163 * T2 - T7
T6 = C164 * T6 - T7
T7 = S2 + T4
T8 = S2 - T4
```

```
      S2 = T7 + T2
      S4 = T7 - T2
      S6 = T8 + T6
      S8 = T8 - T6
      T7 = C165 * (T1 + T5)
      T2 = T7 - C164 * T1
      T4 = T7 - C163 * T5
      T6 = S10 + T3
      T8 = S10 - T3
      S10 = T6 + T2
      S12 = T6 - T2
      S14 = T8 + T4
      S16 = T8 - T4
      X(IP( 2)) = R2 + S10
      X(IP(16)) = R2 - S10
      Y(IP( 2)) = S2 - R10
      Y(IP(16)) = S2 + R10
      X(IP( 3)) = R9 + S13
      X(IP(15)) = R9 - S13
      Y(IP( 3)) = S9 - R13
      Y(IP(15)) = S9 + R13
      X(IP( 4)) = R8 - S16
      X(IP(14)) = R8 + S16
      Y(IP( 4)) = S8 + R16
      Y(IP(14)) = S8 - R16
      X(IP( 6)) = R6 + S14
      X(IP(12)) = R6 - S14
      Y(IP( 6)) = S6 - R14
      Y(IP(12)) = S6 + R14
      X(IP( 7)) = R11 - S15
      X(IP(11)) = R11 + S15
      Y(IP( 7)) = S11 + R15
      Y(IP(11)) = S11 - R15
      X(IP( 8)) = R4 - S12
      X(IP(10)) = R4 + S12
      Y(IP( 8)) = S4 + R12
      Y(IP(10)) = S4 - R12
C
      GOTO 20
C
  20        CONTINUE
  10    CONTINUE
```

```
RETURN
END
```

Appendix 4: Programs for Short FFTs[1]

This appendix will discuss efficient short FFT programs that can be used in both the Cooley-Tukey (Chapter 9) and the Prime Factor FFT algorithms (Chapter 10). Links and references are given to Fortran listings that can be used "as is" or put into the indexed loops of existing programs to give greater efficiency and/or a greater variety of allowed lengths. Special programs have been written for lengths: $N = 2, 3, 4, 5, 7, 8, 9, 11, 13, 16, 17, 19, 25$, etc.

In the early days of the FFT, multiplication was done in software and was, therefore, much slower than an addition. With modem hardware, a floating point multiplication can be done in one clock cycle of the computer, microprocessor, or DSP chip, requiring the same time as an addition. Indeed, in some computers and many DSP chips, both a multiplication and an addition (or accumulation) can be done in one cycle while the indexing and memory access is done in parallel. Most of the algorithms described here are not hardware architecture specific but are designed to minimize both multiplications and additions.

The most basic and often used length FFT (or DFT) is for $N = 2$. In the Cooley Tukey FFT, it is called a "butterfly" and its reason for fame is requiring no multiplications at all, only one complex addition and one complex subtraction and needing only one complex temporary storage location. This is illustrated in Figure 1: The Prime Factor and Winograd Transform Algorithms (Figure 10.1) and code is shown in Figure 2: The Prime Factor and Winograd Transform Algorithms (Figure 10.2). The second most used length is $N = 4$ because it is the only other short length requiring no multiplications and a minimum of additions. All other short FFT require some multiplication but for powers of two, $N = 8$ and $N = 16$ require few enough to be worth special coding for some situations.

Code for other short lengths such as the primes $N = 3, 5, 7, 11, 13, 17$, and 19 and the composites $N = 9$ and 25 are included in the programs for the prime factor algorithm or the WFTA. They are derived using the theory in Chapters 5, 6, and 9. They can also be found in references ... and

If these short FFTs are used as modules in the basic prime factor algorithm (PFA), then the straight forward development used for the modules in Figure 17.12 are used. However if the more complicated indexing use to achieve in-order, in-place calculation used in {xxxxx}

[1] This content is available online at <http://cnx.org/content/m17646/1.4/>.

require different code.

For each of the indicated lengths, the computer code is given in a Connexions module.

They are not in the collection Fast Fourier Transforms[2] as the printed version would be too long. However, one can link to them on-line from the following buttons:

N=2[3]
N=3[4]
N=4[5]
N=5[6]
N=7[7]
N= 8
N= 9
N= 11
N= 13
N= 16
N= 17
N= 19
N= 25

Versions for the in-place, in-order prime factor algorithm {pfa} can be obtained from:

N=2[8]
N=3[9]
N=4[10]
N=5[11]
N=7[12]
N=8[13]
N=9[14]
N=11[15]
N=13[16]

[2]*Fast Fourier Transforms* <http://cnx.org/content/col10550/latest/>
[3]"N=2" <http://cnx.org/content/m17625/latest/>
[4]"N=3" <http://cnx.org/content/m17626/latest/>
[5]"N=4" <http://cnx.org/content/m17627/latest/>
[6]"N=5" <http://cnx.org/content/m17628/latest/>
[7]"N=7" <http://cnx.org/content/m17629/latest/>
[8]"pN=2" <http://cnx.org/content/m17631/latest/>
[9]"pN=3" <http://cnx.org/content/m17632/latest/>
[10]"pN=4" <http://cnx.org/content/m17633/latest/>
[11]"pN=5" <http://cnx.org/content/m17634/latest/>
[12]"pN=7" <http://cnx.org/content/m17635/latest/>
[13]"pN=8" <http://cnx.org/content/m17636/latest/>
[14]"pN=9" <http://cnx.org/content/m17637/latest/>
[15]"N = 11 Winograd FFT module" <http://cnx.org/content/m17377/latest/>
[16]"N = 13 Winograd FFT module" <http://cnx.org/content/m17378/latest/>

N=16[17]
N=17[18]
N=19[19]
N=25[20]

A technical report that describes the length 11, 13, 17, and 19 is in {report 8105} and another technical report that describes a program that will automatically generate a prime length FFT and its flow graph si in {report xxx}.

[17] "N = 16 FFT module" <http://cnx.org/content/m17382/latest/>
[18] "N = 17 Winograd FFT module" <http://cnx.org/content/m17380/latest/>
[19] "N = 19 Winograd FFT module" <http://cnx.org/content/m17381/latest/>
[20] "N = 25 FFT module" <http://cnx.org/content/m17383/latest/>

Bibliography

[1] R. C. Agarwal and C. S. Burrus. Fast digital convolution using fermat transforms. In *Proceedings of the IEEE 25th Annual Southwestern Conference*, page 5388211;543, Houston, April 1973.

[2] R. C. Agarwal and C. S. Burrus. Fast convolution using fermat number transforms with applications to digital filtering. *IEEE Transactions on Acoustics, Speech, and Signal Processing*, ASSP-22(2):87–97, April 1974. Reprinted in Number Theory in DSP, by McClellan and Rader, Prentice-Hall, 1979.

[3] R. C. Agarwal and C. S. Burrus. Fast one-dimensional digital convolution by multidimensional techniques. *IEEE Transactions on Acoustics, Speech, and Signal Processing*, ASSP-22(1):18211;10, February 1974. also in IEEE Press DSP Reprints II, 1979; and Number Theory in DSP, by McClellan and Rader, Prentice-Hall, 1979.

[4] R. C. Agarwal and C. S. Burrus. Number theoretic transforms to implement fast digital convolution. *Proceedings of the IEEE*, 63(4):5508211;560, April 1975. also in IEEE Press DSP Reprints II, 1979.

[5] R. C. Agarwal and C. S. Burrus. Number theoretic transforms to implement fast digital convolution. *Proceedings of the IEEE*, 63(4):5508211;560, April 1975. also in IEEE Press DSP Reprints II, 1979.

[6] R. C. Agarwal and J. W. Cooley. New algorithms for digital convolution. *IEEE Trans. on ASSP*, 25(2):3928211;410, October 1977.

[7] R. C. Agarwal and J. W. Cooley. New algorithms for digital convolution. *IEEE Trans. on ASSP*, 25(2):3928211;410, October 1977.

[8] R. C. Agarwal and J. W. Cooley. New algorithms for digital convolution. *IEEE Trans. on ASSP*, 25(2):3928211;410, October 1977.

[9] R. C. Agarwal and J. W. Cooley. New algorithms for digital convolution. *IEEE Trans. on ASSP*, 25(2):3928211;410, October 1977.

[10] E. Anderson, Z. Bai, C. Bischof, S. Blackford, J. Demmel, J. Dongarra, J. Du Croz, A. Greenbaum, S. Hammarling, A. McKenney, and D. Sorensen. *LAPACK Users'*

Guide. Society for Industrial and Applied Mathematics, Philadelphia, PA, 3rd edition, 1999.

[11] Jrg Arndt. *Algorithms for Programmers: Ideas and Source Code*. http://www.jjj.de/fxt/, Bayreuth, Germany, 2008. FFT book available on-line, 1000 pages, continually updated.

[12] L. Auslander, E. Feig, and S. Winograd. Abelian semi-simple algebras and algorithms for the discrete fourier transform. *Advances in Applied Mathematics*, 5:318211;55, 1984.

[13] S. Bagchi and S. Mitra. *The Nonuniform Discrete Fourier Transform and Its Applications in Signal Processing*. Kluwer Academic, Boston, 1999.

[14] D. H. Bailey. Ffts in external or hierarchical memory. *J. Supercomputing*, 4(1):238211;35, May 1990.

[15] C. W. Barnes. Roundoff8211;noise and overflow in normal digital filters. *IEEE Transactions on Circuit and Systems*, CAS-26:1548211;155, March 1979.

[16] C. W. Barnes and S. Shinnaka. Block shift invariance and block implementation of discrete-time filters. *IEEE Transactions on Circuit and Systems*, CAS-27(4):6678211;672, August 1980.

[17] James K. Beard. An in-place, self-reordering fft. In *Proceedings of the ICASSP-78*, pages 632–633, Tulsa, April 1978.

[18] James K. Beard. *The FFT in the 21st Century: Eigenspace Processing*. Kluwer, Boston, 2003.

[19] James K. Beard. *The FFT in the 21st Century: Eigenspace Processing*. Kluwer, Boston, 2003.

[20] Laszlo A. Belady. A study of replacement algorithms for virtual storage computers. *IBM Systems Journal*, 5(2):78–101, 1966.

[21] G. D. Bergland. A fast fourier transform algorithm for real-valued series. *Comm. ACM*, 11(10):703–710, October 1968.

[22] G. D. Bergland. A radix-8 fast fourier transform subroutine for real-valued series. *IEEE Trans. on Audio an Electrocoustics*, 17:138–144, June 1969.

[23] Th. Beth. *Verfahren der Schnellen Fouriertransformation [Fast Fourier Transform Methods]*. Teubner, 1984.

[24] Guoan Bi and Yan Qiu Chen. Fast dft algorithms for length. *IEEE Transactions on Circuits and Systems 8211; II*, 45(6):6858211;690, June 1998.

[25] R. E. Blahut. *Fast Algorithms for Digital Signal Processing.* Addison-Wesley, Inc., Reading, MA, 1984.

[26] Richard E. Blahut. *Fast Algorithms for Digital Signal Processing.* Addison-Wesley, Reading, Mass., 1985.

[27] Richard E. Blahut. *Fast Algorithms for Digital Signal Processing.* Addison-Wesley, Reading, Mass., 1985.

[28] Richard E. Blahut. *Fast Algorithms for Digital Signal Processing.* Addison-Wesley, Reading, Mass., 1985.

[29] Richard E. Blahut. *Fast Algorithms for Digital Signal Processing.* Addison-Wesley, Reading, Mass., 1985.

[30] Richard E. Blahut. *Fast Algorithms for Digital Signal Processing.* Addison-Wesley, Reading, Mass., 1985.

[31] Richard E. Blahut. *Fast Algorithms for Digital Signal Processing.* Addison-Wesley, Reading, Mass., 1985.

[32] Richard E. Blahut. *Fast Algorithms for Digital Signal Processing.* Addison-Wesley, Reading, Mass., 1985.

[33] Richard E. Blahut. *Algebraic Methods for Signal Processing and Communications Coding.* Springer-Verlag, New York, 1992.

[34] L. I. Bluestein. A linear filtering approach to the computation of discrete fourier transform. *IEEE Transactions on Audio Electroacoustics*, AU-18:4518211;455, December 1970.

[35] Leo I. Bluestein. A linear filtering approach to the computation of the discrete fourier transform. *Northeast Electronics Research and Eng. Meeting Record*, 10:2188211;219, 1968.

[36] R. N. Bracewell. *The Hartley Transform.* Oxford Press, 1986.

[37] William L. Briggs and Van Emden Henson. *The DFT: An Owner's Manual for the Discrete Fourier Transform.* SIAM, Philadelphia, 1995.

[38] E. Oran Brigham. *The Fast Fourier Transform and Its Applications.* Prentice-Hall, Englewood Cliffs, NJ, 1988. Expansion of the 1974 book.

[39] E. Oran Brigham. *The Fast Fourier Transform and Its Applications.* Prentice-Hall, Englewood Cliffs, NJ, 1988. Expansion of the 1974 book.

[40] V. Britanak and K. R. Rao. The fast generalized discrete fourier transforms: A unified approach to the discrete sinusoidal transforms computation. *Signal Processing*, 79:1358211;150, 1999.

[41] G. Bruun. z-transform dft filters and ffts. *IEEE Transactions on ASSP*, 26(1):568211;63, February 1978.

[42] O. Buneman. Stable online creation of sines or cosines of successive angles. *Proc. IEEE*, 75(10):14348211;1435, 1987.

[43] C. S. Burrus. Block implementation of digital filters. *IEEE Transactions on Circuit Theory*, CT-18(6):6978211;701, November 1971.

[44] C. S. Burrus. Block realization of digital filters. *IEEE Transactions on Audio and Electroacoustics*, AU-20(4):2308211;235, October 1972.

[45] C. S. Burrus. Digital filter structures described by distributed arithmetic. *IEEE Transactions on Circuit and Systems*, CAS-24(12):6748211;680, December 1977.

[46] C. S. Burrus. Index mapping for multidimensional formulation of the dft and convolution. *IEEE Transactions on Acoustics, Speech, and Signal Processing*, ASSP-25(3):2398211;242, June 1977.

[47] C. S. Burrus. Index mapping for multidimensional formulation of the dft and convolution. *IEEE Transactions on Acoustics, Speech, and Signal Processing*, ASSP-25(3):2398211;242, June 1977.

[48] C. S. Burrus. Recursive digital filter structures using new high speed convolution algorithms. In *IEEE International Conference on Acoustics, Speech, and Signal Processing*, page 3638211;365, Hartford, CT, May 1977.

[49] C. S. Burrus. Efficient fourier transform and convolution algorithms. In J. S. Lim and A. V. Oppenheim, editors, *Advanced Topics in Signal Processing*, chapter 4, page 1998211;245. Prentice-Hall, Englewood Cliffs, NJ, 1988.

[50] C. S. Burrus. Efficient fourier transform and convolution algorithms. In J. S. Lim and A. V. Oppenheim, editors, *Advanced Topics in Signal Processing*, chapter 4, page 1998211;245. Prentice-Hall, Englewood Cliffs, NJ, 1988.

[51] C. S. Burrus. Unscrambling for fast dft algorithms. *IEEE Transactions on Acoustics, Speech, and Signal Processing*, 36(7):10868211;1087, July 1988.

[52] C. S. Burrus. Goertzel's algorithm. Unpublished notes, ECE Dept., Rice University, 1992.

[53] C. S. Burrus. The quick fourier transform. Unpublished notes, ECE Dept., Rice University, 1992.

[54] C. S. Burrus and R. C. Agarwal. Efficient implementation of recursive digital filters. In *Proceedings of the Seventh Asilomar Conference on Circuits and Systems*, page 2808211;284, Pacific Grove, CA, November, 5 1973.

[55] C. S. Burrus and R. C. Agarwal. The use of number theoretic transforms for convolution. In *Presented at the IEEE Arden House Workshop on Digital Signal Processing*, Harriman, NY, January 1974.

[56] C. S. Burrus and P. W. Eschenbacher. An in8211;place, in8211;order prime factor fft algorithm. *IEEE Transactions on Acoustics, Speech, and Signal Processing*, 29(4):8068211;817, August 1981. Reprinted in it DSP Software, by L.R. Morris, 1983.

[57] C. S. Burrus and P. W. Eschenbacher. An in8211;place, in8211;order prime factor fft algorithm. *IEEE Transactions on Acoustics, Speech, and Signal Processing*, 29(4):8068211;817, August 1981. Reprinted in it DSP Software, by L.R. Morris, 1983.

[58] C. S. Burrus and P. W. Eschenbacher. An in8211;place, in8211;order prime factor fft algorithm. *IEEE Transactions on Acoustics, Speech, and Signal Processing*, 29(4):8068211;817, August 1981. Reprinted in it DSP Software, by L.R. Morris, 1983.

[59] C. S. Burrus and T. W. Parks. *DFT/FFT and Convolution Algorithms*. John Wiley & Sons, New York, 1985.

[60] C. S. Burrus and T. W. Parks. *DFT/FFT and Convolution Algorithms*. John Wiley & Sons, New York, 1985.

[61] C. S. Burrus and T. W. Parks. *DFT/FFT and Convolution Algorithms*. John Wiley & Sons, New York, 1985.

[62] C. S. Burrus and T. W. Parks. *DFT/FFT and Convolution Algorithms*. John Wiley & Sons, New York, 1985.

[63] C. S. Burrus and T. W. Parks. *DFT/FFT and Convolution Algorithms*. John Wiley & Sons, New York, 1985.

[64] C. S. Burrus and T. W. Parks. *DFT/FFT and Convolution Algorithms*. John Wiley & Sons, New York, 1985.

[65] C. S. Burrus and T. W. Parks. *DFT/FFT and Convolution Algorithms*. John Wiley & Sons, New York, 1985.

[66] C. S. Burrus and T. W. Parks. *DFT/FFT and Convolution Algorithms*. John Wiley & Sons, New York, 1985.

[67] C. S. Burrus and T. W. Parks. *DFT/FFT and Convolution Algorithms*. John Wiley & Sons, New York, 1985.

[68] C. S. Burrus and T. W. Parks. *DFT/FFT and Convolution Algorithms.* John Wiley & Sons, New York, 1985.

[69] C. Sidney Burrus, Ramesh A. Gopinath, and Haitao Guo. *Introduction to Wavelets and the Wavelet Transform.* Prentice Hall, Upper Saddle River, NJ, 1998.

[70] C. Sidney Burrus and Ivan W. Selesnick. Fast convolution and filtering. In V. K. Madisetti and D. B. Williams, editors, *The Digital Signal Processing Handbook*, chapter 8. CRC Press, Boca Raton, 1998.

[71] Iain R. Byam. A new fast fourier transform algorithm. Technical report, University of the West Indies (UWI), St. Augustine, Trinidad, June 1999. A short version is in Technical Report PG-99001, ECE Dept., Univ. of the West Indies, Aug. 1999.

[72] P. Brgisser, M. Clausen, and M. A. Shokrollahi. *Algebraic Complexity Theory.* Springer, 1997.

[73] S. C. Chan and K. L. Ho. Direct methods for computing discrete sinusoidal transforms. *IEE Proc. F*, 137(6):4338211;442, 1990.

[74] D. Chanoux. Synthesis of recursive digital filters using the fft. *IEEE Transactions on Audio and Electroacoustics*, AU-18:2118211;212, June 1970.

[75] K. T. Chen. A new record: The largest known prime number. *IEEE Spectrum*, 27(2):47, February 1990.

[76] K. M. Cho and G. C. Temes. Real-factor fft algorithms. In *Proceedings of IEEE ICASSP-78*, pages 634–637, Tulsa, OK, April 1978.

[77] Shuni Chu and C. S. Burrus. A prime factor fft algorithm using distributed arithmetic. *IEEE Transactions on Acoustics, Speech, and Signal Processing*, 30(2):2178211;227, April 1982.

[78] Shuni Chu and C. S. Burrus. A prime factor fft algorithm using distributed arithmetic. *IEEE Transactions on Acoustics, Speech, and Signal Processing*, 30(2):2178211;227, April 1982.

[79] Shuni Chu and C. S. Burrus. A prime factor fft algorithm using distributed arithmetic. *IEEE Transactions on Acoustics, Speech, and Signal Processing*, 30(2):2178211;227, April 1982.

[80] Shuni Chu and C. S. Burrus. A prime factor fft algorithm using distributed arithmetic. *IEEE Transactions on Acoustics, Speech, and Signal Processing*, 30(2):2178211;227, April 1982.

[81] T. A. C. M. Claasen and W. F. G. Mecklenbraker. On stationary linear time8211;varying systems. *IEEE Trans. on Circuits and Systems*, 29(3):1698211;184, March 1982.

[82] DSP Committee, editor. *Digital Signal Processing II, selected reprints.* IEEE Press, New York, 1979.

[83] DSP Committee, editor. *Digital Signal Processing II, selected reprints.* IEEE Press, New York, 1979.

[84] DSP Committee, editor. *Digital Signal Processing II, selected reprints.* IEEE Press, New York, 1979.

[85] DSP Committee, editor. *Programs for Digital Signal Processing.* IEEE Press, New York, 1979.

[86] J. W. Cooley. How the fft gained acceptance. *IEEE Signal Processing Magazine*, 9(1):108211;13, January 1992. Also presented at the ACM Conference on the History of Scientific and Numerical Computation, Princeton, NJ, May 1987 and published in: A History of Scientific Computing, edited by S. G. Nash, Addison-Wesley, 1990, pp. 133-140.

[87] J. W. Cooley. How the fft gained acceptance. *IEEE Signal Processing Magazine*, 9(1):108211;13, January 1992. Also presented at the ACM Conference on the History of Scientific and Numerical Computation, Princeton, NJ, May 1987 and published in: A History of Scientific Computing, edited by S. G. Nash, Addison-Wesley, 1990, pp. 133-140.

[88] J. W. Cooley and J. W. Tukey. An algorithm for the machine calculation of complex fourier series. *Math. Computat.*, 19:2978211;301, 1965.

[89] J. W. Cooley and J. W. Tukey. An algorithm for the machine calculation of complex fourier series. *Math. Computat.*, 19:2978211;301, 1965.

[90] J. W. Cooley and J. W. Tukey. An algorithm for the machine calculation of complex fourier series. *Math. Computat.*, 19:2978211;301, 1965.

[91] J. W. Cooley and J. W. Tukey. An algorithm for the machine calculation of complex fourier series. *Math. Computat.*, 19:2978211;301, 1965.

[92] J. W. Cooley and J. W. Tukey. An algorithm for the machine computation of the complex fourier series. *Math. Computation*, 19:2978211;301, April 1965.

[93] James W. Cooley. The structure of fft algorithms, April 1990. Notes for a Tutorial at IEEE ICASSP-90.

[94] James W. Cooley, Peter A. W. Lewis, and Peter D. Welch. Historical notes on the fast fourier transform. *IEEE Transactions on Audio and Electroacoustics*, 15:2608211;262, June 1967.

[95] James W. Cooley, Peter A. W. Lewis, and Peter D. Welch. Historical notes on the fast fourier transform. *IEEE Transactions on Audio and Electroacoustics*, 15:2608211;262, June 1967.

[96] Thomas H. Cormen, Charles E. Leiserson, and Ronald L. Rivest. *Introduction to Algorithms*. The MIT Press, Cambridge, Massachusetts, 1990.

[97] R. E. Crochiere and L. R. Rabiner. *Multirate Digital Signal Processing*. Prentice-Hall, Englewood Cliffs, NJ, 1983.

[98] Ronald E. Crochiere and Alan V. Oppenheim. Analysis of linear digital networks. *Proc. IEEE*, 63(4):5818211;595, 1975.

[99] Ingrid Daubechies. *Ten Lectures on Wavelets*. SIAM, Philadelphia, PA, 1992. Notes from the 1990 CBMS-NSF Conference on Wavelets and Applications at Lowell, MA.

[100] Murray Dow. Transposing a matrix on a vector computer. *Parallel Computing*, 21(12):1997–2005, 1995.

[101] P. Duhamel. Implementation of 'split-radix' fft algorithms for complex, real, and real-symmetric data. *IEEE Trans. on ASSP*, 34:2858211;295, April 1986. A shorter version appeared in the ICASSP-85 Proceedings, p. 20.6, March 1985.

[102] P. Duhamel. Implementation of 'split-radix' fft algorithms for complex, real, and real-symmetric data. *IEEE Trans. on ASSP*, 34:2858211;295, April 1986. A shorter version appeared in the ICASSP-85 Proceedings, p. 20.6, March 1985.

[103] P. Duhamel. Implementation of 'split-radix' fft algorithms for complex, real, and real-symmetric data. *IEEE Trans. on ASSP*, 34:2858211;295, April 1986. A shorter version appeared in the ICASSP-85 Proceedings, p. 20.6, March 1985.

[104] P. Duhamel. Implementation of 'split-radix' fft algorithms for complex, real, and real-symmetric data. *IEEE Trans. on ASSP*, 34:2858211;295, April 1986. A shorter version appeared in the ICASSP-85 Proceedings, p.20.6, March 1985.

[105] P. Duhamel and H. Hollmann. Split radix fft algorithm. *Electronic Letters*, 20(1):148211;16, January 5 1984.

[106] P. Duhamel and H. Hollmann. Split radix fft algorithm. *Electronic Letters*, 20(1):148211;16, January 5 1984.

[107] P. Duhamel and H. Hollmann. Split-radix fft algorithm. *Electronics Lett.*, 20(1):14–16, 1984.

[108] P. Duhamel and H. Hollmann. Split radix fft algorithm. *Electronic Letters*, 20(1):148211;16, January 5 1984.

[109] P. Duhamel, B. Piron, and J. M. Etcheto. On computing the inverse dft. *IEEE Transactions on ASSP*, 36(2):2858211;286, February 1978.

[110] P. Duhamel and M. Vetterli. Cyclic convolution of real sequences: Hartley versus fourier and new schemes. *Proceedings of the IEEE International Conference on Acoustics, Speech, and Signal Processing (ICASSP-86)*, page 6.5, April 1986.

[111] P. Duhamel and M. Vetterli. Cyclic convolution of real sequences: Hartley versus fourier and new schemes. *Proceedings of the IEEE International Conference on Acoustics, Speech, and Signal Processing (ICASSP-86)*, page 6.5, April 1986.

[112] P. Duhamel and M. Vetterli. Cyclic convolution of real sequences: Hartley versus fourier and new schemes. *Proceedings of the IEEE International Conference on Acoustics, Speech, and Signal Processing (ICASSP-86)*, page 6.5, April 1986.

[113] P. Duhamel and M. Vetterli. Improved fourier and hartley transfrom algorithms, application to cyclic convolution of real data. *IEEE Trans. on ASSP*, 35(6):8188211;824, June 1987.

[114] P. Duhamel and M. Vetterli. Fast fourier transforms: a tutorial review and a state of the art. *Signal Processing*, 19:2598211;299, April 1990.

[115] P. Duhamel and M. Vetterli. Fast fourier transforms: A tutorial review and a state of the art. *Signal Processing*, 19(4):2598211;299, April 1990.

[116] Pierre Duhamel. Algorithms meeting the lower bounds on the multiplicative complexity of length- dfts and their connection with practical algorithms. *IEEE Trans. Acoust., Speech, Signal Processing*, 38(9):1504–1511, 1990.

[117] S. Egner and M. Pschel. Automatic generation of fast discrete signal transforms. *IEEE Transactions on Signal Processing*, 49(9):19928211;2002, 2001.

[118] Sebastian Egner and Markus Pschel. Automatic generation of fast discrete signal transforms. *IEEE Transactions on Signal Processing*, 49(9):19928211;202, 2001.

[119] D. F. Elliott and K. F. Rao. *Fast Transforms: Algorithms, Analyses and Applications*. Academic Press, New York, 1982.

[120] Douglas F. Elliott, editor. *Handbook of Digital Signal Processing*. Academic Press, San Diego, CA, 1987. Chapter 7 on FFT by Elliott.

[121] Douglas F. Elliott, editor. *Handbook of Digital Signal Processing*. Academic Press, San Diego, CA, 1987. Chapter 7 on FFT by Elliott.

[122] Funda Ergn. Testing multivariate linear functions: Overcoming the generator bottleneck. In *Proc. Twenty-Seventh Ann. ACM Symp. Theory of Computing*, page 4078211;416, Las Vegas, Nevada, June 1995.

[123] D. M. W. Evans. A second improved digit8211;reversal permutation algorithm for fast transforms. *IEEE Transactions on Acoustics, Speech, and Signal Processing*, 37(8):12888211;1291, August 1989.

[124] W. E. Ferguson, Jr. A simple derivation of glassman general-n fast fourier transform. *Comput. and Math. with Appls.*, 8(6):4018211;411, 1982. Also, in Report AD-A083811, NTIS, Dec. 1979.

[125] W. E. Ferguson, Jr. A simple derivation of glassman general-n fast fourier transform. *Comput. and Math. with Appls.*, 8(6):4018211;411, 1982. Also, in Report AD-A083811, NTIS, Dec. 1979.

[126] P. A. Franasek and B. Liu. On a class of time8211;varying filters. *IEEE Trans. on Information Theory*, 13:477, 1967.

[127] Franz Franchetti, Stefan Kral, Juergen Lorenz, and Christoph W. Ueberhuber. Domain specific compiler techniques. Manuscript in preparation.

[128] Matteo Frigo. A fast fourier transform compiler. In *Proc. ACM SIGPLAN'99 Conference on Programming Language Design and Implementation (PLDI)*, volume 34, page 1698211;180, Atlanta, Georgia, May 1999. ACM.

[129] Matteo Frigo. A fast fourier transform compiler. In *Proceedings of the 1999 ACM SIGPLAN Conference on Programming Language Design and Implentation*, PLDI-99, Atlanta, May 1999.

[130] Matteo Frigo and Steven G. Johnson. The fastest fourier transform in the west. Technical report MIT-LCS-TR-728, Laboratory for Computer Science, MIT, Cambridge, MA, September 1997.

[131] Matteo Frigo and Steven G. Johnson. Fftw: An adaptive software architecture for the fft. In *Proc. IEEE Int'l Conf. Acoustics, Speech, and Signal Processing*, volume 3, page 13818211;1384, Seattle, WA, May 1998.

[132] Matteo Frigo and Steven G. Johnson. Fftw: An adaptive software architecture for the fft. In *Proceedings of the IEEE International Conference on Acoustics, Speech, and Signal Processing*, volume III, page 13818211;1384, ICASSP-98, Seattle, May 128211;15 1998.

[133] Matteo Frigo and Steven G. Johnson. The fftw web page. http://www.fftw.org/, 2003.

[134] Matteo Frigo and Steven G. Johnson. The design and implementation of fftw3. *Proc. IEEE*, 93(2):2168211;231, 2005.

[135] Matteo Frigo and Steven G. Johnson. The design and implementtion of fftw. *Proceedings of the IEEE*, 93(2):2168211;231, February 2005.

[136] Matteo Frigo and Steven G. Johnson. The design and implementtion of fftw. *Proceedings of the IEEE*, 93(2):2168211;231, February 2005.

[137] Matteo Frigo, Charles E. Leiserson, Harald Prokop, and Sridhar Ramachandran. Cache-oblivious algorithms. In *Proc. 40th Ann. Symp. Foundations of Computer Science (FOCS '99)*, New York, USA, October 1999.

[138] Paul A. Fuhrman. *A Polynomial Approach to Linear Algebra.* Springer Verlag, New York, 1996.

[139] W. M. Gentleman and G. Sande. Fast fourier transforms8212;for fun and profit. *Proc. AFIPS*, 29:5638211;578, 1966.

[140] S. P. Ghanekar, S. Tantaratana, and L. E. Franks. High-precision multiplier8211;free fir filter realization with periodically time8211;varying coefficients. In *Paper Summaries for the 1992 DSP Workshop*, page 3.3.1, Starved Rock Lodge, Utica, Ill., 1992.

[141] S. P. Ghanekar, S. Tantaratana, and L. E. Franks. A class of high-precision multiplier8211;free fir filter realizations with periodically time8211;varying coefficients. *IEEE Transactions on Signal Processing*, 43(4):8228211;830, 1995.

[142] Sachin Ghanekar, Sawasd Tantaratana, and Lewis E. Franks. Implementation of recursive filters using highly quantized periodically time8211;varying coefficients. In *Proceedings of the ICASSP-91*, page 16258211;1628, Toronto, Canada, May 1991.

[143] J. A. Glassman. A generalization of the fast fourier transform. *IEEE Transactions on Computers*, C-19(2):1058211;116, Feburary 1970.

[144] G. Goertzel. An algorithm for the evaluation of finite trigonometric series. *Amer. Math. Monthly*, 65:34–35, January 1958.

[145] B. Gold and K. L. Jordan. A note on digital filter synthesis. *Proceedings of the IEEE*, 56:17178211;1718, October 1968.

[146] B. Gold and C. M. Rader. *Digital Processing of Signals.* McGraw-Hill, New York, 1969.

[147] B. Gold and C. M. Rader. *Digital Processing of Signals.* McGraw-Hill, New York, 1969.

[148] I. J. Good. Interaction algorithm and practical fourier analysis. *J. Royal Statist. Soc., B*, 20:3618211;372, 1958. Addendum: vol. 22, 1960, pp 3728211;375.

[149] I. J. Good. Interaction algorithm and practical fourier analysis. *J. Royal Statist. Soc., B*, 20:3618211;372, 1958. Addendum: vol. 22, 1960, pp 3728211;375.

[150] I. J. Good. Interaction algorithm and practical fourier analysis. *J. Royal Statist. Soc., B*, 20:3618211;372, 1958. Addendum: vol. 22, 1960, pp 3728211;375.

[151] R. A. Gopinath and C. S. Burrus. Wavelet transforms and filter banks. In Charles K. Chui, editor, *Wavelets: A Tutorial in Theory and Applications*, page 6038211;655. Academic Press, San Diego, CA, 1992. Volume 2 in the series: Wavelet Analysis and its Applications.

[152] Ramesh A. Gopinath. *Wavelets and Filter Banks 8211; New Results and Applications*. Ph. d. thesis, Rice University, Houston, Tx, August 1992.

[153] John Granata, Michael Conner, and Richard Tolimieri. Recursive fast algorithms and the role of the tensor product. *IEEE Transactions on Signal Processing*, 40(12):29218211;2930, December 1992.

[154] John Granata, Michael Conner, and Richard Tolimieri. The tensor product: A mathematical programming language for ffts. *IEEE Signal Processing Magazine*, 9(1):408211;48, January 1992.

[155] H. Guo, G. A. Sitton, and C. S. Burrus. The quick discrete fourier transform. In *Proceedings of the IEEE International Conference on Acoustics, Speech, and Signal Processing*, page III:4458211;448, IEEE ICASSP-94, Adelaide, Australia, April 198211;22 1994.

[156] H. Guo, G. A. Sitton, and C. S. Burrus. The quick discrete fourier transform. In *Proceedings of the IEEE International Conference on Acoustics, Speech, and Signal Processing*, page III:4458211;448, IEEE ICASSP-94, Adelaide, Australia, April 198211;22 1994.

[157] H. Guo, G. A. Sitton, and C. S. Burrus. The quick discrete fourier transform. In *Proceedings of the IEEE International Conference on Acoustics, Speech, and Signal Processing*, page III:4458211;448, IEEE ICASSP-94, Adelaide, Australia, April 198211;22 1994.

[158] H. Guo, G. A. Sitton, and C. S. Burrus. The quick fourier transform: an fft based on symmetries. *IEEE Transactions on Signal Processing*, 46(2):3358211;341, February 1998.

[159] H. Guo, G. A. Sitton, and C. S. Burrus. The quick fourier transform: an fft based on symmetries. *IEEE Transactions on Signal Processing*, 46(2):3358211;341, February 1998.

[160] H. Guo, G. A. Sitton, and C. S. Burrus. The quick fourier transform: an fft based on symmetries. *IEEE Transactions on Signal Processing*, 46(2):3358211;341, February 1998.

[161] Haitao Guo and C. Sidney Burrus. Approximate fft via the discrete wavelet transform. In *Proceedings of SPIE Conference 2825*, Denver, August 68211;9 1996.

[162] Haitao Guo and C. Sidney Burrus. Approximate fft via the discrete wavelet transform. In *Proceedings of SPIE Conference 2825*, Denver, August 68211;9 1996.

[163] Haitao Guo and C. Sidney Burrus. Wavelet transform based fast approximate fourier transform. In *Proceedings of the IEEE International Conference on Acoustics, Speech, and Signal Processing*, volume 3, page III:19738211;1976, IEEE ICASSP-97, Munich, April 218211;24 1997.

[164] Haitao Guo and C. Sidney Burrus. Wavelet transform based fast approximate fourier transform. In *Proceedings of the IEEE International Conference on Acoustics, Speech, and Signal Processing*, volume 3, page III:19738211;1976, IEEE ICASSP-97, Munich, April 218211;24 1997.

[165] G. H. Hardy and E. M. Wright. *An Introduction to the Theory of Numbers.* Oxford, London, fourth edition, 1938, 1960.

[166] Markus Hegland. A self-sorting in-place fast fourier transform algorithm suitable for vector and parallel processing. *Numerische Mathematik*, 68(4):5078211;547, 1994.

[167] Markus Hegland and W. W. Wheeler. Linear bijections and the fast fourier transform. *Applicable Algebra in Engineering, Communication and Computing*, 8(2):1438211;163, 1997.

[168] M. T. Heideman. private communication, 1985.

[169] M. T. Heideman. *Fast Algorithms for the DFT and Convolution with Constrained Inputs and Restricted Outputs.* Ph. d. thesis, Rice University, May 1986.

[170] M. T. Heideman. *Multiplicative Complexity, Convolution, and the DFT.* Springer8211;Verlag, 1988.

[171] M. T. Heideman and C. S. Burrus. On the number of multiplications necessary to compute a length- dft. *IEEE Transactions on Acoustics, Speech, and Signal Processing*, 34(1):918211;95, February 1986.

[172] M. T. Heideman and C. S. Burrus. On the number of multiplications necessary to compute a length- dft. *IEEE Transactions on Acoustics, Speech, and Signal Processing*, 34(1):918211;95, February 1986.

[173] M. T. Heideman, D. H. Johnson, and C. S. Burrus. Gauss and the history of the fast fourier transform. *IEEE ASSP Magazine*, 1(4):14–21, 1984.

[174] M. T. Heideman, D. H. Johnson, and C. S. Burrus. Gauss and the history of the fft. *IEEE Acoustics, Speech, and Signal Processing Magazine*, 1(4):148211;21, October 1984. also in Archive for History of Exact Sciences, 1985.

[175] M. T. Heideman, D. H. Johnson, and C. S. Burrus. Gauss and the history of the fft. *IEEE Acoustics, Speech, and Signal Processing Magazine*, 1(4):148211;21, October 1984. also in Archive for History of Exact Sciences, 1985.

[176] M. T. Heideman, D. H. Johnson, and C. S. Burrus. Gauss and the history of the fft. *Archive for History of Exact Sciences*, 34(3):2658211;277, 1985.

[177] M. T. Heideman, H. W. Johnson, and C. S. Burrus. Prime factor fft algorithms for real8211;valued series. In *Proceedings of the IEEE International Conference on Acoustics, Speech, and Signal Processing*, page 28A.7.18211;4, San Diego, CA, March 1984.

[178] M. T. Heideman, H. W. Johnson, and C. S. Burrus. Prime factor fft algorithms for real8211;valued series. In *Proceedings of the IEEE International Conference on Acoustics, Speech, and Signal Processing*, page 28A.7.18211;4, San Diego, CA, March 1984.

[179] M. T. Heideman, H. W. Johnson, and C. S. Burrus. Prime factor fft algorithms for real8211;valued series. In *Proceedings of the IEEE International Conference on Acoustics, Speech, and Signal Processing*, page 28A.7.18211;4, San Diego, CA, March 1984.

[180] Michael T. Heideman and C. Sidney Burrus. On the number of multiplications necessary to compute a length- dft. *IEEE Trans. Acoust., Speech, Signal Processing*, 34(1):91–95, 1986.

[181] H. D. Helms. Fast fourier transform method of computing difference equations and simulating filters. *IEEE Trans. on Audio and Electroacoustics*, AU-15:858211;90, June 1967.

[182] N. J. Higham. The accuracy of floating-point summation. *SIAM J. Sci. Comput.*, 14(4):7838211;799, July 1993.

[183] W. K. Hocking. Performing fourier transforms on extremely long data streams. *Computers in Physics*, 3(1):598211;65, January 1989.

[184] Jia-Wei Hong and H. T. Kung. I/o complexity: the red-blue pebbling game. In *Proc. Thirteenth Ann. ACM Symp. Theory of Computing*, page 3268211;333, Milwaukee, 1981.

[185] A. N. Hossen, U. Heute, O. V. Shentov, and S. K. Mitra. Subband dft 8211; part ii: Accuracy, complexity, and applications. *Signal Processing*, 41:2798211;295, 1995.

[186] Jechang Jeong and William J. Williams. A fast recursive bit-reversal algorithm. In *Proceedings of the ICASSP-90*, page 15118211;1514, Albuquerque, NM, April 1990.

[187] H. W. Johnson and C. S. Burrus. Large dft modules: N = 11, 13, 17, 19, and 25. Technical report 8105, Department of Electrical Engineering, Rice University, Houston, TX 772518211;1892, 1981.

[188] H. W. Johnson and C. S. Burrus. Large dft modules: N = 11, 13, 17, 19, and 25. Technical report 8105, Department of Electrical Engineering, Rice University, Houston, TX 772518211;1892, 1981.

[189] H. W. Johnson and C. S. Burrus. Large dft modules: N = 11, 13, 17, 19, and 25. Technical report 8105, Department of Electrical Engineering, Rice University, Houston, TX 772518211;1892, 1981.

[190] H. W. Johnson and C. S. Burrus. The design of optimal dft algorithms using dynamic programming. In *Proceedings of the IEEE International Conference on Acoustics, Speech, and Signal Processing*, page 208211;23, Paris, May 1982.

[191] H. W. Johnson and C. S. Burrus. The design of optimal dft algorithms using dynamic programming. *IEEE Transactions on Acoustics, Speech, and Signal Processing*, 31(2):3788211;387, April 1983.

[192] H. W. Johnson and C. S. Burrus. An in-order, in-place radix-2 fft. In *Proceedings of the IEEE International Conference on Acoustics, Speech, and Signal Processing*, page 28A.2.18211;4, San Diego, CA, March 1984.

[193] H. W. Johnson and C. S. Burrus. An in-order, in-place radix-2 fft. In *Proceedings of the IEEE International Conference on Acoustics, Speech, and Signal Processing*, page 28A.2.18211;4, San Diego, CA, March 1984.

[194] H. W. Johnson and C. S. Burrus. An in-order, in-place radix-2 fft. In *Proceedings of the IEEE International Conference on Acoustics, Speech, and Signal Processing*, page 28A.2.18211;4, San Diego, CA, March 1984.

[195] H. W. Johnson and C. S. Burrus. An in-place in-order radix-2 fft. In *Proc. IEEE Int'l Conf. Acoustics, Speech, and Signal Processing*, page 28A.2.18211;4, 1984.

[196] H. W. Johnson and C. S. Burrus. An in-place, in-order radix-2 fft. In *ICASSP-84 Proceedings*, page 28A.2, March 1984.

[197] Howard W. Johnson and C. S. Burrus. On the structure of efficient dft algorithms. *IEEE Transactions on Acoustics, Speech, and Signal Processing*, 33(1):2488211;254, February 1985.

[198] Howard W. Johnson and C. S. Burrus. On the structure of efficient dft algorithms. *IEEE Transactions on Acoustics, Speech, and Signal Processing*, 33(1):2488211;254, February 1985.

[199] Howard W. Johnson and C. S. Burrus. On the structure of efficient dft algorithms. *IEEE Transactions on Acoustics, Speech, and Signal Processing*, 33(1):2488211;254, February 1985.

[200] J. Johnson, R. W. Johnson, D. Rodriguez, and R. Tolimieri. A methodology for designing, modifying, and implementing fourier transform algorithms on various architectures. *Circuits, Systems and Signal Processing*, 9(4):4498211;500, 1990.

[201] Steven G. Johnson and Matteo Frigo. A modified split-radix fft with fewer arithmetic operations. *IEEE Transactions on Signal Processing*, 55(1):1118211;119, January 2007.

[202] Steven G. Johnson and Matteo Frigo. A modified split-radix fft with fewer arithmetic operations. *IEEE Trans. Signal Processing*, 55(1):1118211;119, 2007.

[203] Steven G. Johnson and Matteo Frigo. A modified split-radix fft with fewer arithmetic operations. *IEEE Transactions on Signal Processing*, 55(1):1118211;119, January 2007.

[204] Steven G. Johnson and J. D. Joannopoulos. Block-iterative frequency-domain methods for maxwell's equations in a planewave basis. *Optics Express*, 8(3):1738211;190, 2001.

[205] Douglas L. Jones. *The DFT, FFT, and Practical Spectral Analysis.* Connexions, February 2007. http://cnx.org/content/col10281/1.2/.

[206] Douglas L. Jones. *The DFT, FFT, and Practical Spectral Analysis.* Connexions, February 2007. http://cnx.org/content/col10281/1.2/.

[207] Alan H. Karp. Bit reversal on uniprocessors. *SIAM Rev.*, 38(1):18211;26, 1996.

[208] Donald E. Knuth. *The Art of Computer Programming, Vol. 2, Seminumerical Algorithms.* Addison-Wesley, Reading, MA, third edition, 1997.

[209] Donald E. Knuth. *The Art of Computer Programming, Vol. 2, Seminumerical Algorithms.* Addison-Wesley, Reading, MA, third edition, 1997.

[210] Donald E. Knuth. *Fundamental Algorithms*, volume 1 of *The Art of Computer Programming.* Addison-Wesley, 3nd edition, 1997.

[211] John F. Kohne. A quick fourier transform algorithm. Technical report TR-1723, Naval Electronics Laboratory Center, July 1980.

[212] D. P. Kolba and T. W. Parks. A prime factor fft algorithm using high speed convolution. *IEEE Trans. on ASSP*, 25:2818211;294, August 1977. also in.

[213] D. P. Kolba and T. W. Parks. A prime factor fft algorithm using high speed convolution. *IEEE Trans. on ASSP*, 25:2818211;294, August 1977. also in.

[214] D. P. Kolba and T. W. Parks. A prime factor fft algorithm using high speed convolution. *IEEE Trans. on ASSP*, 25:2818211;294, August 1977. also in.

[215] H. Krishna, B. Krishna, K.-Y. Lin, and J.-D. Sun. *Computational Number Theory and Digital Signal Processing*. CRC Press, Boca Raton, FL, 1994.

[216] Z. Li, H. V. Sorensen, and C. S. Burrus. Fft and convolution algorithms for dsp microprocessors. In *Proceedings of the IEEE International Conference on Acoustics, Speech, and Signal Processing*, page 2848211;292, Tokyo, Japan, April 1986.

[217] J. S. Lim and A. V. Oppenheim. *Advanced Topics in Signal Processing*. Prentice-Hall, Englewood Cliffs, NJ, 1988.

[218] J. S. Lim and A. V. Oppenheim. *Advanced Topics in Signal Processing*. Prentice-Hall, Englewood Cliffs, NJ, 1988.

[219] Jae S. Lim and A. V. Oppenheim. *Advanced Topics in Signal Processing*, chapter 4. Prentice-Hall, 1988.

[220] C. M. Loeffler and C. S. Burrus. Equivalence of block filter representations. In *Proceedings of the 1981 IEEE International Symposium on Circuits and Systems*, pages 546–550, Chicago, IL, April 1981.

[221] C. M. Loeffler and C. S. Burrus. Periodically time8211;varying bandwidth compressor. In *Proceedings of the IEEE International Symposium on Circuits and Systems*, page 6638211;665, Rome, Italy, May 1982.

[222] C. M. Loeffler and C. S. Burrus. Optimal design of periodically time varying and multirate digital filters. *IEEE Transactions on Acoustics, Speech, and Signal Processing*, ASSP-32(5):991–924, October 1984.

[223] Chao Lu, James W. Cooley, and Richard Tolimieri. Fft algorithms for prime transform sizes and their implementations on vax, ibm3090vf, and ibm rs/6000. *IEEE Transactions on Signal Processing*, 41(2):6388211;648, February 1993.

[224] D. P-K. Lun and W-C. Siu. An analysis for the realization of an in-place and in-order prime factor algorithm. *IEEE Transactions on Signal Processing*, 41(7):23628211;2370, July 1993.

[225] T. Lundy and J. Van Buskirk. A new matrix approach to real ffts and convolutions of length. *Computing*, 80(1):238211;45, 2007.

[226] J. D. Markel. Fft pruning. *IEEE Trans on Audio and Electroacoustics*, 19(4):3058211;311, June 1971.

[227] J. B. Martens. Recursive cyclotomic factorization 8211; a new algorithm for calculating the discrete fourier transform. *IEEE Trans. on ASSP*, 32(4):7508211;762, August 1984.

[228] J. B. Martens. Recursive cyclotomic factorization 8211; a new algorithm for calculating the discrete fourier transform. *IEEE Trans. on ASSP*, 32(4):7508211;762, August 1984.

[229] J. B. Martens. Recursive cyclotomic factorization 8211; a new algorithm for calculating the discrete fourier transform. *IEEE Trans. on ASSP*, 32(4):7508211;762, August 1984.

[230] J. B. Martens. Recursive cyclotomic factorization8212;a new algorithm for calculating the discrete fourier transform. *IEEE Trans. Acoust., Speech, Signal Processing*, 32(4):750–761, 1984.

[231] D. Maslen and D. Rockmore. Generalized ffts 8211; a survey of some recent results. In *Proceedings of IMACS Workshop in Groups and Computation*, volume 28, page 1828211;238, 1995.

[232] J. H. McClellan and C. M. Rader. *Number Theory in Digital Signal Processing*. Prentice-Hall, Englewood Cliffs, NJ, 1979.

[233] J. H. McClellan and C. M. Rader. *Number Theory in Digital Signal Processing*. Prentice-Hall, Englewood Cliffs, NJ, 1979.

[234] J. H. McClellan and C. M. Rader. *Number Theory in Digital Signal Processing*. Prentice-Hall, Englewood Cliffs, NJ, 1979.

[235] J. H. McClellan and C. M. Rader. *Number Theory in Digital Signal Processing*. Prentice-Hall, Englewood Cliffs, NJ, 1979.

[236] J. H. McClellan and C. M. Rader. *Number Theory in Digital Signal Processing*. Prentice-Hall, Englewood Cliffs, NJ, 1979.

[237] J. H. McClellan and C. M. Rader. *Number Theory in Digital Signal Processing*. Prentice-Hall, Englewood Cliffs, NJ, 1979.

[238] J. H. McClellan and C. M. Rader. *Number Theory in Digital Signal Processing*. Prentice-Hall, Inc., Englewood Cliffs, NJ, 1979.

[239] J. H. McClellan and C. M. Rader. *Number Theory in Digital Signal Processing*. Prentice-Hall, Englewood Cliffs, NJ, 1979.

[240] J. W. Meek and A. S. Veletsos. Fast convolution for recursive digital filters. *IEEE Transactions on Audio and Electroacoustics*, AU-20:938211;94, March 1972.

[241] R. Meyer, R. Reng, and K. Schwarz. Convolution algorithms on dsp processors. In *Proceedings of the ICASSP-91*, page 21938211;2196, Toronto, Canada, May 1991.

[242] R. Meyer and K. Schwarz. Fft implementation on dsp-chips, Sept. 18 1990. preprint.

[243] R. Meyer, K. Schwarz, and H. W. Schuessler. Fft implementation on dsp-chips 8212; theory and practice. In *Proceedings of the ICASSP-90*, page 15038211;1506, Albuquerque, NM, April 1990.

[244] R. A. Meyer and C. S. Burrus. A unified analysis of multirate and periodically time varying digital filters. *IEEE Transactions on Circuits and Systems*, CAS-22(3):1628211;168, March 1975.

[245] R. A. Meyer and C. S. Burrus. Design and implementation of multirate digital filters. *IEEE Transactions on Acoustics, Speech, and Signal Processing*, ASSP-24(1):538211;58, February 1976.

[246] S. K. Mitra and R. Gransekaran. A note on block implementation of iir digital filters. *IEEE Transactions on Circuit and Systems*, CAS-24(7), July 1977.

[247] S. K. Mitra and R. Gransekaran. Block implementation of recursive digital filters 8211; new structures and properties. *IEEE Transactions on Circuit and Systems*, CAS-25(4):2008211;207, April 1978.

[248] Jacques Morgenstern. Note on a lower bound of the linear complexity of the fast fourier transform. 20(2):305–306, 1973.

[249] L. R. Morris. *Digital Signal Processing Software.* DSPSW, Inc., Toronto, Canada, 1982, 1983.

[250] L. R. Morris. *Digital Signal Processing Software.* DSPSW, Inc., Toronto, Canada, 1982, 1983.

[251] L. R. Morris. *Digital Signal Processing Software.* DSPSW, Inc., Toronto, Canada, 1982, 1983.

[252] Douglas G. Myers. *Digital Signal Processing, Efficient Convolution and Fourier Transform Techniques.* Prentice-Hall, Sydney, Australia, 1990.

[253] Douglas G. Myers. *Digital Signal Processing, Efficient Convolution and Fourier Transform Techniques.* Prentice-Hall, Sydney, Australia, 1990.

[254] Douglas G. Myers. *Digital Signal Processing, Efficient Convolution and Fourier Transform Techniques.* Prentice-Hall, Sydney, Australia, 1990.

[255] Kenji Nakayama. An improved fast fourier transform algorithm using mixed frequency and time decimations. *IEEE Trans. Acoust., Speech, Signal Processing*, 36(2):2908211;292, 1988.

[256] P. J. Nicholson. Algebraic theory of finite fourier transforms. *Journal of Computer and System Sciences*, 5:5248211;547, 1971.

[257] P. J. Nicholson. Algebraic theory of finite fourier transforms. *Journal of Computer and System Sciences*, 5(2):5248211;547, February 1971.

[258] Ivan Niven and H. S. Zuckerman. *An Introduction to the Theory of Numbers.* John Wiley & Sons, New York, fourth edition, 1980.

[259] Ivan Niven and H. S. Zuckerman. *An Introduction to the Theory of Numbers.* John Wiley & Sons, New York, fourth edition, 1980.

[260] Ivan Niven and H. S. Zuckerman. *An Introduction to the Theory of Numbers.* John Wiley & Sons, New York, fourth edition, 1980.

[261] H. J. Nussbaumer. *Fast Fourier Transform and Convolution Algorithms.* Springer-Verlag, Heidelberg, Germany, second edition, 1981, 1982.

[262] H. J. Nussbaumer. *Fast Fourier Transform and Convolution Algorithms.* Springer-Verlag, Heidelberg, Germany, second edition, 1981, 1982.

[263] H. J. Nussbaumer. *Fast Fourier Transform and Convolution Algorithms.* Springer-Verlag, Heidelberg, Germany, second edition, 1981, 1982.

[264] H. J. Nussbaumer. *Fast Fourier Transform and Convolution Algorithms.* Springer-Verlag, Heidelberg, Germany, second edition, 1981, 1982.

[265] H. J. Nussbaumer. *Fast Fourier Transform and Convolution Algorithms.* Springer-Verlag, Heidelberg, Germany, second edition, 1981, 1982.

[266] H. J. Nussbaumer. *Fast Fourier Transform and Convolution Algorithms.* Springer-Verlag, Heidelberg, Germany, second edition, 1981, 1982.

[267] H. J. Nussbaumer. *Fast Fourier Transform and Convolution Algorithms.* Springer-Verlag, Heidelberg, Germany, second edition, 1981, 1982.

[268] H. J. Nussbaumer. *Fast Fourier Transformation and Convolution Algorithms.* Springer, 2nd edition, 1982.

[269] A. V. Oppenheim and R. W. Schafer. *Discrete-Time Signal Processing.* Prentice-Hall, Englewood Cliffs, NJ, 1989.

[270] A. V. Oppenheim and R. W. Schafer. *Discrete-Time Signal Processing.* Prentice-Hall, Englewood Cliffs, NJ, 1989.

[271] A. V. Oppenheim and R. W. Schafer. *Discrete-Time Signal Processing.* Prentice-Hall, Englewood Cliffs, NJ, second edition, 1999. Earlier editions in 1975 and 1989.

[272] A. V. Oppenheim and R. W. Schafer. *Discrete-Time Signal Processing.* Prentice-Hall, Englewood Cliffs, NJ, second edition, 1999. Earlier editions in 1975 and 1989.

[273] A. V. Oppenheim and R. W. Schafer. *Discrete-Time Signal Processing.* Prentice-Hall, Englewood Cliffs, NJ, second edition, 1999. Earlier editions in 1975 and 1989.

[274] A. V. Oppenheim and R. W. Schafer. *Discrete-Time Signal Processing.* Prentice-Hall, Englewood Cliffs, NJ, second edition, 1999. Earlier editions in 1975 and 1989.

[275] A. V. Oppenheim and R. W. Schafer. *Discrete-Time Signal Processing.* Prentice-Hall, Englewood Cliffs, NJ, second edition, 1999. Earlier editions in 1975 and 1989.

[276] A. V. Oppenheim and R. W. Schafer. *Discrete-Time Signal Processing.* Prentice-Hall, Englewood Cliffs, NJ, second edition, 1999. Earlier editions in 1975 and 1989.

[277] A. V. Oppenheim, R. W. Schafer, and J. R. Buck. *Discrete-Time Signal Processing.* Prentice Hall, 2nd edition, 1999.

[278] A. V. Oppenheim, R. W. Schafer, and J. R. Buck. *Discrete-Time Signal Processing.* Prentice-Hall, Upper Saddle River, NJ, 2nd edition, 1999.

[279] Oystein Ore. *Number Theory and Its History.* McGraw-Hill, New York, 1948.

[280] Victor Ya. Pan. The trade-off between the additive complexity and the asyncronicity of linear and bilinear algorithms. *Information Proc. Lett.*, 22:118211;14, 1986.

[281] Christos H. Papadimitriou. Optimality of the fast fourier transform. 26(1):95–102, 1979.

[282] T. W. Parks and C. S. Burrus. *Digital Filter Design.* John Wiley & Sons, New York, 1987.

[283] T. W. Parks and C. S. Burrus. *Digital Filter Design.* John Wiley & Sons, New York, 1987.

[284] T. W. Parsons. A winograd-fourier transform algorithm for real-valued data. *IEEE Trans. on ASSP*, 27:398–402, August 1979.

[285] F. Perez and T. Takaoka. A prime factor fft algorithm implementation using a program generation technique. *IEEE Transactions on Acoustics, Speech and Signal Processing*, 35:12218211;1223, August 1987.

[286] I. Pitas and C. S. Burrus. Time and error analysis of digital convolution by rectangular transforms. *Signal Processing*, 5(2):1538211;162, March 1983.

[287] I. Pitas and C. S. Burrus. Time and error analysis of digital convolution by rectangular transforms. *Signal Processing*, 5(2):1538211;162, March 1983.

[288] J. M. Pollard. The fast fourier transform in a finite field. *Mathematics of Computation*, 25(114):3658211;374, April 1971.

[289] Miodrag Popovi263; and Dragutin 352;evi263;. A new look at the comparison of the fast hartley and fourier transforms. *IEEE Transactions on Signal Processing*, 42(8):21788211;2182, August 1994.

[290] W. H. Press, B. P. Flannery, S. A. Teukolsky, and W. T. Vetterling. *Numerical Recipes in C: The Art of Scientific Computing.* Cambridge Univ. Press, New York, NY, 2nd edition, 1992.

[291] M. Pschel and J. M. F. Moura. Algebraic signal processing theory. available at http://arxiv.org/abs/cs.IT/0612077.

[292] M. Pschel and J. M. F. Moura. The algebraic approach to the discrete cosine and sine transforms and their fast algorithms. *SIAM Journal of Computing*, 32(5):12808211;1316, 2003.

[293] M. Pschel and J. M. F. Moura. Algebraic signal processing theory: 1-d space. *IEEE Transactions on Signal Processing*, 56(8):3586–3599, 2008.

[294] M. Pschel and J. M. F. Moura. Algebraic signal processing theory: Cooley-tukey type algorithms for dcts and dsts. *IEEE Transactions on Signal Processing*, 56(4):1502–1521, 2008. a longer version is available at http://arxiv.org/abs/cs.IT/0702025.

[295] M. Pschel and J. M. F. Moura. Algebraic signal processing theory: Foundation and 1-d time. *IEEE Transactions on Signal Processing*, 56(8):3572–3585, 2008.

[296] Markus Pschel, Jos[U+FFFD] F. Moura, Jeremy R. Johnson, David Padua, Manuela M. Veloso, Bryan W. Singer, Jianxin Xiong, Franz Franchetti, Aca Ga269;i263;, Yevgen Voronenko, Kang Chen, Robert W. Johnson, and Nicholas Rizzolo. Spiral: Code generation for dsp transforms. *Proc. IEEE*, 93(2):232–275, 2005.

[297] Z. Qian, C. Lu, M. An, and R. Tolimieri. Self-sorting in-place fft algorithm with minimum working space. *IEEE Trans. Acoust., Speech, Signal Processing*, 42(10):28358211;2836, 1994.

[298] L. R. Rabiner and B. Gold. *Theory and Application of Digital Signal Processing.* Prentice-Hall, Englewood Cliffs, NJ, 1975.

[299] L. R. Rabiner and B. Gold. *Theory and Application of Digital Signal Processing.* Prentice-Hall, Englewood Cliffs, NJ, 1975.

[300] L. R. Rabiner and B. Gold. *Theory and Application of Digital Signal Processing.* Prentice-Hall, Englewood Cliffs, NJ, 1975.

[301] L. R. Rabiner and C. M. Rader, editors. *Digital Signal Processing, selected reprints.* IEEE Press, New York, 1972.

[302] L. R. Rabiner and C. M. Rader, editors. *Digital Signal Processing, selected reprints.* IEEE Press, New York, 1972.

[303] L. R. Rabiner and C. M. Rader, editors. *Digital Signal Processing, selected reprints.* IEEE Press, New York, 1972.

[304] Lawrence Rabiner. The chirp z-transform algorithm: a lesson in serendipity. *IEEE Signal Processing Magazine*, 24:1188211;119, March 2004.

[305] Lawrence R. Rabiner, Ronald W. Schafer, and Charles M. Rader. The chirp -transform algorithm. *IEEE Trans. Audio Electroacoustics*, 17(2):868211;92, 1969.

[306] L.R. Rabiner, R.W. Schafer, and C.M. Rader. The chirp z-transform algorithm. *IEEE Transactions on Audio Electroacoustics*, AU-17:868211;92, June 1969.

[307] L.R. Rabiner, R.W. Schafer, and C.M. Rader. The chirp z-transform algorithm. *IEEE Transactions on Audio Electroacoustics*, AU-17:868211;92, June 1969.

[308] C. M. Rader. Discrete fourier transforms when the number of data samples is prime. *Proceedings of the IEEE*, 56:11078211;1108, June 1968.

[309] C. M. Rader. Discrete fourier transforms when the number of data samples is prime. *Proc. IEEE*, 56:11078211;1108, June 1968.

[310] C. M. Rader and N. M. Brenner. A new principle for fast fourier transformation. *IEEE Transactions on Acoustics, Speech, and Signal Processing*, ASSP-24(3):264–266, June 1976.

[311] Charles M. Rader. Discrete convolution via mersenne transforms. *IEEE Transactions on Computers*, 21(12):12698211;1273, December 1972.

[312] Charles M. Rader. Number theoretic convolution. In *IEEE Signal Processing Workshop*, Arden House, Harriman, NY, January 1972.

[313] Charles M. Rader and N. M. Brenner. A new principle for fast fourier transformation. *IEEE Trans. Acoust., Speech, Signal Processing*, 24:264–265, 1976.

[314] K. R. Rao and P. Yip. *Discrete Cosine Transform: Algorithms, Advantages, Applications.* Academic Press, San Diego, CA, 1990.

[315] J. M. Rius and R. De Porrata-D[U+FFFD]. New fft bit-reversal algorithm. *IEEE Transactions on Signal Processing*, 43(4):9918211;994, April 1995.

[316] Christian Roche. A split8211;radix partial input/output fast fourier transform algorithm. *IEEE Transactions on Signal Processing*, 40(5):12738211;1276, May 1992.

[317] D. Rockmore. Some applications of generalized fft's. In *Proceedings of DIMACS Workshop in Groups and Computation*, volume 28, page 3298211;370, 1995.

[318] J. J. Rodr[U+FFFD]ez. An improved fft digit8211;reversal algorithm. *IEEE Transactions on Acoustics, Speech, and Signal Processing*, 37(8):12988211;1300, August 1989.

[319] J.H. Rothweiler. Implementation of the in-order prime factor fft algorithm. *IEEE TRANS. ON ASSP*, 30:105–107, February 1982.

[320] J.H. Rothweiler. Implementation of the in-order prime factor fft algorithm. *IEEE TRANS. ON ASSP*, 30:105–107, February 1982.

[321] Petr Rsel. Timing of some bit reversal algorithms. *Signal Processing*, 18(4):4258211;433, December 1989.

[322] Ali Saidi. Decimation-in-time-frequency fft algorithm. In *Proc. IEEE Int'l Conf. Acoustics, Speech, and Signal Processing*, volume 3, page 4538211;456, 1994.

[323] Ali Saidi. Decimation-in-time-frequency fft algorithm. In *Proceedings of the IEEE International Conference on Acoustics, Speech, and Signal Processing*, volume 3, page III:4538211;456, IEEE ICASSP-94, Adelaide, Australia, April 198211;22 1994.

[324] Ali Saidi. Decimation-in-time-frequency fft algorithm, 1996. manuscript.

[325] G. Sande. Fast fourier transform - a gobally complex algorithm with locally real implementations. *Proc. 4th Annual Princeton Conference on Information Sciences and Systems*, pages 136–142, March 1970.

[326] James C. Schatzman. Accuracy of the discrete fourier transform and the fast fourier transform. *SIAM J. Scientific Computing*, 17(5):11508211;1166, 1996.

[327] James C. Schatzman. Index mapping for the fast fourier transform. *IEEE Transactions on Signal Processing*, 44(3):7178211;719, March 1996.

[328] Manfred R. Schroeder. *Number Theory in Science and Comminication.* Springer8211;Verlag, Berlin, second edition, 1984, 1986.

[329] I. W. Selesnick and C. S. Burrus. Automating the design of prime length fft programs. In *Proceedings of the IEEE International Symposium on Circuits and Systems*, volume 1, page 1338211;136, ISCAS-92, San Diego, CA, May 1992.

[330] I. W. Selesnick and C. S. Burrus. Automating the design of prime length fft programs. In *Proceedings of the IEEE International Symposium on Circuits and Systems*, volume 1, page 1338211;136, ISCAS-92, San Diego, CA, May 1992.

[331] I. W. Selesnick and C. S. Burrus. Multidimensional mapping techniques for convolution. In *Proceedings of the IEEE International Conference on Signal Processing*, volume III, pages III–2888211;291, IEEE ICASSP-93, Minneapolis, April 1993.

[332] I. W. Selesnick and C. S. Burrus. Multidimensional mapping techniques for convolution. In *Proceedings of the IEEE International Conference on Signal Processing*, volume III, pages III–2888211;291, IEEE ICASSP-93, Minneapolis, April 1993.

[333] I. W. Selesnick and C. S. Burrus. Extending winograd's small convolution algorithm to longer lengths. In *Proceedings of the IEEE International Symposium on Circuits and Systems*, volume 2, page 2.4498211;452, IEEE ISCAS-94, London, June 30 1994.

[334] I. W. Selesnick and C. S. Burrus. Extending winograd's small convolution algorithm to longer lengths. In *Proceedings of the IEEE International Symposium on Circuits and Systems*, volume 2, page 2.4498211;452, IEEE ISCAS-94, London, June 30 1994.

[335] Ivan W. Selesnick and C. Sidney Burrus. Automatic generation of prime length fft programs. *IEEE Transactions on Signal Processing*, 44(1):148211;24, January 1996.

[336] Ivan W. Selesnick and C. Sidney Burrus. Automatic generation of prime length fft programs. *IEEE Transactions on Signal Processing*, 44(1):148211;24, January 1996.

[337] R. G. Shenoy, Daniel Burnside, and T. W. Parks. Linear periodic systems and multirate filter design. Technical report GEO-002-92-16b, Schlumberger8211;Doll Research Note, September 1992.

[338] R. G. Shenoy, T. W. Parks, and Daniel Burnside. Fourier analysis of linear periodic systems and multirate filter design. In *Paper Summaries for the 1992 DSP Workshop*, page 2.4.1, Starved Rock Lodge, Utica, Ill., 1992.

[339] O. V. Shentov, S. K. Mitra, U. Heute, and A. N. Hossen. Subband dft 8211; part i: Definition, interpretation and extensions. *Signal Processing*, 41:2618211;278, 1995.

[340] R.C. Singleton. An algorithm for computing the mixed radix fast fourier transform. *IEEE Transactions on Audio and Electroacoustics*, AU-17:93–103, June 1969.

[341] Richard C. Singleton. On computing the fast fourier transform. *Comm. ACM*, 10:6478211;654, 1967.

[342] G. A. Sitton. The qft algorithm, 1985.

[343] G. A. Sitton. The qft algorithm, 1985.

[344] G. A. Sitton. The qdft: An efficient method for short prime length dfts, June 1991.

[345] Winthrop W. Smith and Joanne M. Smith. *Handbook of Real-Time Fast Fourier Transforms*. IEEE Press, New York, 1995.

[346] H. V. Sorensen and C. S. Burrus. Efficient computation of the short-time fft. In *Proceedings of the IEEE International Conference on Acoustics, Speech, and Signal Processing*, pages 1894–1897, New York, April 1988.

[347] H. V. Sorensen and C. S. Burrus. Efficient computation of the dft with only a subset of input or output points. *IEEE Transactions on Signal Processing*, 41(3):11848211;1200, March 1993.

[348] H. V. Sorensen and C. S. Burrus. Fast dft and convolution algorithms. In Sanjit K. Mitra and James F. Kaiser, editors, *Handbook for Digital Signal Processing*, chapter 8. John Wiley & Sons, New York, 1993.

[349] H. V. Sorensen, C. S. Burrus, and D. L. Jones. A new efficient algorithm for computing a few dft points. In *Proceedings of the IEEE International Symposium on Circuits and Systems*, page 19158211;1918, Espoo, Finland, June 1988.

[350] H. V. Sorensen, M. T. Heideman, and C. S. Burrus. On computing the split8211;radix fft. *IEEE Transactions on Acoustics, Speech, and Signal Processing*, 34(1):1528211;156, February 1986.

[351] H. V. Sorensen, M. T. Heideman, and C. S. Burrus. On computing the split8211;radix fft. *IEEE Transactions on Acoustics, Speech, and Signal Processing*, 34(1):1528211;156, February 1986.

[352] H. V. Sorensen, D. L. Jones, C. S. Burrus, and M. T. Heideman. On computing the discrete hartley transform. *IEEE Transactions on Acoustics, Speech, and Signal Processing*, 33(5):12318211;1238, October 1985.

[353] H. V. Sorensen, D. L. Jones, C. S. Burrus, and M. T. Heideman. On computing the discrete hartley transform. *IEEE Transactions on Acoustics, Speech, and Signal Processing*, 33(5):12318211;1238, October 1985.

[354] H. V. Sorensen, D. L. Jones, C. S. Burrus, and M. T. Heideman. On computing the discrete hartley transform. *IEEE Transactions on Acoustics, Speech, and Signal Processing*, 33(5):12318211;1238, October 1985.

[355] H. V. Sorensen, D. L. Jones, C. S. Burrus, and M. T. Heideman. On computing the discrete hartley transform. *IEEE Transactions on Acoustics, Speech, and Signal Processing*, 33(5):12318211;1238, October 1985.

[356] H. V. Sorensen, D. L. Jones, M. T. Heideman, and C. S. Burrus. Real valued fast fourier transform algorithms. *IEEE Transactions on Acoustics, Speech, and Signal Processing*, 35(6):8498211;863, June 1987.

[357] H. V. Sorensen, D. L. Jones, M. T. Heideman, and C. S. Burrus. Real valued fast fourier transform algorithms. *IEEE Transactions on Acoustics, Speech, and Signal Processing*, 35(6):8498211;863, June 1987.

[358] H. V. Sorensen, D. L. Jones, M. T. Heideman, and C. S. Burrus. Real valued fast fourier transform algorithms. *IEEE Transactions on Acoustics, Speech, and Signal Processing*, 35(6):8498211;863, June 1987.

[359] H. V. Sorensen, D. L. Jones, M. T. Heideman, and C. S. Burrus. Real valued fast fourier transform algorithms. *IEEE Transactions on Acoustics, Speech, and Signal Processing*, 35(6):8498211;863, June 1987.

[360] H. V. Sorensen, D. L. Jones, M. T. Heideman, and C. S. Burrus. Real valued fast fourier transform algorithms. *IEEE Transactions on Acoustics, Speech, and Signal Processing*, 35(6):8498211;863, June 1987.

[361] H. V. Sorensen, C. A. Katz, and C. S. Burrus. Efficient fft algorithms for dsp processors using tensor product decomposition. In *Proceedings of the IEEE International Conference on Acoustics, Speech, and Signal Processing*, page 15078211;1510, Albuquerque, NM, April 1990.

[362] Henrik V. Sorensen, C. Sidney Burrus, and Michael T. Heideman. *Fast Fourier Transform Database.* PWS Publishing, Boston, 1995.

[363] Henrik V. Sorensen, C. Sidney Burrus, and Michael T. Heideman. *Fast Fourier Transform Database.* PWS Publishing, Boston, 1995.

[364] R. Stasi324;ski. Prime factor dft algorithms for new small-n dft modules. *IEE Proceedings, Part G*, 134(3):1178211;126, 1987.

[365] R. Stasi324;ski. Extending sizes of effective convolution algorithms. *Electronics Letters*, 26(19):16028211;1604, 1990.

[366] Ryszard Stasi324;ski. The techniques of the generalized fast fourier transform algorithm. *IEEE Transactions on Signal Processing*, 39(5):10588211;1069, May 1991.

[367] T. G. Stockham. High speed convolution and correlation. *Proc. AFIPS Spring Joint Computer Conference*, 28:2298211;233, 1966.

[368] T. G. Stockham. High speed convolution and correlation. In *AFIPS Conf. Proc.*, volume 28, page 2298211;233. 1966 Spring Joint Computer Conference, 1966.

[369] R. Storn. On the bruun algorithm and its inverse. *Frequenz*, 46:1108211;116, 1992. a German journal.

[370] D. Sundararajan, M. O. Ahamad, and M. N. S. Swamy. A fast fft bit-reversal algorithm. *IEEE Transactions on Circuits and Systems, II*, 41(10):7018211;703, October 1994.

[371] D. Sundararajan, M. O. Ahmad, and M. N. S. Swamy. Fast computations of the discrete fourier transform of real data. *IEEE Transactions on Signal Processing*, 45(8):20108211;2022, August 1997.

[372] P. N. Swarztrauber. Vectorizing the ffts. *Parallel Computations*, page 518211;83, 1982. G. Rodrigue ed.

[373] Manfred Tasche and Hansmartin Zeuner. *Handbook of Analytic-Computational Methods in Applied Mathematics*, chapter 8, page 3578211;406. CRC Press, Boca Raton, FL, 2000.

[374] Manfred Tasche and Hansmartin Zeuner. Improved roundoff error analysis for precomputed twiddle factors. *J. Comput. Anal. Appl.*, 4(1):18211;18, 2002.

[375] C. Temperton. Self-sorting in-place fast fourier transforms. *SIAM J. Scientific and Statistical Computing*, 12(4):8088211;823, 1991.

[376] Clive Temperton. A new set of minimum-add small-n rotated dft modules. *Journal of Computational Physics*, 75:1908211;198, 1988.

[377] Clive Temperton. A self-sorting in-place prime factor real/half-complex fft algorithm. *Journal of Computational Physics*, 75:1998211;216, 1988.

[378] Clive Temperton. Nesting strategies for prime factor fft algorithms. *Journal of Computational Physics*, 82(2):2478211;268, June 1989.

[379] Clive Temperton. Self-sorting in-place fast fourier transforms. *SIAM Journal of Sci. Statist. Comput.*, 12(4):8088211;823, 1991.

[380] Clive Temperton. A generalized prime factor fft algorithm for any. *SIAM Journal of Scientific and Statistical Computing*, 13:6768211;686, May 1992.

[381] R. Tolimieri, M. An, and C. Lu. *Algorithms for Discrete Fourier Transforms and Convolution*. Springer, 2nd edition, 1997.

[382] Richard Tolimieri, Myoung An, and Chao Lu. *Algorithms for Discrete Fourier Transform and Convolution*. Springer-Verlag, New York, second edition, 1989, 1997.

[383] Richard Tolimieri, Myoung An, and Chao Lu. *Algorithms for Discrete Fourier Transform and Convolution*. Springer-Verlag, New York, second edition, 1989, 1997.

[384] Richard Tolimieri, Myoung An, and Chao Lu. *Mathematics of Multidimensional Fourier Transform Algorithms*. Springer-Verlag, New York, second edition, 1993, 1997.

[385] Richard Tolimieri, Myoung An, and Chao Lu. *Mathematics of Multidimensional Fourier Transform Algorithms*. Springer-Verlag, New York, second edition, 1993, 1997.

[386] P. R. Uniyal. Transforming real-valued sequences: Fast fourier versus fast hartley transform algorithms. *IEEE Transactions on Signal Processing*, 42(11):32498211;3254, November 1994.

[387] P. P. Vaidyanathan. *Multirate Systems and Filter Banks*. Prentice-Hall, Englewood Cliffs, NJ, 1992.

[388] C. Van Loan. *Computational Framework of the Fast Fourier Transform*. Siam, 1992.

[389] Charles van Loan. *Computational Frameworks for the Fast Fourier Transform.* SIAM, Philadelphia, 1992.

[390] Charles Van Loan. *Matrix Frameworks for the Fast Fourier Transform.* SIAM, Philadelphia, PA, 1992.

[391] M. Vetterli and H. J. Nussbaumer. Simple fft and dct algorithms with reduced number of operations. *Signal Processing*, 6(4):267–278, 1984.

[392] Martin Vetterli and P. Duhamel. Split-radix algorithms for length- dft's. *IEEE Trans. on ASSP*, 37(1):578211;64, January 1989. Also, ICASSP-88 Proceedings, pp.14158211;1418, April 1988.

[393] Martin Vetterli and H. J. Nussbaumer. Simple fft and dct algorithms with reduced number of operations. *Signal Processing*, 6(4):2678211;278, August 1984.

[394] Martin Vetterli and H. J. Nussbaumer. Simple fft and dct algorithms with reduced number of operations. *Signal Processing*, 6(4):2678211;278, August 1984.

[395] Martin Vetterli and H. J. Nussbaumer. Simple fft and dct algorithms with reduced number of operations. *Signal Processing*, 6(4):2678211;278, August 1984.

[396] H. B. Voelcker and E. E. Hartquist. Digital filtering via block recursion. *IEEE Transactions on Audio and Electroacoustics*, AU-18:1698211;176, June 1970.

[397] Y. Voronenko and M. Pschel. Algebraic signal processing theory: Cooley-tukey type algorithms for real dfts. *IEEE Transactions on Signal Processing.* to appear.

[398] Y. Voronenko and M. Pschel. Algebraic derivation of general radix cooley-tukey algorithms for the real discrete fourier transform. In *Proc. International Conference on Acoustics, Speech, and Signal Processing (ICASSP)*, volume 3, pages 876–879, 2006.

[399] A. R. V[U+FFFD]onyi-K[U+FFFD]. A recursive fast fourier transformation algorithm. *IEEE Transactions on Circuits and Systems, II*, 42(9):6148211;616, September 1995.

[400] J. S. Walker. A new bit reversal algorithm. *IEEE Transactions on Acoustics, Speech, and Signal Processing*, 38(8):14728211;1473, August 1990.

[401] Fang Ming Wang and P. Yip. Fast prime factor decomposition algorithms for a family of discrete trigonometric transforms. *Circuits, Systems, and Signal Processing*, 8(4):4018211;419, 1989.

[402] S. A. White. Applications of distributed arithmetic to digital signal processing. *IEEE ASSP Magazine*, 6(3):48211;19, July 1989.

[403] S. Winograd. On computing the discrete fourier transform. *Proc. National Academy of Sciences, USA*, 73:10068211;1006, April 1976.

[404] S. Winograd. On computing the discrete fourier transform. *Proc. National Academy of Sciences, USA*, 73:10068211;1006, April 1976.

[405] S. Winograd. On computing the discrete fourier transform. *Proc. National Academy of Sciences, USA*, 73:10068211;1006, April 1976.

[406] S. Winograd. On computing the discrete fourier transform. *Proc. National Academy of Sciences, USA*, 73:10068211;1006, April 1976.

[407] S. Winograd. On computing the discrete fourier transform. *Proc. National Academy of Sciences, USA*, 73:10068211;1006, April 1976.

[408] S. Winograd. On computing the discrete fourier transform. *Mathematics of Computation*, 32:1758211;199, January 1978.

[409] S. Winograd. On computing the discrete fourier transform. *Mathematics of Computation*, 32:1758211;199, 1978.

[410] S. Winograd. On computing the discrete fourier transform. *Mathematics of Computation*, 32:1758211;199, January 1978.

[411] S. Winograd. On computing the discrete fourier transform. *Math. Computation*, 32(1):1758211;199, January 1978.

[412] S. Winograd. On computing the discrete fourier transform. *Mathematics of Computation*, 32:1758211;199, January 1978.

[413] S. Winograd. On the multiplicative complexity of the discrete fourier transform. *Advances in Mathematics*, 32(2):838211;117, May 1979. also in.

[414] S. Winograd. On the multiplicative complexity of the discrete fourier transform. *Advances in Mathematics*, 32:838211;117, 1979.

[415] S. Winograd. On the multiplicative complexity of the discrete fourier transform. *Advances in Mathematics*, 32(2):838211;117, May 1979. also in.

[416] S. Winograd. *Arithmetic Complexity of Computation.* SIAM CBMS-NSF Series, No. 33. SIAM, Philadelphia, 1980.

[417] S. Winograd. *Arithmetic Complexity of Computation.* Siam, 1980.

[418] S. Winograd. *Arithmetic Complexity of Computation.* SIAM CBMS-NSF Series, No.33. SIAM, Philadelphia, 1980.

[419] S. Winograd. Signal processing and complexity of computation. In *Proceedings of the IEEE International Conference on Acoustics, Speech, and Signal Processing*, page 13818211;1384, ICASSP-80, Denver, April 1980.

[420] Jianxin Xiong, David Padua, and Jeremy Johnson. Spl: A language and compiler for dsp algorithms. In *Proc. ACM SIGPLAN'01 Conf. Programming Language Design and Implementation (PLDI)*, pages 298–308, 2001.

[421] R. Yavne. An economical method for calculating the discrete fourier transform. In *Fall Joint Computer Conference, AFIPS Conference Proceedings*, volume 33 part 1, page 1158211;125, 1968.

[422] R. Yavne. An economical method for calculating the discrete fourier transform. In *Proc. AFIPS Fall Joint Computer Conf.*, volume 33, pages 115–125, 1968.

[423] R. Yavne. An economical method for calculating the discrete fourier transform. In *Fall Joint Computer Conference, AFIPS Conference Proceedings*, volume 33 part 1, page 1158211;125, 1968.

[424] Angelo A. Yong. A better fft bit-reversal algorithm without tables. *IEEE Transactions on Signal Processing*, 39(10):23658211;2367, October 1991.

[425] L. A. Zadeh. Frequency analysis of variable networks. *Proceeding of the IRE*, 38(3):2918211;299, 1950.

[426] Y. Zalcstein. A note on fast convolution. *IEEE Transactions on Computers*, C-20:665, June 1971.

[427] Jan Zeman and Allen G. Lindgren. Fast digital filters with low round8211;off noise. *IEEE Transactions on Circuit and Systems*, CAS-28:7168211;723, July 1981.

[428] Jan Zeman and Allen G. Lindgren. Fast state8211;space decimator with very low round8211;off noise. *Signal Processing*, 3(4):3778211;388, October 1981.

Index of Keywords and Terms

Keywords are listed by the section with that keyword (page numbers are in parentheses). Keywords do not necessarily appear in the text of the page. They are merely associated with that section. *Ex.* apples, § 1.1 (1) **Terms** are referenced by the page they appear on. *Ex.* apples, 1

Attributions

Collection: *Fast Fourier Transforms*
Edited by: C. Sidney Burrus
URL: http://cnx.org/content/col10550/1.22/
License: http://creativecommons.org/licenses/by/3.0/

Module: "Preface: Fast Fourier Transforms"
By: C. Sidney Burrus
URL: http://cnx.org/content/m16324/1.10/
Pages: 1-3
Copyright: C. Sidney Burrus
License: http://creativecommons.org/licenses/by/3.0/

Module: "Introduction: Fast Fourier Transforms"
By: C. Sidney Burrus
URL: http://cnx.org/content/m16325/1.10/
Pages: 5-6
Copyright: C. Sidney Burrus
License: http://creativecommons.org/licenses/by/2.0/

Module: "Multidimensional Index Mapping"
By: C. Sidney Burrus
URL: http://cnx.org/content/m16326/1.12/
Pages: 7-19
Copyright: C. Sidney Burrus
License: http://creativecommons.org/licenses/by/3.0/

Module: "Polynomial Description of Signals"
By: C. Sidney Burrus
URL: http://cnx.org/content/m16327/1.8/
Pages: 21-25
Copyright: C. Sidney Burrus
License: http://creativecommons.org/licenses/by/2.0/

Module: "The DFT as Convolution or Filtering"
By: C. Sidney Burrus
URL: http://cnx.org/content/m16328/1.9/
Pages: 27-37
Copyright: C. Sidney Burrus
License: http://creativecommons.org/licenses/by/2.0/

Module: "Factoring the Signal Processing Operators"
By: C. Sidney Burrus
URL: http://cnx.org/content/m16330/1.8/
Pages: 39-41
Copyright: C. Sidney Burrus
License: http://creativecommons.org/licenses/by/2.0/

Module: "Winograd's Short DFT Algorithms"
By: C. Sidney Burrus, Ivan Selesnick
URL: http://cnx.org/content/m16333/1.14/
Pages: 43-63
Copyright: C. Sidney Burrus, Ivan Selesnick
License: http://creativecommons.org/licenses/by/2.0/

Module: "DFT and FFT: An Algebraic View"
By: Markus Pueschel
URL: http://cnx.org/content/m16331/1.14/
Pages: 65-77
Copyright: Markus Pueschel
License: http://creativecommons.org/licenses/by/2.0/

Module: "The Cooley-Tukey Fast Fourier Transform Algorithm"
By: C. Sidney Burrus
URL: http://cnx.org/content/m16334/1.13/
Pages: 79-96
Copyright: C. Sidney Burrus
License: http://creativecommons.org/licenses/by/2.0/

Module: "The Prime Factor and Winograd Fourier Transform Algorithms"
By: C. Sidney Burrus
URL: http://cnx.org/content/m16335/1.9/
Pages: 97-108
Copyright: C. Sidney Burrus
License: http://creativecommons.org/licenses/by/2.0/

Module: "Implementing FFTs in Practice"
By: Steven G. Johnson, Matteo Frigo
URL: http://cnx.org/content/m16336/1.15/
Pages: 109-135
Copyright: Steven G. Johnson
License: http://creativecommons.org/licenses/by/3.0/

Module: "Algorithms for Data with Restrictions"
By: C. Sidney Burrus
URL: http://cnx.org/content/m16338/1.7/
Pages: 137-138
Copyright: C. Sidney Burrus
License: http://creativecommons.org/licenses/by/2.0/

Module: "Convolution Algorithms"
By: C. Sidney Burrus
URL: http://cnx.org/content/m16339/1.10/
Pages: 139-152
Copyright: C. Sidney Burrus
License: http://creativecommons.org/licenses/by/2.0/

Module: "Comments: Fast Fourier Transforms"
By: C. Sidney Burrus
URL: http://cnx.org/content/m16434/1.8/
Pages: 153-155
Copyright: C. Sidney Burrus
License: http://creativecommons.org/licenses/by/2.0/

Module: "Conclusions: Fast Fourier Transforms"
By: C. Sidney Burrus
URL: http://cnx.org/content/m16340/1.7/
Pages: 157-158
Copyright: C. Sidney Burrus
License: http://creativecommons.org/licenses/by/2.0/

Module: "Appendix 1: FFT Flowgraphs"
By: C. Sidney Burrus
URL: http://cnx.org/content/m16352/1.11/
Pages: 159-164
Copyright: C. Sidney Burrus
License: http://creativecommons.org/licenses/by/2.0/

Module: "Appendix 2: Operation Counts for General Length FFT"
By: C. Sidney Burrus
URL: http://cnx.org/content/m16353/1.8/
Pages: 165-166
Copyright: C. Sidney Burrus
License: http://creativecommons.org/licenses/by/2.0/

Module: "Appendix 3: FFT Computer Programs"
By: C. Sidney Burrus
URL: http://cnx.org/content/m17397/1.5/
Pages: 167-206

Module: "Appendix 4: Programs for Short FFTs"
By: C. Sidney Burrus
URL: http://cnx.org/content/m17646/1.4/
Pages: 207-209

Lightning Source UK Ltd.
Milton Keynes UK
UKHW051932060223
416581UK00008B/566

CW00687357

Hooked on LURE FISHING

Tips and new techniques for over 60 sea and freshwater species

Dominic Garnett & Andy Mytton

With underwater photography by Jack Perks
and original illustrations by Neil Sutherland

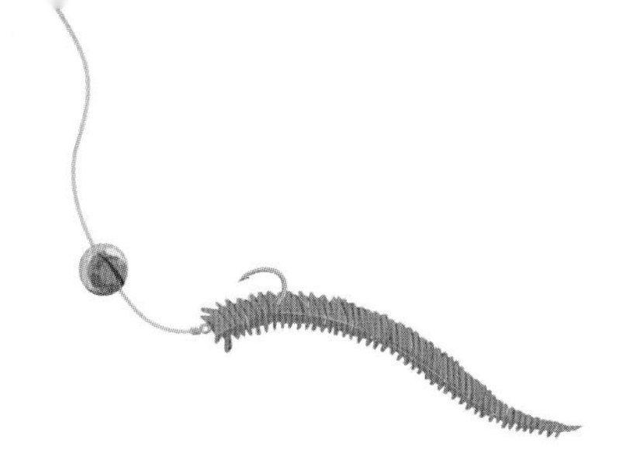

First published in Great Britain by Merlin Unwin Books Ltd, 2019

Merlin Unwin Books Ltd
Palmers House
7 Corve Street
Ludlow
Shropshire SY8 1DB
UK

www.merlinunwin.co.uk

ISBN 978-1-910723-92-0

Designed and set in Bembo 11.5pt by Jo Dovey
Printed by 1010 Printing International Ltd

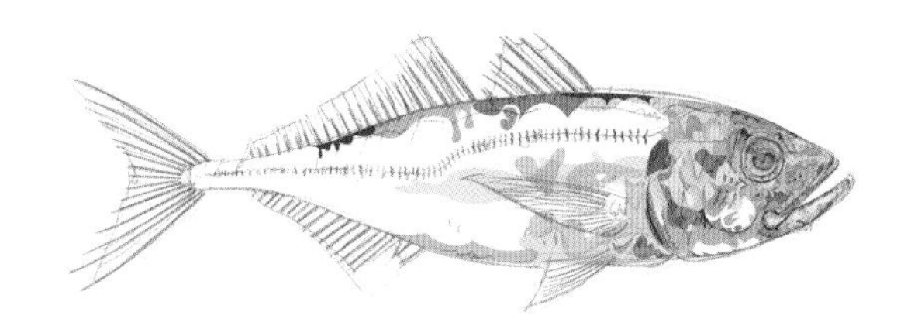

Contents

Introduction

It's one of the oldest clichés going that 'nothing is ever new in angling'. Looking at the vast changes in lure fishing over my mere forty years on the planet, this saying itself is out of date. Take into account the endless variety of new ideas, presentations, better and more affordable tackle, not to mention an entire galaxy of strange lures, and it is nothing short of a revolution. In fact, this very book started out under the slightly pretentious working title of 'Lure Fishing Revolution', but that's another story.

Back in 2015, when I first made the trip to Dorset to meet Andy Mytton, little did I know what was in store. Having cast spinners, plugs and soft baits at perch, pike and trout ever since I was a little kid on holiday, I considered myself a relatively experienced lure angler. What I was to witness next, however, made me feel like a first timer all over again. Those early trips in search of flounder, wrasse and all manner of fierce little critters were mind-blowing. Not so long ago, many of these fish didn't seem very catchable by accident, let alone by design, when lure fishing. I began scribbling notes frantically. Andy's technical know-how and fresh ideas took up whole notebooks and the volume you have in your hands was perhaps inevitable.

A sense of childlike wonder best describes the journey as we went from wide open beaches to tiny rockpools to tempt fish of all descriptions. And what an absolute joy to do so with light tackle! Specialised kit has never been better value and can magnify the most modest fish into a bold fighter. On a more sobering note, in an era when less than 5% of shore-caught sea fish weigh over a kilo, this has to be a better way than to maximise enjoyment than the traditional broom handle rod.

Not that our exploits need be reserved to the sea, or indeed to UK waters. Modern ultralight techniques have been a huge hit on the canals and rivers I've fished for years. They have also translated perfectly to settings as diverse as Amsterdam and the Greek island of Zante, while I even managed to sneak a travel rod on honeymoon in America. At this point I should point out that I'm still a married man – and perhaps the ultimate endorsement of 'fun sized' modern lure fishing is that even my non-angling wife caught the bug on our road trip.

Things got even more interesting as the net widened, further plans hatched and we assembled a little team. The magical photography and brilliant insight of underwater filmmaker Jack Perks was especially exciting, as were the beautiful diagrams and designs of Neil Sutherland. Added to this came a vast array of insights and knowledge learned from our shared network of lure anglers – from species hunting aces like Scott

Fishing should be about enjoyment first. Everything is relative as The General, Dom's six inch plastic model, demonstrates. We forget this at our peril.

Hutchison and Will Pender, to characters like Maurice Minchinton, still avidly lure fishing and learning new tricks in his 70s.

Fishing should be about enjoyment first. Everything is relative, as The General *(pictured below left)* demonstrates. We forget this at our peril.

Indeed, the current community of anglers and ideas in lure fishing is something to be treasured. I can only hope we build on this spirit of openness, enjoyment and cooperation. It is refreshing and much needed in fishing. We can achieve so much more together and I would like to thank everyone who has supported, helped and inspired us in the making of this book.

With the project growing into something bigger than I could have imagined over the three years, the hardest part until now has been keeping a lid on it all! In fact, the book has been the polar opposite of today's reveal-everything-the-very-minute-it-happens social media world. We've been itching to share so many pictures and insights over the last three years.

Whatever your level of lure angling, I hope you will find ideas to stimulate and inspire you in these pages. If I can convey just a fraction of the immense fun and sense of discovery we had in the making of this book, I'm sure you'll enjoy it. And if that is the case, don't guard your sport jealously. Please join the Angling Trust to help conserve sea stocks and share the enjoyment by taking your friends fishing this year.

Dominic Garnett
June 2019

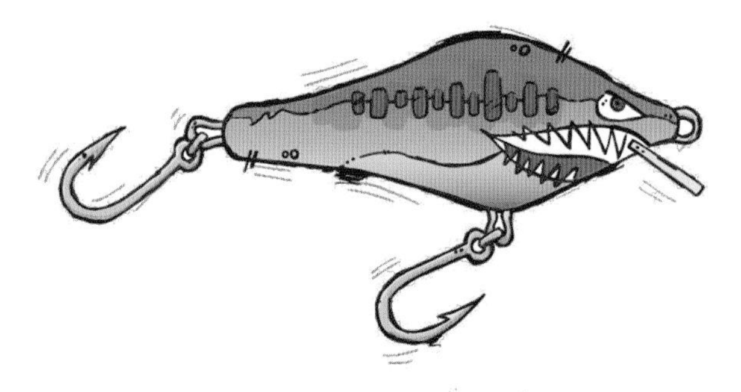

Acknowledgements

Our heartfelt thanks and a big shout out to:

Will Pender, Richard Salter, Scott Hutchison, Ben Bassett, Maurice Minchinton, Merlin, Karen and Jo at Merlin Unwin Books, Paulina Garnett, Shelly and Esmé, Kevin Wilmot and all at *Angling Times*, Richard Cake and Mike at Chesil Bait 'n' Tackle, Steve Partner, Dan 'Esox' Fois, Garrett Fallon, John Cheyne and all at the Angling Trust, Antonius Fischt and John Wheeler. Special thanks also to Jo Dovey who designed this book.

Additional fish species guide images, with thanks, by Scott Hutchison, Richard Salter, Will Pender and Ben Bassett. The photo of the zander (page 161) and the European street fishing pictures courtesy of Antonius Fischt.

This book is dedicated to
lure fishing nuts, species hunters,
adventurers and big kids everywhere

TACKLE

One of the great joys of lure fishing is the minimum of tackle required, allowing the angler to stay mobile and completely focussed. If you're used to a ton of kit, it is a liberating feeling to set out with just a rod, a net and a few bits and pieces in the pocket.

For beginners, or those exploring new ground in their fishing, a perfectly usable lure fishing setup is very affordable indeed. Fishing tackle, however, can be dangerously addictive. Lures, which we'll deal with a little later, are especially collectable. So are rods and reels (and most of us probably own too much tackle!) So, our aim here is to guide you through the essentials, along with more advanced items we've found especially useful.

Rods

A vast array of lure fishing rods await the enthusiast with money to spend. With the growth of methods and brands, the angler has never had better choice or value for money. However, within lure fishing you will find some very specialised tools indeed. When it comes to fishing with light lures, there is no 'jack of all trades', so it's important to pick with care.

So, regardless of things like sexy colours, fittings, carbon content and branding, most important of all is whether the rod is suited to your style of fishing. If possible, set it up with a reel and have a play

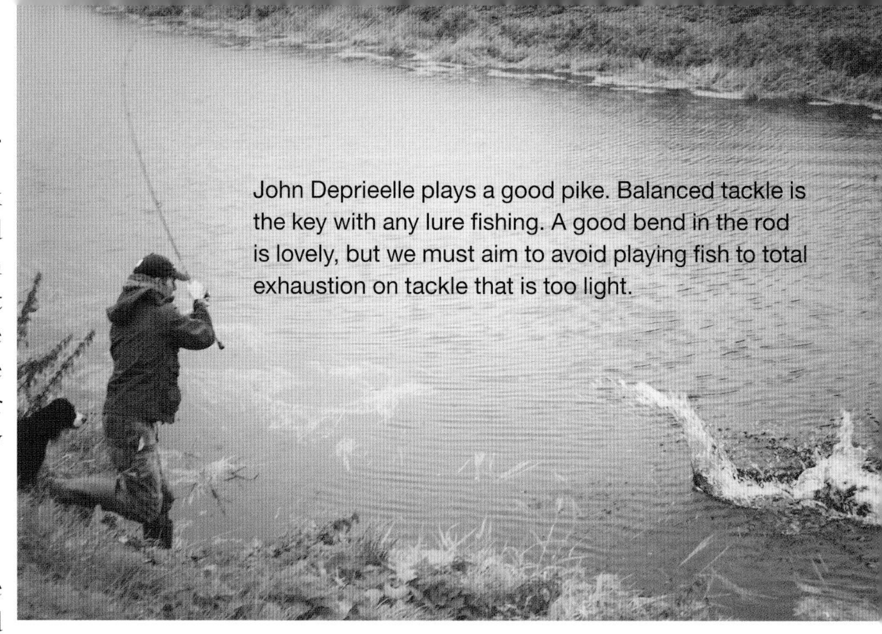

John Deprieelle plays a good pike. Balanced tackle is the key with any lure fishing. A good bend in the rod is lovely, but we must aim to avoid playing fish to total exhaustion on tackle that is too light.

before you buy it, or seek opinions from friends and reviewers. After all, even rods with similar weight ratings can behave quite differently. This is where your friendly guide or tackle shop staff are really worth their salt!

Weight ratings

Lure fishing rods all come with a recommended casting weight in grams and/or ounces. It's vital that you get a good general fit with the typical lures you use, because this will help you not only with casting, but imparting movement, and 'feeling' what your lure is doing. For example, a rod rated 1-5g would be suitable for tiny lures and mini species hunting; a rod casting 7-28g would be capable of throwing much bigger lures longer distances for bass or pike. Where possible ask tackle dealers and friends too, as plenty of rods are rather lighter or heavier actioned than their numbers claim.

Which length rod?

Another consideration for lure rods is overall length. There is no 'best length', because this depends on your style of fishing! In general, shorter rods of 6-8 feet stow away neatly and are ideal for the smaller end of the lure spectrum. They also don't get in the way on boats or in tight areas where a long blank is a nuisance. There is also a good case for shorter rods in situations when fish need to be played hard, for example when battling wrasse from the rocks, since a 'shortie' has less flex and more grunt at close range.

Light, responsive and easy on the eye, the modern lure fishing rod is a thing of beauty. With so many variations, however, it pays to choose with care!

There are other situations where a longer rod is preferable, however. On venues that are overgrown or it's tricky to access the water, the extra 'reach' of a rod of nine to ten feet can be an advantage. Longer rods are also better for distance casts, as there is more spring and energy stored in a longer blank. In fact, if there's one type of rod that's rather under-represented, it's the super-light blank in lengths of nine feet and above.

Tip sections: Tubular or solid?

Lure anglers will often debate the merits of tip sections, in other words the final foot or two of the rod. A responsive, sensitive tip is vital in both imparting life into your artificials and feeling for bites, especially when using smaller lures. It must still be crisp and rigid enough to set the hook when you strike, however.

Just to confuse things a little more, rod tips will also be either tubular (i.e. hollow) or solid carbon (often marked as T or S in the model number). This is a commonly misunderstood area, which is why it's important to choose carefully!

Generally speaking, tubular tips are firmer and therefore ideal for working small crankbaits, darting jig heads and other lures where you want to impart more action. They're also better for setting hooks in tougher-mouthed customers, such as zander and pike. Logic also suggests they are better suited to lures that use treble hooks, as more force is required where the pressure is spread across multiple points.

Solid tips, on the other hand, are rather softer, with more 'give'. This helps for protecting very light traces and playing fish with softer mouths such as scad and perch, which are more likely to come adrift if a rod is overly severe. This softer feel also helps with mini species, which can move further with tiny scented lures without feeling too much spring or resistance. Hence a softer tip can help bites 'develop'.

If you were to fish more crankbaits/ hard baits with treble hooks it would be advisable to fish a slower rod that absorbs the head shakes of fish. The faster the rod the more chance the fish has got of levering itself off.

A striking test

TIP

How powerful is your rod when it comes to setting the hook in a fish?

This can vary a lot, but a quick test can help. Start by placing a reel on the rod and set up a lure with a barbless hook attached to the end. Now, hold the lure and ask your friend or tackle shop assistant to lift the rod slowly, imparting roughly a 1ft movement into the tip of the rod. If it feels that you would need to strike a lot more than this to penetrate your hand or a fish's mouth, then you have quite a soft tipped rod.

Generally speaking this rod would be ideal for fish that engulf the lure and are generally softer mouthed, such as scad, mackerel and perch. If you feel that with 1ft movement there is already plenty of pressure to set the hook, then you have a fast or extra fast rod with a firm tip. This would be ideal for tougher-mouthed species such as zander.

How many sections?

Another issue of practicality, rods come in various numbers of sections. At one end of the scale, one-piece rods are quite popular, especially in shorter lengths. These are great for leaving ready set up and have very smooth actions.

Most lure rods are multi-section, however, with two piece being the most common. An even number of sections makes sense, too, because you can then stow away the rod set-up and ready to go by separating it into two equal halves (a velcro 'rod-wrap' is a cheap and sensible investment to avoid mishaps).

Modern travel rods will pack down to under 2ft (45cm) and are fantastic for the nomadic angler.

However, travel rods are another option that have improved hugely in quality lately. These are a godsend for the angler who needs to stow their rod inside a rucksack or suitcase. Because of all the joints involved, the action won't be quite as smooth as a one or two piece rod, but newer models are impressive.

Last of all, we also have telescopic rods. Highly portable, but unless you invest some extra money, the cheaper rods can be fairly dreadful in performance terms! Like-for-like, you are probably better off buying a multi-piece rod in four or more sections that will sneak into a suitcase in a protective tube.

GETTING A GRIP

How do you prefer to hold the rod when fishing? It's a personal thing, but the key is that you are comfortable and in control. There are a few different grips you could try:

Standard 'Thumb on top' grip

The basic grip which most anglers use. For a little more control over your lures, you could also position your thumb directly on top of the rod. This position can be handy for making nice, straight casts, with the thumb helping to align each shot with your target.

Index finger grip

A variation of the basic grip, you could also try placing your index finger on top of the rod handle when retrieving. Best suited to lighter rods, this can help when flicking out accurate casts, or adding small jigging, shaking or lifting motions when moving the lure.

Trigger grip / pen hold

As the title suggests, the rod is held like a pen. Ideal with feather-light modern tackle and arguably the best grip of all to impart small lifts, taps and shudders into smaller lures. However, some anglers find it reduces leverage for striking into tougher-mouthed fish.

Reels

Whichever school of lure fishing you choose, a quality, dependable reel is a must. Perfectly usable models are now available for a modest sum, but as with rods you should select with care. Pick your reel from a reputable maker, with a reliable drag system and, preferably, at least one spare spool, which will give you greater flexibility by allowing you to pack two different lines should you want to go lighter or heavier. A smooth drag is important for landing fish, because if this jolts and jams it can put undue stress on your leader knot or give the fish sufficient leverage to throw the hook.

Setting the drag

TIP

One of the most common errors in lure fishing (and any fishing!) is not setting the reel's drag carefully. For those too shy to ask, this is the mechanism that determines how strong a pull it takes for the reel to release line. Too loose, and you might not set the hook or stop a fish reaching that nasty snag; too tight and you might get broken or 'bump' off a softer-mouthed fish.

So, it's down to situation and experience, but always check before you start fishing. Trust us, you don't want to be adjusting it when there's a big fish on the line! In general terms though, you'll want to set it a bit tighter for stronger tackle and species that like snags or have tough mouths (pike and wrasse both love snags, have tough lips and take few prisoners). For smaller species and the likes of softer-mouthed scad, however, a looser drag prevents hook pulls and breaks. And, if you are on very light line (say 4lb braid) and a big bass hits, you are going to need some give if you are to land that fish without getting broken off!

The two main types of reel are the classic **fixed spool** and the **multiplier** (smaller, ergonomically designed models of multipliers are known as a **baitcasters**). For the majority of light lure fishing, however, a fixed spool reel is all you need because these are better suited to casting very small artificials. If you do a lot of boat fishing or casting larger lures, however, a baitcaster could be worth the investment.

How to use and tune a baitcaster

So, you've chosen a baitcaster reel and loaded it with line; first cast and you immediately get a tangle! No sniggering, we've all done it. Why is this, you wonder? More importantly, how can it be avoided?

Understanding how to load your rod on the cast and tune the baitcasting reel to your needs is the key to avoiding tangles and wasted time. In short, baitcasters can be superb, but it's advisable to practise and get a feel for them, ideally before going fishing.

So, let's tune up! The first step is to find the spool tension knob, which is usually located above the reel handle/drag star: this applies resistance on the spool to prevent an 'overrun' (i.e. the spool turning too rapidly and throwing line everywhere) as the lure hits the water. Setting the tension is simple: attach your chosen lure, tighten the knob and release the spool. Now, gradually loosen the tension so that the lure falls slowly to the ground. To start off, you want the spool to stop turning when the lure hits the floor. As you become more experienced, however, you can loosen this further so the spool turns more freely and you can get more distance. However, this also means you'll need to slow things down yourself, by 'feathering' the cast with your thumb on the turning spool to prevent an overrun as the lure hits the water.

Baitcasting reels, with their braking systems revealed: Centrifugal (L) and Magnetic (R)

Above all, it's important that your reel should balance with your chosen rod. It mustn't feel too big or small, or seem too heavy or light. The safest way to pick wisely is to actually try it out for yourself at a tackle shop, attaching it to your chosen rod. Ideally, the central point of balance should be just above the reel, where you grip the rod handle.

Multiplier reels are growing in popularity, too, especially the modern **baitcasters**. These are essentially a compact, ergonomically designed evolution of the old school, round-bodied multiplier, which tend to be reserved for heavier lines and boat fishing these days. The new breed of baitcaster reels are thought to offer more precision when casting, especially when using special trick shots such as the skip cast (a low delivery, to skim the lure beneath cover). Widely used in the USA, their popularity is steadily growing in Europe, too. They take a little getting used to, but have some advantages of their own, including greater 'cranking' power (as one turn of the reel is equivalent to several wraps of line on the reel, hence the old-fashioned name 'multiplier'). Baitcasters also come into their own when fishing vertically from a boat, because they are easier to use with one hand, even for the angler who is also steering or using a fish finder.

Braking systems explained

Besides controlling things with your thumb, the braking systems found on modern reels are also a big help. To get the hang of these, the first step is to identify what type of braking system your reel has. It will be one of three: a centrifugal brake, magnetic brake or an electric brake.

Centrifugal brakes are small brake blocks that put tension on the spool and slow it down. The more blocks in place, the quicker the spool will slow down. If you're getting overruns at the start of your cast then it is advisable to engage more blocks. However, more blocks also means less casting distance.

It's all about adjusting things to your set-up. For instance, if you are casting into a headwind or using a lure with more resistance, such as a spinnerbait, you will lose momentum much faster than if you were using a streamlined lure or casting with the wind behind you. The trouble is, your reel spool does not know when it's time to let rip or to slow things down. It's thus a happy balance between turning freely enough to hit the distance you need, but not so freely that you get an overrun and risk the dreaded 'birds nest'!

Magnetic brakes are a series of magnets that sit under the side plate of the reel; they can often be added or removed as required. A magnetic brake works in a similar way to brake blocks – they slow the spool at the start of the cast and as the lure travels through the air. The benefit of the magnetic brake is that you can control it using the fine tuning dial at the side of the reel.

Electric brakes are new to the market in the UK. Shimano's DC braking system is a prime example of this system, which could revolutionise lure fishing with baitcasting reels. These newer model reels have an inbuilt electric circuit board which adjusts the brake to prevent overruns automatically. Clever stuff!

Beware of saltwater!

TIP

For all sea fishing, you must wash off your rod, reel and any lures used after each trip. Yes, it's a chore, but saltwater is ruthless stuff, capable of corroding just about any metal fitting. Perhaps the easiest way to wash off your gear is in a shower unit, giving everything a good rinse. Pay particular attention to rod rings, reel seat and the reel itself. Always screw down your drag and make sure your handle is tight when washing, to prevent ingress!

If your reel ever gets fully submerged in saltwater then wash immediately when home, taking the spool off and leaving to dry. Modern reels are leaning towards removing the anti-reverse switch, to protect the core of the reel, as these reels are less susceptible to water ingress into the body of the reel and are better for saltwater use. High-end reels are also worth treating with a service every so often.

Hooks and lures can also suffer, so do wash off any that you used. One useful tip here is to keep a small container purely for lures you've tried on the day, so that you know which ones need a wash when you return home.

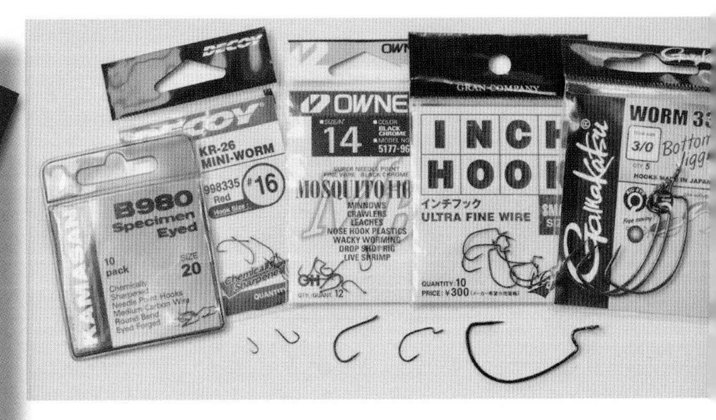

A wide range of hooks will be required by the species hunter. Tiny models are ideal for smaller mouths, while big fish and snaggy areas might call for tough, weedless patterns.

Hooks

While many lures come with hooks incorporated, a selection of hook patterns will also be required for many of the methods detailed in this book. Particularly useful for the species hunter are small barbless and micro barbed hooks, right down to sizes 20 or even smaller for micro species. Other especially useful designs include drop shot hooks (those purpose-made for the method are spot on) and weedless patterns for special presentations such as the Texas Rig. For greater detail, you'll find suggested sizes and advice in our sections on species and rigs.

Line

It may not be the sexiest item to buy, but good quality fishing line is essential for any fishing. It is your direct connection to both the lure you're using and the fish you catch. Strange, therefore, that anglers buy expensive rods and reels but look for the cheapest lines!

Braided Line (often just called 'braid' by anglers) is the most popular choice for 90% of modern lure fishing. With zero stretch it is incredibly sensitive, allowing you to feel even tiny bites or impart delicate movements to a lure with the merest shudder. Compared with standard fishing line, it's very thin. This makes longer casting easier, and you can get a lot of it even on quite a small reel. However, this fineness can make lighter braids quite delicate around rocks and sharp underwater objects. This is part of the reason we tend to use a final couple of feet or so of tougher trace material, such as fluorocarbon.

Which braids are best? Most braided lines on the UK market are either flat or round in profile, composed of either 4 or 8 strands of woven material (hence called 4 or 8 carrier). Generally, round profile braids sink through the water quicker, while flat will 'hang' more, slowing the rate of your lure. 4 strand lines tend to be flat, with a larger diameter than 8 strand braids. They are tougher therefore, but will achieve a little less distance on the cast than 8 strand braids, which usually have a rounder profile and sink faster.

For finesse fishing, invest in the best line you can: a good 8 strand braid with a diameter of 0.5-0.8 PE is ideal for lighter lines (roughly 5lb to 8lb breaking strain).

Where distance and a fast sink rate is required, use an 8 strand line with a round profile. Where a slower fall rate and more abrasion resistance are required, look for a tougher 4 strand line.

TIP

Diameter counts

The diameter of fishing line is often overlooked, but can significantly alter the presentation and sink rate of your lure. The larger the diameter, the slower a line will fall and the more it will 'hang' in the water: think of it as a kind of 'parachute effect'! The same can be said about performance above the water: the larger the line diameter, the more drag and the less efficiently you can cast; in a crosswind this can also cause you to lose contact with your lure. Also remember that lures will sink more slowly in salt than freshwater!

We recommend an 8 strand for fine LRF (Light Rock Fishing) around clean ground, lake and canal fishing and 4 strand for rock fishing or river fishing.

Monofilament or 'mono' is rapidly falling out of favour in lure fishing, but does still have its uses. For one thing, it is cheap. On cost, mono is perhaps

Check and discard TIP

Keep checking the last few feet of your reel line, as this is where most wear and tear occurs. If you find any abrasion, damage or kinks, do yourself a favour and cut and discard a short section. This quick 'freshening up' could well save you a lost fish later on! To recycle old mono or braid: see **www.anglers-nlrs.co.uk**

Braided line tends to last longer than mono, but if your line is getting old or tired, don't wait for problems to occur before replacing. It can be costly filling entire spools, so another money saving tip is to 'pad' these out with some old line or tape before adding expensive new braid. Much of the time you only really need 50-80 metres of line.

better for the beginner than braid. Mono tangles less easily than braid. Once you're comfortable casting small lures, though, old fashioned mono doesn't really come close to the performance of braid.

Mono is probably best for fishing crankbaits: its elasticity helps when playing fish or fishing around snags.

Fluorocarbon is not often used as a main reel line, but regularly used in lure fishing to make a 'trace' (the final few inches or feet before the lure). Fluorocarbon is tougher and more abrasion resistant than braid or mono, hence a short length of fluorocarbon is a good insurance policy when fishing around rocks or obstructions. It is also quite unobtrusive in water – although not as 'invisible' as tackle companies like to claim! Last but not least, fluorocarbon line has very little stretch and sinks quickly, making it useful for some other applications too. Occasionally

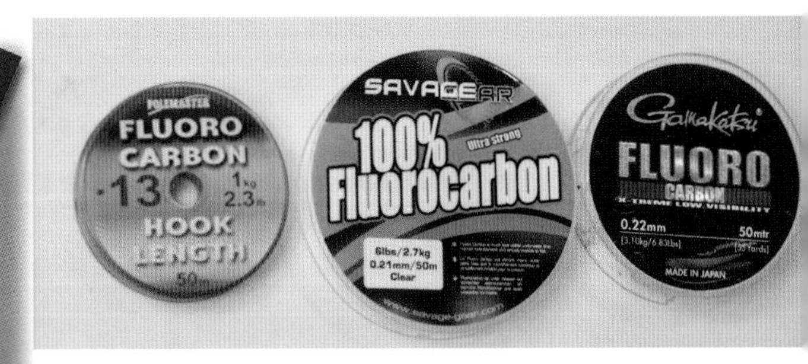

it is used as a reel line; and indeed, it tends to tangle less and transmits bites better in windy conditions, as it is heavier, cutting through any breeze and tending to sink faster than braided mainline.

See p222 for a selection of useful knots.

Drab or bright coloured lines?

One common source of confusion over fishing line is the colour. Does a brightly coloured line put the fish off? In most cases, the answer is no and brightly coloured line has a few advantages. Firstly, it's easier to see where your lure is travelling or pick out those suspicious movements that could signal a bite. Sometimes you will see the line move, even though you feel nothing. However, it's a personal choice and some freshwater anglers tackling clear venues might be wary of using a very vividly coloured line in pursuit of sharp-sighted fish such as trout.

Different braids for different jobs; bright colours can be easier for night fishing and untangling – while interval markings are also popular these days (top spool).

Wire is a specialist trace material aimed at toothy fish such as pike. It is stiffer and far more visible than standard mono or fluorocarbon line. There are many kinds now on the market, including supple and fine grades of wire that offer some 'insurance' against teeth without the stiffness that might kill the action of smaller lures. Look for fine wolfram and titanium wire for ultimate finesse presentation.

Swivels, snap links and other fittings are also worth discussing briefly as we talk traces. Snap links come in various guises and are a handy way of changing lures quickly. Look for strong, compact models that match the sizes of your typical lures, to avoid a clumsy presentation. Rig rings and swivels are a simple means to connect two lines together, such as a mainline to your trace. In some circumstances, for example when connecting regular fishing lines to wire or very fine braid, a ring or swivel can create a safer connection and eliminate the risk of a harder trace material cutting through a softer one.

Shot and other small weights are also important for some lure fishing presentations, especially with drop shotting and split shot rigging (*see p41 and 44*). A selection of medium to large split shot is a wise investment – although some prefer cylindrical styl weights that ping off the line less easily. Also handy are a few cone and drilled bullet weights, along with drop shot sinkers; we'll show their use as we discuss rigs and presentations. Cone and drop shot weights come in a variety of materials including brass, lead and tungsten. The differing densities of the metals mean you get different diameters for the same weight. They also feedback differently on the tip.

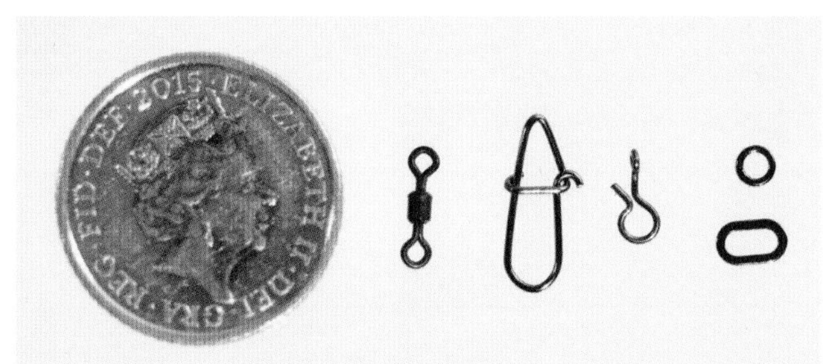

Tiny fittings and trace components
Left: *starting from nearest to the coin*, swivel, snap link, fly snap, circular and oval rig/leader rings

Below: *from left*, 3 sizes of olivette, round and oval split shot, 2 cylindrical 'styl' weights, 2 drilled bullets in oval and cone designs

OTHER TACKLE ITEMS & ESSENTIALS

Before we deal with lures, jig heads and other terminal tackle, we should also consider the other items at the angler's disposal. Some of the following are 'optional' or specific to a particular style of fishing. Others, however, could be regarded as essential, or fall into what we would call the damned-bloody-useful category!

1. Tackle bags and alternatives: A rucksack or other wearable bag is a compact way to carry a lot of tackle. Modern purpose-made fishing bags are excellent, with integrated lure boxes and accessory slots. If you wade, a multi-pocket game angler's fly vest is a great alternative.

2, 3 & 4: Lure boxes and containers come in various guises. Quality multi-compartment boxes are water-tight (to avoid rusting) and compact, to keep lures and components like sinkers and snap links safe and orderly. Although more costly, special slotted containers and even fly boxes are excellent for storing items such as jig heads.

5. A rubber sealed container ensures that modern hybrid lures don't dry out, or stink out your other things! A small airtight food container is ideal.

6. Trace wallet: There's nothing worse than lines, traces and packets of hooks escaping everywhere in your bag. A trace wallet with various pockets is a handy solution to carry all these items.

7. Essential tools include compact, sharp scissors to cut braid, long nosed pliers or forceps to remove hooks (these must be 1ft long for pike and larger fish with teeth!) Buy quality models with firm grip; you'll need them for fish like wrasse! Meanwhile, a coarse angler's disgorger is better for tiny hooks and mini species, while a hook sharpener is also useful, as are split ring pliers to open up components and change hooks on lures.

8. Head torch. An absolute must, not only for night fishing, but setting up and packing away in the dark. If you can afford it, go for a better quality model with an adjustable beam.

9. Landing net and handle: these are essential most of the time if you hope to land any larger fish. Look for soft, fish-friendly mesh. Rubber mesh is arguably better still and snags fewer hooks. For all but the easiest spots, an extra long extendable handle is a godsend. Specialist nets can go anything up to 6m and are a must for those who fish piers and high walls. Be a little generous with the size of your net, because while a small fish will fit inside a large one, the reverse isn't true!

10. Unhooking mat, plus scales or measure. These days, an unhooking mat is seen as vital to protect your catch from damage on the bank (and to avoid being crucified on social media!) You may also want to record the weight or length. Here, a special measure has been glued into an unhooking mat – which can be rolled neatly for compact storage.

Waders

Although perhaps not essential, waders can be extremely useful. Some waters and particular spots simply cannot be reached without them; although you must always wade with great care when it comes to strong currents or deep, sticky mud! Chest waders are the most versatile and come in breathable or neoprene material. Breathable is more expensive but better for summer fishing. Neoprenes are very warm for winter or coastal fishing – but can be sweaty and unpleasant in hot weather.

Cheaper waders come with built in boots; the posher 'stocking foot' versions take special boots. 'Cleated' or standard grip soles are fine for mud, sand and gravel. But if you regularly wade in fast rivers or other places with slippery rocks, felt soles and studded versions are the answer. Stay safe when wading, one secure step at a time. It's good practice to wear a belt around your waist which will trap air in your waders and keep you buoyant should you fall in! A wading staff is a good investment if you wade anywhere with currents.

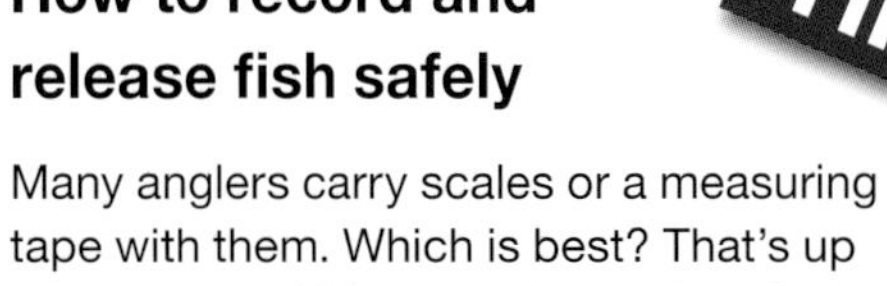

How to record and release fish safely

TIP

Many anglers carry scales or a measuring tape with them. Which is best? That's up to you. In the UK, we tend to weigh. Spring or digital scales are simple enough to use; the quickest way is weigh the fish in the net head and then deduct the weight of the net afterwards. A soft weigh sling is better for large fish like pike. Measuring fish tends to be quicker and kinder, however, and is the method preferred in competitions and by most European anglers. Fish are measured from the nose to the 'fork' of the tail (the notch where it goes in).

Whichever way you record a catch, be quick and kind. Always use wet hands and if you want a picture, consider briefly retaining it in a submerged landing net while you arrange the shot. These are precious living creatures, not just a social media post, so always be quick, careful and well-organised.

LURES

From beautifully-crafted traditional plugs to outlandish-looking soft plastics, anglers find lures compelling and collectable. More than mere tools of the trade, they are the opium of anglers. The variety on offer is bewildering and once the bug really bites, most of us amass more than we'll ever need. Out of this morass of lures, our authors aim to break things down into simpler categories in this section of the book.

When it comes to species hunting using light tackle, we are dealing with lures of up to an ounce, typically, but often a great deal smaller. With the right tackle, those of just half a gram can be cast and worked effectively. Not only do tiny lures and light tackle provide fantastic sport, they also allow us to target a much wider range of fish than the usual suspects we were raised on.

A word of warning, however: buying lures is addictive! Nor are they always cheap. Just as with baits and rigs, they go in and out of fashion, while companies claim their lures are 'irresistible' to fish or better than all the others. Take such claims with a pinch of salt. You will inevitably find your favourites, so follow your nose and spend carefully, because lures are designed to appeal to anglers as well as fish. There is nothing wrong with this in one sense, because confidence is key if we are to convince the fish our lures are worth eating! So, where do we start?

Metals

Although this is a book about the new wave of light lure fishing, we can think of no better place to begin than classic metal lures and their modern variants. Today's lure anglers often simply categorise all these lures as **metals.**

Above: A selection of metal lures. These include slow jigging spoons in wide and slender designs (middle rows), vibe baits (bottom left) and spinners (bottom right). Generally speaking, the heavier and more slender the design, the faster they will sink. You'll notice that some modern metals have extra 'twin assist' hooks *(top right)*. These can be removed for catch and release fishing.

These are where lure fishing started for many of us (including both authors), casting cheap spinners and spoons for obliging fish like mackerel and small perch.

Nor were we the first by any stretch. The simple flash and throb of metal lures have been effective for thousands of years; early spoons were cut from copper, sometimes from the curved sides of metal cooking pots. Earlier models were made of bone. The simple yet subtle blade or leaf-shaped metals we use today are the nearest we get to this early blueprint, yet still work. You might also hear the term 'pirks' by sea anglers; these tend to be much bigger beasts for deep sea fishing.

The best modern metals provide a concentrated weight in a small package for excellent casting, along with a seductive wobble and flash when retrieved. Simple they might be, but a fluttering metal is hard to fish badly and they win their place in any lure box. Indeed, metals remain among the longest casting lures and the best at cutting through strong winds. So what are the options?

Spinners are another distinctive old school favourite metal lure, which add further vibration by means of a revolving 'blade' around a central body. They are also among the easiest to use, and a straight cast-it-out and crank-it-in will work. Models such as the classic Mepps have been standard fare for decades, catching most predators, and there is something fish still find hypnotic about that spinning blade.

Perhaps the main drawback with traditional spinners is line twist, caused by constantly revolving blades and bodies. This can be countered by swivels or special weights, but the best solution is simply to move with the times and use modern 'in-line' or fixed body spinners, which are infinitely better. These newer models tend to feature much heavier bodies that cast better and won't constantly rotate and mess your line up. Major win! There are definitely days when bladed lures will outfish modern soft plastics, too. In murky water, for instance, or for fish hunting small, silvery fry or fish in the mood to chase, the spinner is a winner.

Other lures come in many guises, from classic spoons to various modern designs. We pretty much term them all as simply '**metals**' however. Along with general fit metals like the good old Toby or Kuusamo Professor, the new breed of lures is getting broader than ever. For the ultra light species hunter, the Japanese lure market is especially interesting, and many of these exceptional quality lures are making their way to the UK and Europe now. Various descriptively named metals include High-Speed, Micro, High Pitch, Long Fall and Slow Pitch, but the most common approach seen in the UK is the so-called 'slow jigging technique' which we discuss further in our rigs and presentations chapter (*see p35*).

Made of less dense metals such as nickel or zinc, the newer breed of top quality Japanese lures are compact and heavy enough for long casts, but will fall slowly and tantalisingly, keeping your lure in the 'zone' (the particular depth where the fish are holding) rather than bombing through too quickly. Why exactly is this important? Well, for one thing, we are not in tropical seas. Our cooler seas and generally slower fish species respond excellently to slower presentations.

Metals might be old hat in one sense, then, but are a good example of how lure fishing is evolving and how we are bringing new tricks and twists to an 'old' concept. Besides mid-water sport, another underused trick is to drag or hop a small metal more patiently across the bottom, where flatfish and other ambush merchants are sure to be drawn to the flash and disturbance. Some fantastic results can be had doing this, and fishing a small metal 'tipped' with a piece of scented lure such as Isome.

Loopy movement TIP

For most metals, it's good to encourage a nice open wobble, rather than constricting their movement with tight knots and stiff lines. For this reason, it's often good to attach them via a nice round-ended snaplink or split ring. If going direct to the leader, try a non-slip loop knot. Quality braided main line is also a must.

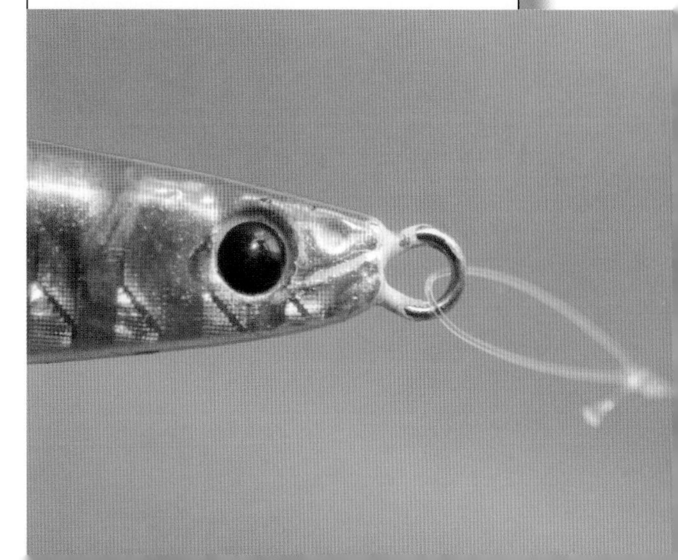

TIP

Straight and steady?

When it comes to catching predators in out-and-out 'chase mode', you can sometimes be 'too clever' with retrieves. Fish like mackerel and pollack are not bright creatures by any stretch, and throwing in lots of needless erratic twitches and pauses can actually lead to these fish missing the target!

Cheap, not cheerful: When bargain lures are a false economy...

We've all done it; bought a bargain or second-hand lure of dubious quality. Just occasionally, the new addition to our collection will be as great as we thought; but buying cheap always comes with risks! Garish paint jobs are one thing, but the biggest area in which cheapskate companies cut corners is with hooks. Low quality models are an accident waiting to happen should you hook a bigger fish. So, if you must grab that bargain, consider replacing the hooks with better quality versions.

Be wary of cheap lures. Hooks are often suspect on bargain models, while this cheap jointed plug literally split in two!

Hard baits

Besides all the newer soft plastic lures, classic hard baits made of wood or plastic still catch plenty of fish. These old school lures will work everywhere from the surface to deep down, kicking and wriggling to get the attention of the fish. The best are beautifully made and addictively collectible.

Plugs (aka **'Crankbaits'**) are also much loved by lure collectors for their traditional charm and craftsmanship. Legend has it that James Heddon accidentally made the first in the 19th century when the piece of wood he'd been whittling was tossed into a mill pond; as it wobbled in the current a bass grabbed it and the rest is history. Whether or not this is a true story or Heddon had been drinking, the idea caught on.

Some classic plugs are still hand cut from wood and feature a little 'lip' or protruding diving vane, which helps the lure dive and wiggle when retrieved. Modern versions are more often made of tough, durable plastic. Some are one-piece, others jointed in one or more places to provide even more wiggle. They are worth their place in any lure box too, especially for the more athletic predators that like to chase, such as bass, pollack, pike and trout.

Even simpler than plugs are **Jerkbaits**. These have no pronounced lip or vane to make them wiggle, but glide or twitch from side to side when the angler imparts movement. Usually this is done with a series of short, pronounced taps or 'jerks' on the rod tip, hence the name. Most commonly used for pike, there are also much smaller versions around. Some are beautifully made by hand and they work tantalisingly well for

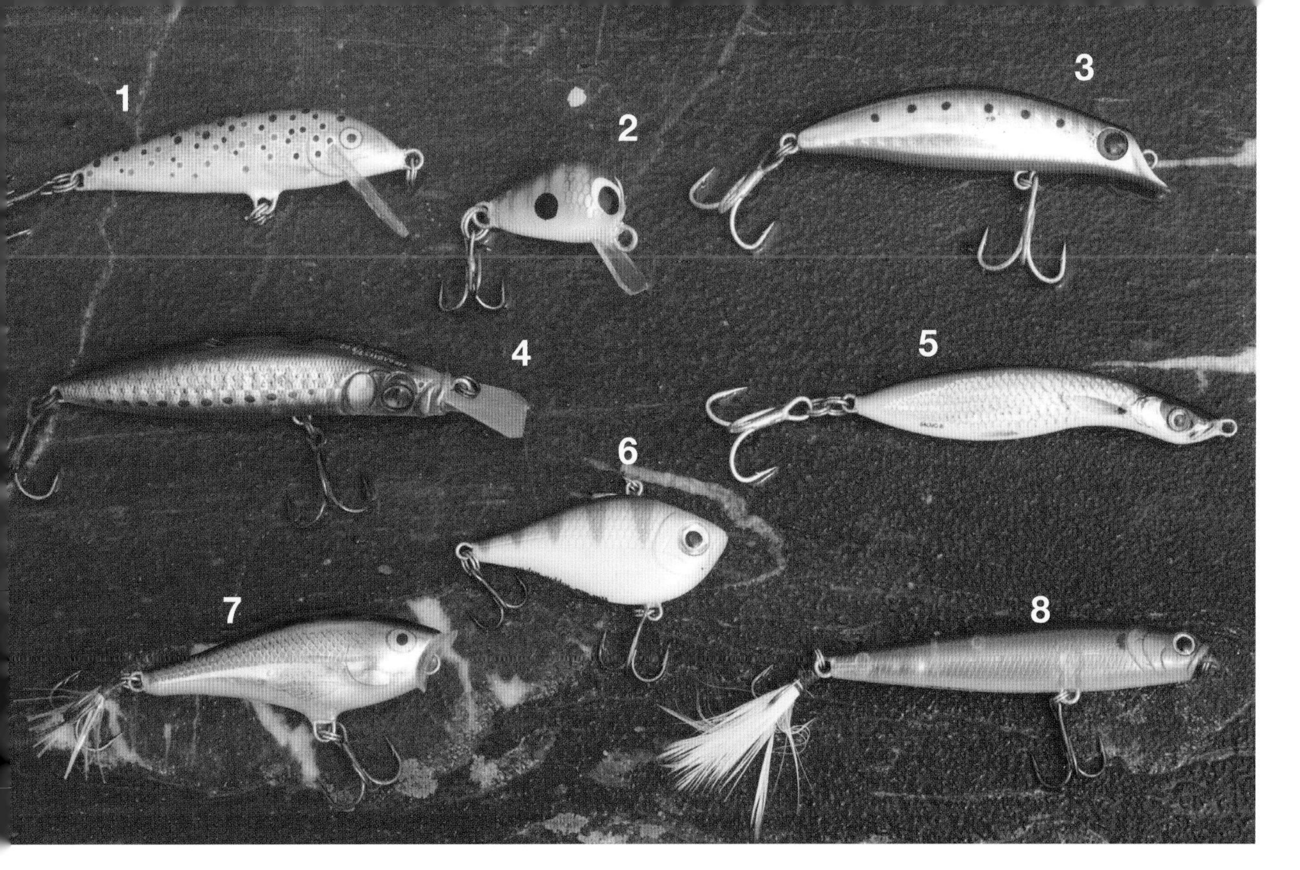

Above: Some typical lighter hard baits:
1: Floating Rapala (brown trout) 2: Mark Houghton Handmade 3: Ima Sasuke Shad 4: Megabass Zonk 5: Salmo Wave 6: Rapala Rippin Rap 7: Rapala Skitterpop 8: Lucky Craft Gunfish

a variety of predators that prefer dinner in the upper layers of the water.

Top water lures such as surface poppers and prop baits are also worth mentioning while we round up traditional hard lures. As the name suggests, such artificials don't dive at all, but are designed to glide, pop or splash their way across the surface. Their application is more limited to those predators prepared to crash into a fleeing victim at the surface. They are an exhilarating choice for bass or chub, for instance.

Right: Hard baits from surface to well-sunk (top to bottom): Surface popper, shallow runner, diving plug, deep diving plug. Note the bigger the vane or 'lip', and the closer it is positioned to the horizontal, the more steeply the lure will dive.

'Match the hatch' or stand out?

There are so many lures these days it can be tough to choose one. A few simple questions are perhaps a better place to start than the latest trends, however:

- What are the fish actually eating?
- What depth is this taking place?
- Are you looking at shy fish in clear water, or a situation where the fish need waking up or a lure that stands out?

If you can make a rough match of prey, this is often a good starting point. If the water is swarming with shiny fry, for example, a small, pearl or silver lure would make perfect sense. Of course, the way you make your lure behave can also be critical to success, especially with fish such as bass and trout, which will key in to a particular prey source. Predators looking for bullheads or blennies in stony waters, for example, will expect dinner to be small, dark and appear close to the bottom. A fleeing bleak, on the other hand, will be found just an inch or two under the surface of the water. Can you give a passable match? In most cases, about right will do. Fish are not brain surgeons, after all, but getting the outline, speed and positioning of your lure right can all help. So will a bit of thought as to 'natural' behaviour: most prey items don't swim straight at predators or land on their heads!

While the ideal scenario is to match a favoured prey item, however, all predatory fish are curious and most will react to quite a wide range of lures. Few fish won't find something resembling a worm attractive, while those with territorial instincts might also attack a lure that resembles nothing natural whatsoever. Why else would a spinnerbait or a rattling pink jerkbait work? Do these 'irritate' the fish?

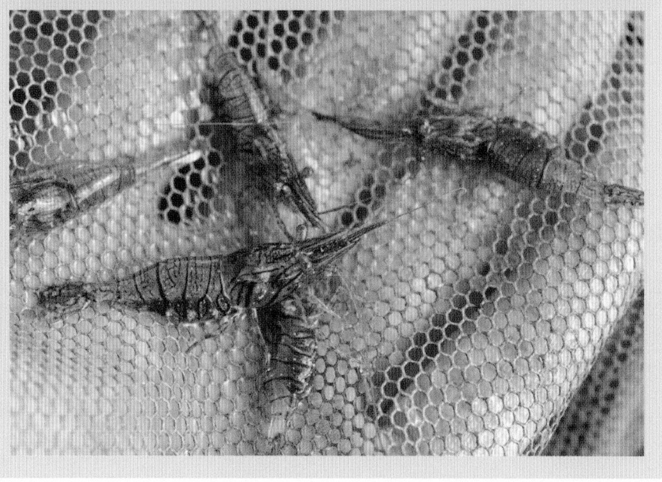

Imitative lure fishing starts with simple questions: what are the fish eating? Should I mimic nature or provoke?

Above: A hook sharpener is one of the wisest investments you'll ever make. Only a minority of anglers bother with this simple task, but it could save you a lot of lost fish!

Lure maintenance

One more neglected aspect of lure fishing is the need to look after your artificials. We've seen some horror stories in our time and, yes, discovered some rusty pieces of junk in our own tackle boxes! Water, teeth, rocks and other wear and tear will take its toll on all lures, but hooks are the single component that suffer most. A hook sharpener is a must, then, but from time to time you should also consider replacing hooks on the most well-used lures.

Do also rinse artificials in tap water after any trip to the salt. In short, it's a little extra faff to take care of lures, but neglecting them is a recipe for missed bites and lost fish.

TIP

Should I remove the barbs on hooks?

These days a lot of anglers and fishing clubs insist on single, barbless hooks, believing they are kinder to the fish. It's true that trebles can make a mess of smaller mouths, while any barbed hook can easily get stuck in nets, clothing and unlucky fingers!

Some anglers think that barbless hooks come adrift easier. This can be true when you lose tension for a second (in a weedy swim, for example) or for predators with softer mouths or head-shaking antics (perch are a classic for coming adrift!)

So, what is the best answer? 'Bumped' hooks seem a good compromise. By using pliers or forceps, aim to crush your barbs down a little and remove their harshness but not squash the barb totally flat. This way you have a hook that should stay put when it matters, but is kinder on your catch.

VARIOUS POPULAR WAYS TO RETRIEVE LURES

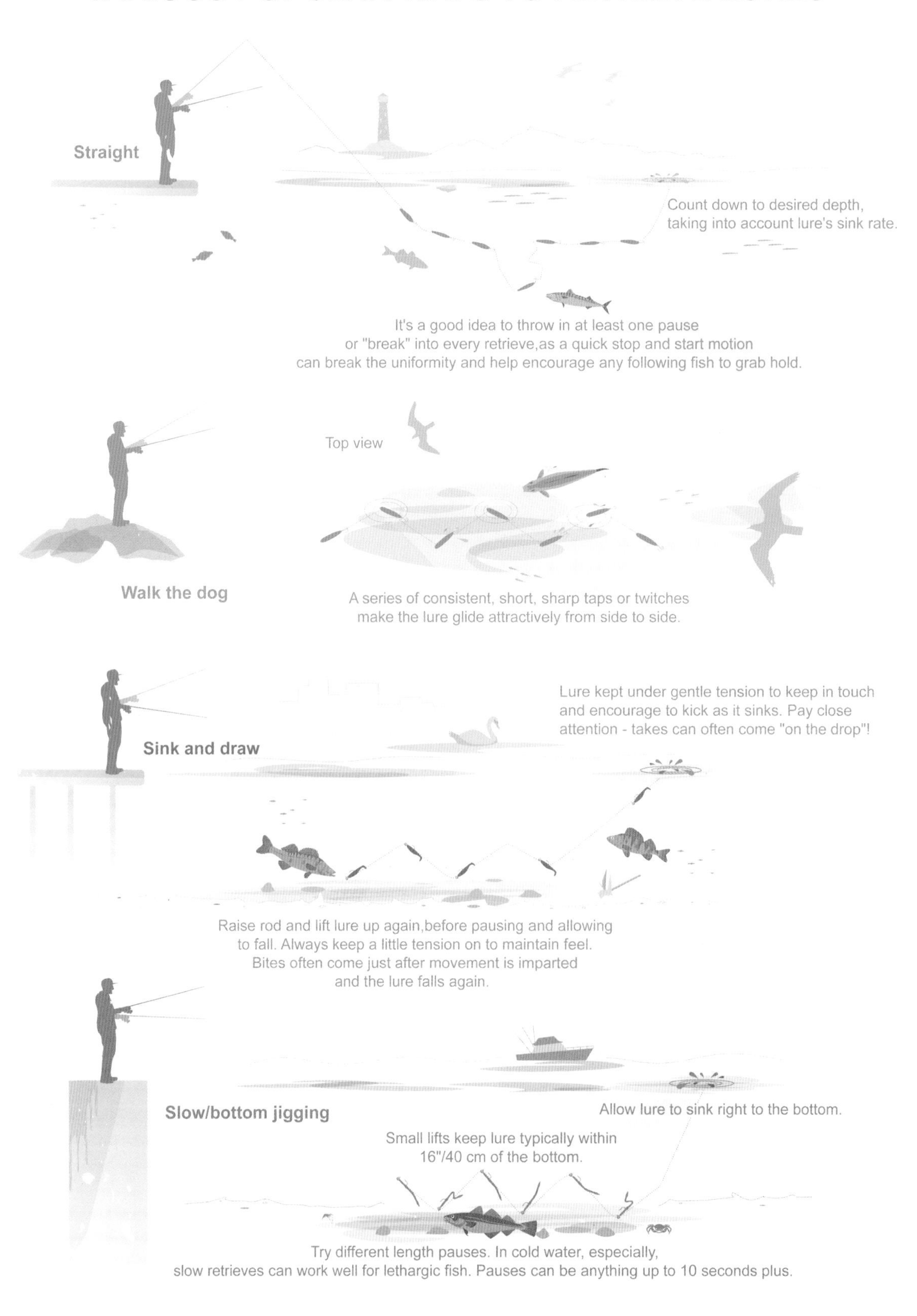

Jigs and soft baits

Lures made of soft plastic are now available in a vast range of shapes and sizes, totally changing the sport in the process. Flexible, attractive and cheap compared to hand-crafted traditionals, these creations have become the mainstay for most modern lure anglers.

There are various ways to present the new generation of soft plastic lures. Perhaps the simplest are the range of 'ready to go' options in which the lures can be used straight out of the packet. Sometimes known as **swimbaits**, these are easy to use, if less versatile in their applications. A more popular option is to add your own weight and hook, however, and the easiest way to do this is with a **jig head**.

Jig heads

The jig head is an essential piece of terminal tackle for light lure fishing. They come in all shapes and sizes. Indeed, if we took the far extremes, these would range from 0.5g to 200g and with hooks sizes from size 14 through to a size 10/0. For our purposes however, light lure fishing focuses on the smaller sizes of jig head and weights of less than an ounce (30g) and often as little as just a gram.

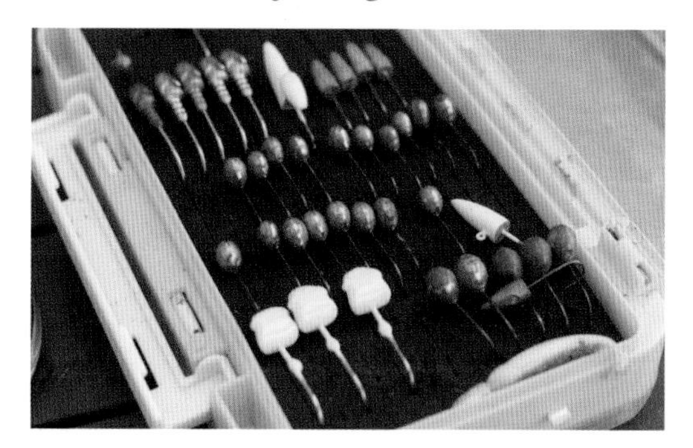

Above: A selection of jig heads for smaller species. A slotted foam or fly box is ideal for storage.

Jig heads are mostly sold separately to lure bodies, allowing you to select a style, weight and hook to suit your target species, choice of soft lure and method of fishing. This means we can mix and match components a great deal, customising our approach to suit our own fishing and the conditions. For different types of jig head and the various presentations that can be achieved, *see p36.*

Which size and type of hook?

Hooks vary vastly on jig heads, but regardless of the type or size you pick, quality is vital; avoid cheap and suspect products at all costs. Our choice might range from a tiny jig head for small rockfish, right through to a 50g model for targeting zander on a deep, windy reservoir.

Pay close attention first to the thickness or 'gauge' of metal. A finer hook will penetrate a fish's mouth with less force than a thicker gauge; but your hook should also suit the size and strength of fish you are targeting. A thin hook found on an LRF jig head may not be best suited to fish with bony or hard mouths such as pike or zander, for example. On other jig heads you'll find out-turned points to help hook the fish; indeed some anglers believe a slightly out-turned hook point will catch hold in the top of a fish's mouth a lot easier than an in-turned point.

You'll also find that hooks tend to be designed for either saltwater or freshwater use. Saltwater patterns are designed to resist corrosion. That's not to say you can't necessarily use freshwater hooks at sea, but always rinse after fishing with them as they will degrade quicker. You might also find that finer hooks turn or bend out of shape too easily in rough or rocky ground

Different Types of Soft Lure

In the past two decades or so, the range of soft plastic lures has increased hugely. Bearing a curious resemblance to the jelly sweets loved by children, they tend to have the same mesmerising effect on anglers. They have a huge variety of sizes, colours and swimming actions, too. However, we can break these lures down into a handful of basic categories.

Above: A selection of soft plastic fish lures, from paddle tails *(top two layers and bottom right)* to pin tails *(bottom left)*.

Worms, grubs and soft stickbaits

The original soft lure, plastic worms have been an established favourite with American bass anglers for decades. These days, many variants exist, from tiny little 'grubs' to foot long eels and so-called 'stick baits' that would be a mouthful for the greediest predator. While we term most of these lures as 'worms', they can make a passable impression of anything from a leech in smaller sizes, right up to small eels and slender bait fish. Most fish find a worm profile attractive too; and we will also deal with flavoured lures and artificial baits separately, towards the end of this chapter.

Shads and pintails

Next to worm-style lures, there is an endless variety of lures we might categorise as 'fish' imitations. They vary widely, from realistic sandeels, needlefish and gudgeon, to more generic shapes. Perhaps the classic design is the shad. The name derives from America, where shad of various kinds are common prey. But for most anglers, these lures are just a good all-round preyfish outline rather than an exact species. Most such lures have prominent tails that

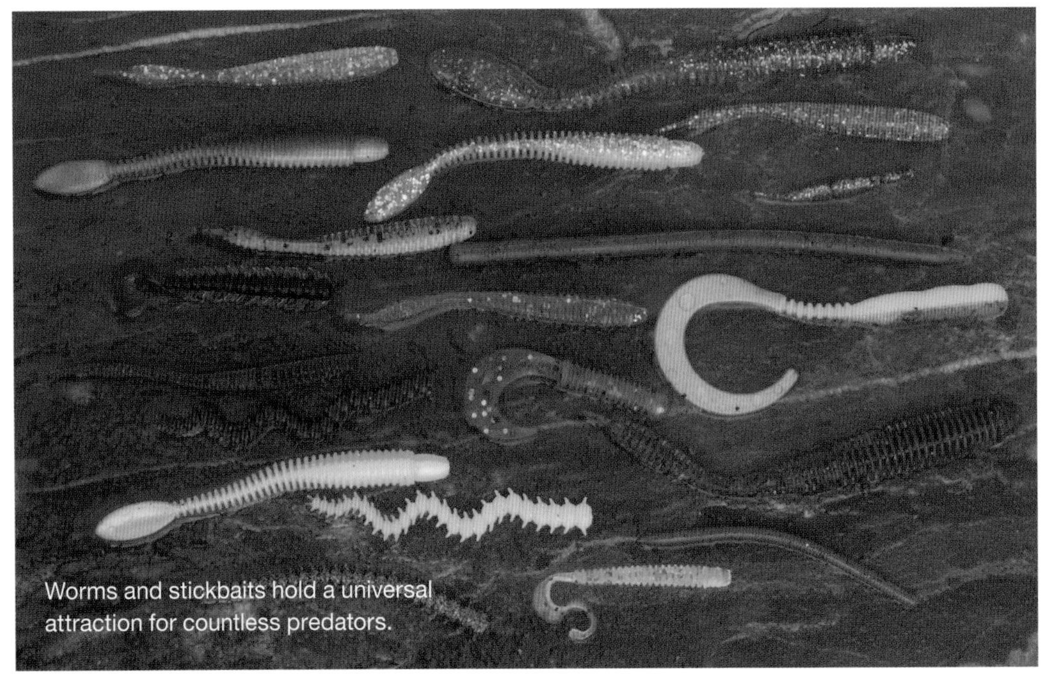

Worms and stickbaits hold a universal attraction for countless predators.

Above: Crayfish, shrimps, crabs: you name it, there's a soft plastic to copy it.

wiggle effortlessly, even at quite slow speeds, which earn them their other name paddle tails. Indeed, they are quite hard to fish badly, whether you jig them along the bottom or simply reel in steadily.

Next, we also have a variety of smaller minnows and more slender lures popular with many predators. Some of these have the classic paddle tail, while others are referred to as **pintails**. The name denotes a more slender profile, and rather than a kicking movement, they tend to dart, flit and tremble. Because they are so slender at the back, even the very tiniest twitch or shudder will make them kick, making them ideal for either drop shot or light vertical jigging presentations. These lures don't kick steadily on a straight retrieve like a paddle tail, and tend to suit a twitchy, stop-start motion better.

Crayfish, crabs & creatures

If we broaden the scope of our lures to cover animals such as prawns, crayfish, crabs and others, the variety of lures gets even wider. Again, if in doubt, a match the hatch policy is the best place to start. Many of our freshwater fisheries now have issues with signal crayfish, for example, a problem that can be turned into an opportunity for those targeting large perch and chub.

For sea fishing, the range of prey is vast, but a majority of our smaller marine predators will eat creatures such as shrimp, prawns, crabs and baby lobsters. It can take something of a leap of faith to imitate them successfully, but suitable lures can be excellent. Just use your head and try to take your cue from nature, because crabs or crayfish don't go flying about in mid-water. They tend to creep more slowly around cover, perhaps darting away when threatened.

Above: A selection of creature imitations, from marine worms to crabs. Edible lures are a great tool for today's lure angler. A kind of hybrid between a lure and a real bait, they work when other methods fail. Unlike some of the rubber shads and worms, many of the modern artificial baits like Gulp! and Isome are also biodegradable, which is another benefit for those conscious of plastic pollution or in spots where lure losses are high.

Flavoured and scented specials

Straddling the boundary between plastic lures and real bait, a new generation of hybrid lures have been a revelation for today's angler. The likes of Berkley Gulp!, Isome worms and others are all very effective fish-catchers. Most types imitate marine worms, although crabs and other creatures also work.

Do these still count as true lures? Perhaps it is less 'pure' lure fishing when our artificial is so potently scented that it will be taken completely static. Nevertheless, these new school artificial baits are deadly, not to mention a fantastic 'get out of jail free' card when conditions make fishing really difficult.

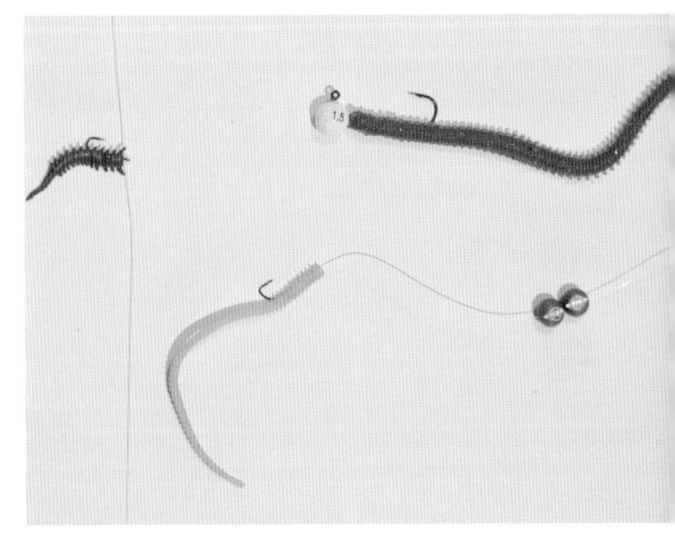

Above: The new breed of flavoured lures are highly versatile, suiting various presentations

A pollack homes in on a simple jigged soft plastic, a simple yet very effective presentation.

RIGS & PRESENTATIONS

With so many ways to fish and species to catch, the art of lure fishing has never been more varied. It can be as simple or complex as you like. Many of the common, more obliging fish are ideal for kids and complete novices to catch, and yet there are techniques to challenge the most experienced of anglers.

Lure fishing is never static (quite literally!) but always progressing. Methods constantly evolve, hence we cannot hope to cover every single foot of water here.

This chapter should provide a useful grounding in some of the key rigs, presentations and styles of fishing, however, including some of the most interesting recent developments.

Jig Presentations

Once upon a time, a jighead was just a lump of lead moulded around hook. Today, a phenomenal variety of jig heads help the angler achieve all manner of special presentations. More on these shortly, but first, let's look at the basics of rigging up correctly.

1. How to rig a jig

The combinations of jig head and soft plastic body are endless, but whatever your choice, start with reasonable size match. Neither component should dwarf the other – and it is important that the hook is big enough and the point will stand proud of the lure *(see top left overleaf)*.

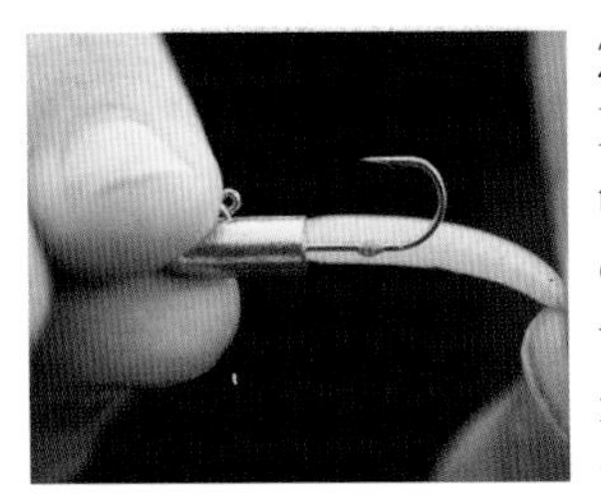

2. Now to measure up. Hold the components together to get an idea of where the hook point will emerge. It's a good idea to gently score the desired 'exit point' gently or perhaps count to a certain 'rib' on a contoured lure.

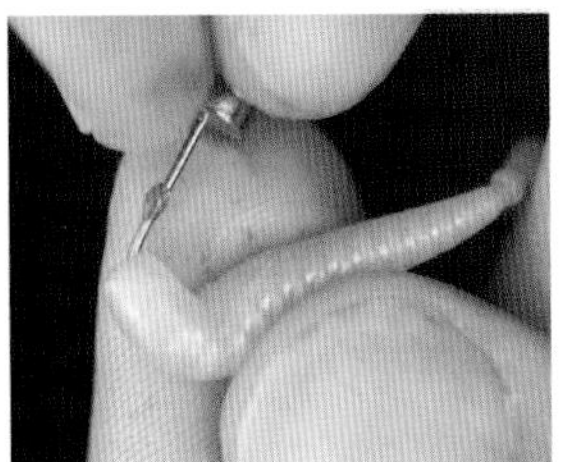

3. Now insert the hook point directly down the lure from the top, as straight as you can, before feeding it through the lure body. You may well need to bend it to reach your exit point cleanly. The process is rather like threading a worm onto a hook.

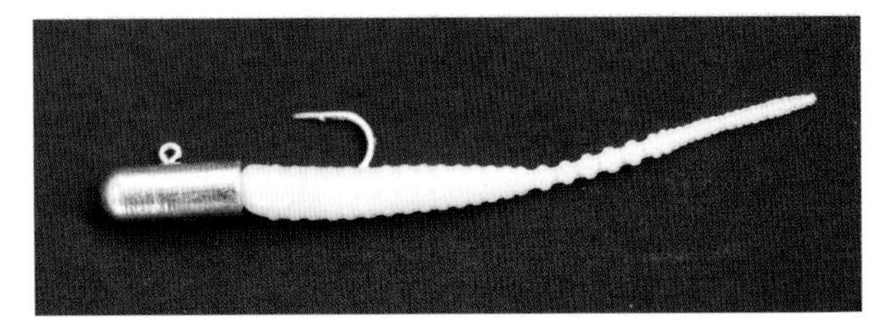

4. Rigged and ready. The hook point is well exposed; the soft plastic body is nice and straight, without being bunched up or uneven. Note that there is also plenty of body and tail beyond the hook, for free movement (it's not desirable to have the a hook running most of the length of the body because this restricts things).

Jig head types and presentations

While you could catch plenty of fish on the standard ball-shaped jig head, there is a wide choice of jig heads available on the market today. These can help us achieve various effects, depending on our target species, venue and other considerations. Here are some of the most common types clearly explained.

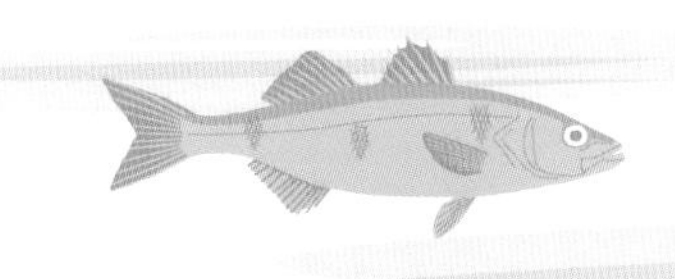

A continuous series of short twitches while reeling will cause this design of jig head to dart and flutter forwards, like a fleeing prey fish.

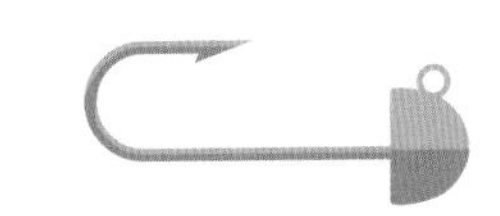

STAND UP JIG HEAD

Upright position at rest is highly attractive to foraging fish. A floating plastic lure accentuates this effect.

BLADED JIG HEAD

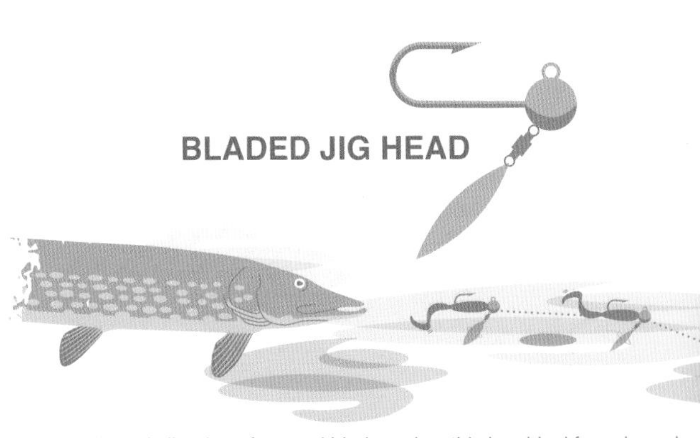

The additional vibration of a metal blade makes this lure ideal for coloured wa or lethargic fish. Will work with steady or stop-start retrieves alike.

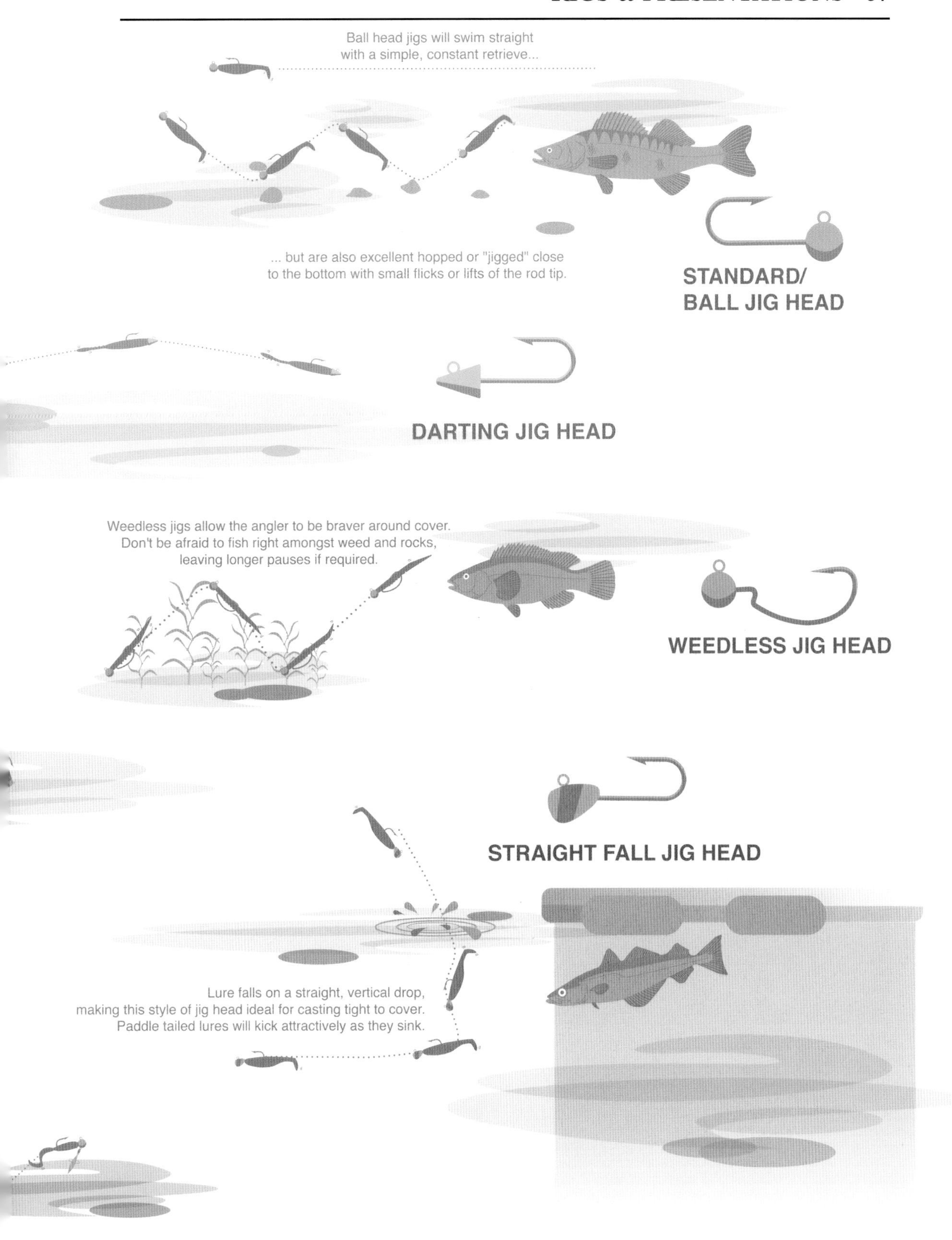
Ball head jigs will swim straight
with a simple, constant retrieve...
... but are also excellent hopped or "jigged" close
to the bottom with small flicks or lifts of the rod tip.
STANDARD/
BALL JIG HEAD
DARTING JIG HEAD
Weedless jigs allow the angler to be braver around cover.
Don't be afraid to fish right amongst weed and rocks,
leaving longer pauses if required.
WEEDLESS JIG HEAD
STRAIGHT FALL JIG HEAD
Lure falls on a straight, vertical drop,
making this style of jig head ideal for casting tight to cover.
Paddle tailed lures will kick attractively as they sink.

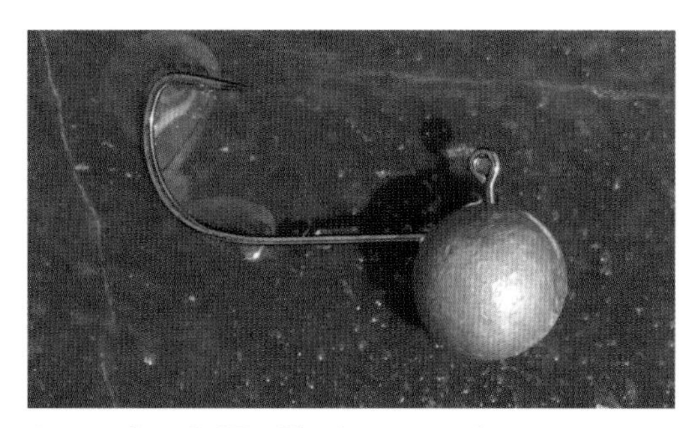

Standard/Ball Jig Head

The most common jig head type, the ball head, is the most versatile of the lot. It will sink fairly straight and can be used to work all layers of the water. It's often the first choice when working deep as the round profile does not dig into the bottom and is less likely to jam in between rocks and snags.

There are a wide variety of ball jig heads, from plain to painted models with eyes, with various hook types. You will also find a fair variety of eye positions (also called 'line ties', in other words where your line attaches to the lure); the position of these can make a big difference to how your jig behaves. 90 degrees to the hook shank is the most common eye position but you can also find hooks with the eye or line tie at 45-60 degrees, which will glide on the drop. Jigs with a line tie at 60 degrees or so are less likely to pick up weed on the retrieve.

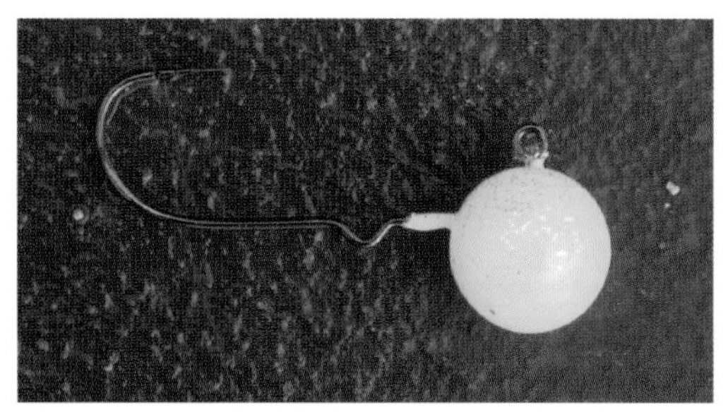

Above: A standard jig head in high-vis. Useful for coloured water and inquisitive species.

It's also worth noting that standard ball jig heads will swim straight. This is useful for fish such as pollack, garfish and others that prefer a constant retrieve. For active fish hunting in the mid to upper layers then, you might look to use a ball head with a 45-60 degree line tie. In this situation, a gliding head cast toward feeding fish will help tempt them to follow and bite. For fish closer to the bottom, the standard 90 degree angle is more suitable.

Darting Jig Head

With a more elongated or pointed shape,

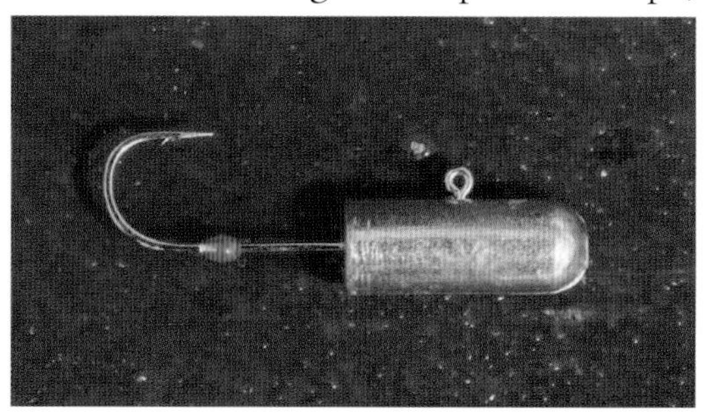

darting jig heads are designed to weave and dart forward vigorously. Small taps or short sharp turns on the reel make these jig heads swim and dart forward erratically, a great way to mimic fleeing baitfish. You can experiment with retrieves. Straight and quick will work, but a series of little taps on the rod tip can be even more attractive, producing an action not dissimilar to the gliding motion of a jerkbait.

Darting jig heads work best for active, chasing fish, where their gliding forward movement tends to trigger more bites than a standard jig head. We also suspect they can help convert more follows to bites, however, when the fish are fickle.

Weedless Jig Head

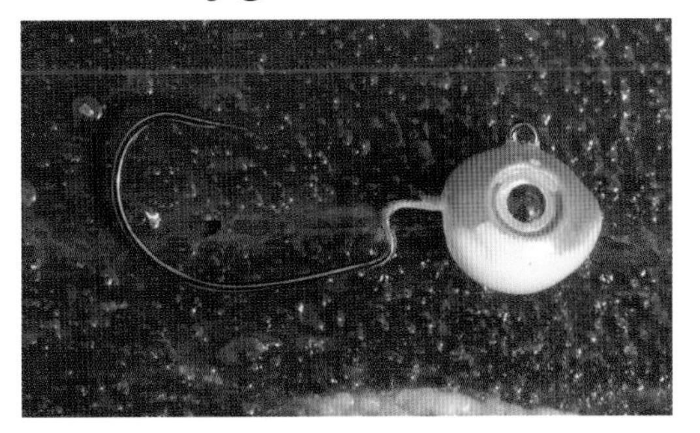

A simple way to make your lure more snag proof, weedless jig heads are highly valued by specialists who fish in difficult spots where weed is thick or losses are high. They're not always widely sold, but highly useful if you can find them. The offset hook design allows you to mask the hook point within a suitable soft lure to reduce the risk of snagging. Note that in spite of the 'weedless' moniker, these designs are also superb for rocky ground. We'll explore weedless rigging further later in this chapter.

Stand Up Jig Head

No, not a jig head that tells jokes at a comedy club, but one designed for working the bottom. The steeper shape will present a soft plastic lure in an attractive, more upright position so that the fish can see and take it more easily. Whether this mimics prey in a natural feeding position (i.e. like a small fish or creature standing on its head to feed or burrow into the bottom) or simply makes your lure easier to grab, it's a trick worth having in your locker. Buoyant lures will accentuate this effect further and the Ned Rig is a classic example of this *(see page 53)*.

Straight Fall Jig head

A clever concept for anyone who likes to fish tight to structure, this type of jig head will fall perfectly straight (most lures will naturally tend to fall back towards the angler a little). Hence this type of jig head is perfect for casting close to walls, posts and other features, as the shape will keep it close to the structure as it falls.

Bladed Jig Head

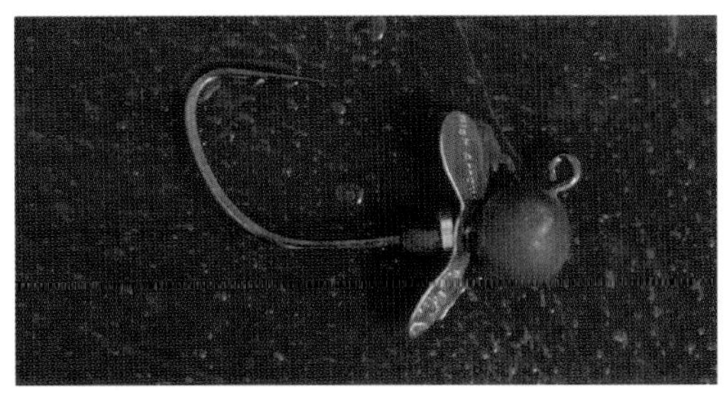

Some jig heads add a further fish-attracting twist with the addition of a flashing blade. This could be a detached, spinner-style blade, or indeed a rotating 'collar' (as pictured). This not only adds vibration, but tends to make the lure fish a little slower. Very useful when bites are hard won, or in murky waters where fish find it hard to locate your artificial by sight alone.

Tungsten Jig Heads

By volume, tungsten gives the most weight of all the materials used for jig heads, allowing a high mass even with small-sized jig heads. Tungsten is also harder than lead and seems to give better feel or 'feedback' to an angler working the bottom. So why aren't we all using tungsten? The short answer is the high cost. Even so, for the competition angler or anyone who needs to present a very small jig at greater depths, it is worth carrying a few.

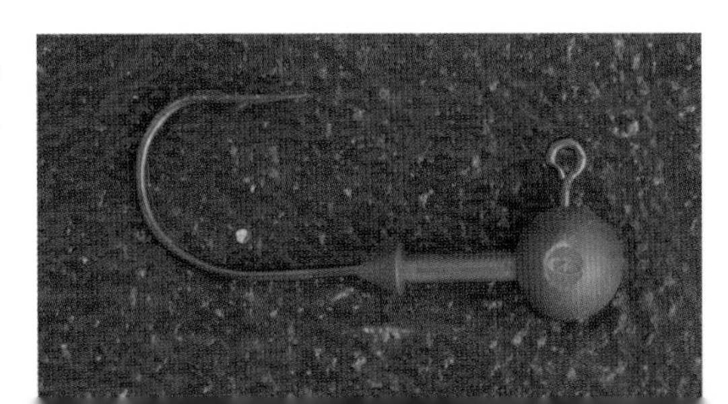

Faithful retainers

TIP

With a bit of persuasion, most of our soft plastic lures can be rigged snugly in place on a jig head. However, prolonged use and bites can sometimes create annoying slippage. Some jig heads have a special retaining spike, collar or blob of epoxy to combat this. However, another great way to stop soft lures sliding down the hook shank is to cut a little section of rubber (elastic band is ideal) and slide this on to keep things secure, as pictured on the large lure below.

to fish. Scandinavian anglers fish them baited – but they'll also work well with tiny soft plastics and scented lures such as Isome.

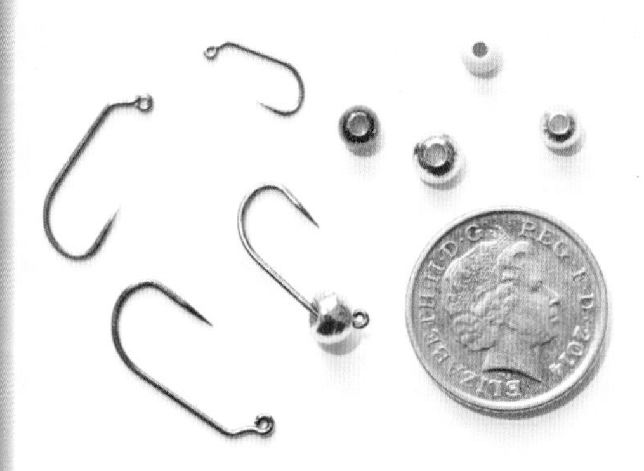

Alternatively, fly fishers' jig hooks and tungsten beads (*above*) are superb for creating tiny jig heads from sizes 8 right down to 20. These come in various sizes and colours at a very fair price, and offer a high mass in a very neat little package.

Ice jigs and other jig heads

Although not always easy to find, ice jigs and other small, specialist jig heads can be another useful addition to your lure box. With very light tackle, lures of just half a gram and upwards can be used, and the typically bright designs are very attractive

This canal zander was perfectly hooked on a drop shotted shad.

Drop Shot Rigs

An excellent method to present smaller soft plastic lures with great control and finesse, drop shotting has really come to the fore in recent years. In a nutshell, rather than using a lure that has its own weight or a jig head, drop shotting involves a separate weight and the lure presented some distance away, sitting at a right angle to your trace line.

Originating in 1970s America, it was used for largemouth bass and other lake fish. It is a method associated with fishing around heavy cover such as jetties, docks or sunken trees, particularly when a very slow and precise presentation retrieve is required. Of course, because our weight is totally separate to the artificial, the method also has the advantage that a tiny lure can be used even at great depths or in difficult conditions where conventional methods would make this impossible.

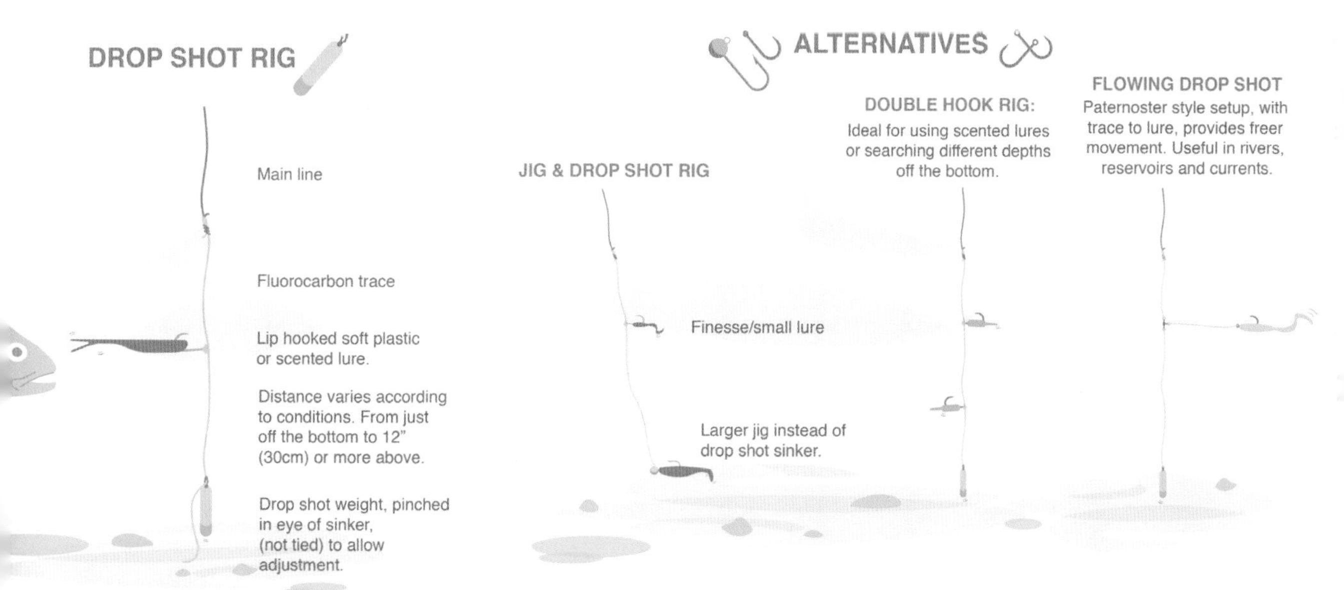

Main line and traces: Strengths vary according to setting and species. Most often, braided mainline is attached to a fluorocarbon trace of 2-3ft (60-90cm), via a secure knot or rig ring.

Hook: The best drop shot hooks have a wide gape profile and are strong yet fine and deadly sharp. Many companies now produce dedicated hook patterns for the method. Sizes will vary from quite large hooks (1/0 to 2) for deep water and larger fish, right through to sizes 10-16 for tiny lures and mini species. The hook can be tied using a Palomar knot, but we much prefer the special 'drop shot knot' (*see p222*).

Lures: Any number of fish, worm or creature imitations can be used. Ultra flexible lures that will quiver with the slightest touch and collapse on the strike are best. Note how the lure is not threaded onto the hook, but simply nicked through the head. This ensures that the virtually the whole of the lure can move freely, while the hook point is well exposed.

Weights: Another clever aspect of the technique, special drop shot sinkers have a narrow 'eye' at the top, which the end of your trace can be squeezed into without having to tie a knot. This has several benefits. First, it allows you to change the depth your lure fishes, or quickly switch to a heavier or lighter weight in seconds. Handy, for instance, if the fish move higher in the water or the current picks up. If you get snagged, the weight will also tend to pull free so you can still get your lure back (and possibly still land a hooked fish that was stuck!) However, you can also use standard sinkers and simply tie your weight on in conventional style. This might be necessary, for example, if you fish from a boat and need a much heavier weight of say an ounce or two, rather than mere grams.

Typical drop shot sinkers (*above*) come in either cylinder or ball type designs. The elongated standard shape is good for vertical fishing and won't easily get stuck or roll into nooks and crannies. A rounded weight is better when you want to bounce across the bottom a little more actively, or perhaps move with the current on a river.

Double drop shot rig: If it's not too snaggy, you could double your chances and perhaps try two different lure sizes and colours, where rules permit.

Jig and drop shot rig: Another cunning variation is to swap the sinker for a lure! This works best with a standard jig, allowing you to fish a large, heavy lure and a smaller, more subtle offering in the same cast.

Flowing drop shot presentation Besides presenting the lure tight against the trace, you could also incorporate a longer length of line, much like a paternoster rig. A 3-way swivel is the easiest way to set up. This allows more movement that can be ideal for flowing water, or indeed an alternative presentation for tricky fish that have become wary of the regular presentation. The length of snood can be adjusted to suit the situation from 1ft to as much as 12ft!

How do you fish it?

Drop shotting is typically used for fishing close in, often right under the rod tip. It's particularly good when you are directly above a structure such as a pier or wall. However, it is also possible to cast over short distances and retrieve your lure in a slightly more active manner. It's ideal, for example, to tempt fish from beneath a wall or overhanging bush on the far bank of a canal or river. As the distance from weight to angler increases, however, the angle of the line will also change quite a lot. The longer the cast, the closer your lure will be to the bottom. So, when fishing further out, you might need to position your lure 60cm from the sinker,

A small black bream homes in on a drop shotted shad.

in order to fish it 30cm off the bottom, for example. Besides bank fishing, drop shot is excellent from a boat, too, either on the spot, or slowly drifting to cover likely water.

Spotting bites

Drop shotting tends to demand a slower and more patient approach than traditional lure fishing. Tiny twitches and plucks are sufficient, and it's easier to impart too much action rather than too little! When fishing right under the rod tip, the idea is to move the lure but not the weight. Some anglers even incorporate a little pole elastic into their rigs, next to the weight, to allow more play on the lure while keeping the sinker static on the deck.

A rod with a firm yet sensitive tip is a must for this style of fishing, in order to feel for bites and hit them crisply. Takes can vary greatly; on occasion, fish will hit hard and virtually hook themselves. More often, however, you'll get softer plucks and pulls, or notice the line starting to move away. Be ready to strike instantly!

It takes more patience than other lure methods, however; you might loiter in a particular hotspot for a long time if fish are present. It also teaches us that lures don't need to be raced in; on the contrary it can take fish several seconds or even minutes to react.

Allowing tiny hooks and lures to be used, the split shot rig is superb for miniature predators.

Split Shot Rig

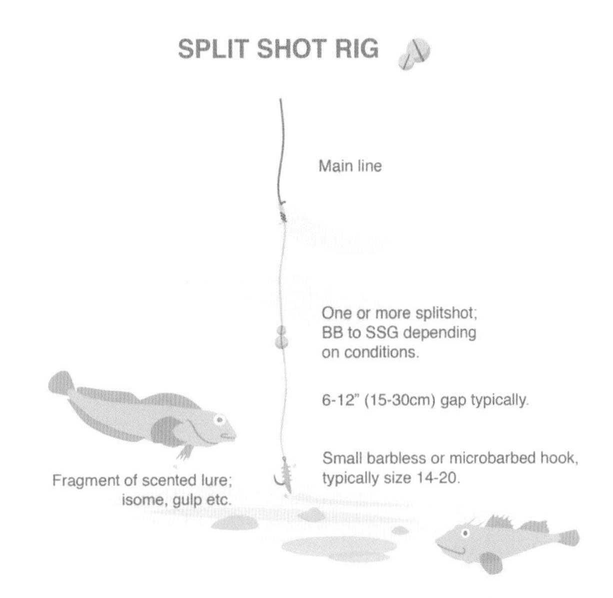

This simplest of rigs is cheap, versatile and very easy to set up. For the species hunter, it is incredibly useful for the smaller predator fish, too. Owing to the lightness of the tackle, it is only effective for close range fishing. Indeed, this is the set-up for searching right beneath a harbour wall or pier for smaller fish.

Main line and trace: Split shot rigs tend to use light mainlines, usually braid in the 4-8lb (2-4kg) bracket. Traces tend to be fine mono or fluorocarbon, in breaking strains of 3-4lb (1.5-2kg), to allow the use of small hooks and fine lures.

Split shot and other small weights: these form our critical mass for casting, helping the lure to sink to the desired depth and register bites. Usually, a couple of BB or AA weights will suffice, but you could go up to SSG size depending on conditions. An olivette or small bullet can also be used as a tidy alternative. A distance of 6-12ft (15-30cm) from weight to hook tends to be optimum, allowing the lure to move more naturally, wafting in any current. You can alter this to suit: too close and you may scare your quarry; too far and you might miss your bite.

Hook: Depending on the target species, we might use anything from a tiny size 20-24 hook, right up to perhaps a 10 or 12. Aside from specialist patterns, coarse fishing hooks are often excellent, especially small barbless carp patterns.

Lures: You can use any tiny soft plastic, or a little sliver taken off a larger lure. However the most effective offering of all is probably a slender, scented worm-style lure like Isome and Gulp! You can hook tiny fragments at one end, so that they waft enticingly. For larger pieces, though, threading up the hook and along the leader is more secure, to prevent fish stripping your lure clean off, it also helps with catching short biting fish.

How do you fish it?

With small lures and diminutive target species, only tiny movements tend to be required. In saltwater, the tide will usually wash your lure around with little need for any dramatic lifts or twitches. Get a feel for where the bottom is and then try gently lifting, dropping, or often simply 'hanging' the lure just off the bottom. You can also try fishing in midwater next to structures such as walls, floating platforms and pier legs. Finally, you might also try casting and jigging slowly across the bottom for the likes of small flatfish and others.

Spotting and hitting bites is a case of timing. A fine, solid-tipped rod is best.

Brighten up or tone down your shot?

For split shot rigs, some anglers like coloured split shot for particularly curious or aggressive fish. However, this also runs the risk that the quarry will attack the weight rather than your lure! When we observed fish underwater, they commonly seemed to attack traditional, metallic traditional shot, too. One solution to dull shiny weights is to soak them in vinegar.

Usually it's wise to ignore the tiniest shivers and nibbles, letting the bite develop. Once the rod tip pulls and holds, moves more dramatically or the line moves off, strike quickly to set the hook.

Weedless & Texas Rigging

Given that so many predators love weedy or craggy places, a snag-proof presentation can be extremely useful for lure fishing. With a little trickery, we can partially bury our hook so it will avoid the ugly stuff on bottom, yet still catch the mouth of a biting fish. There are different ways to achieve this. Some lures are bought as 'weedless'; the remainder you'll need you to rig yourself.

Key to all these presentations are special weedless hooks, designed to be concealed inside a soft plastic lure. You can sometimes buy special jig heads of this type, but another way is to buy weedless hooks and combine these with different weights for more versatility.

The right way to rig a weedless lure

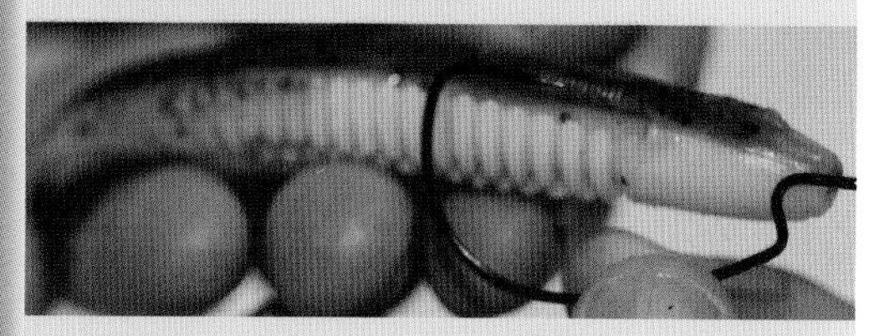

1. Select a soft plastic lure along with a hook that is a reasonable sized match. The best soft plastics for the job are either worm style designs or special slotted lures. Some are purpose-made for the job and are slotted or contoured to conceal the hook yet collapse on the take so you connect with the fish. Hold your chosen hook and measure up – taking note of where it should pass through the back (you can even make a little scratch or count the lines/ribs).

2. Next carefully thread the hook through the 'nose' of the lure cleanly, like this.

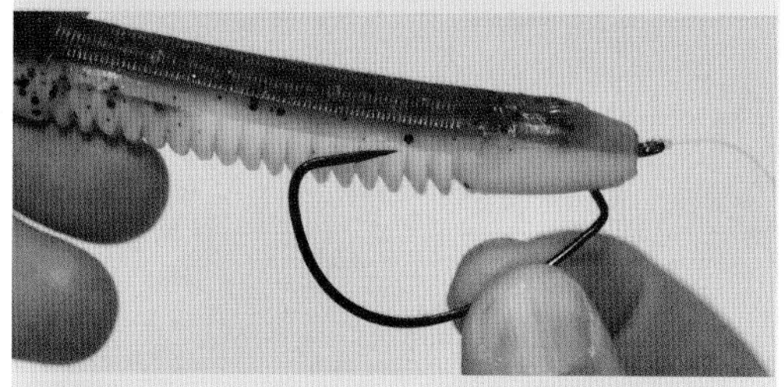

3. Pass the hook right through until you reach the special 'notch' at the front the weedless hook. You will need to twist the hook round so that the point faces upwards as you do this. Note, if you seem to get issues with the lure slipping out of position with softer materials, you can also thread on a tiny section of rubber or plastic to keep things snug (see our tip on page 40)

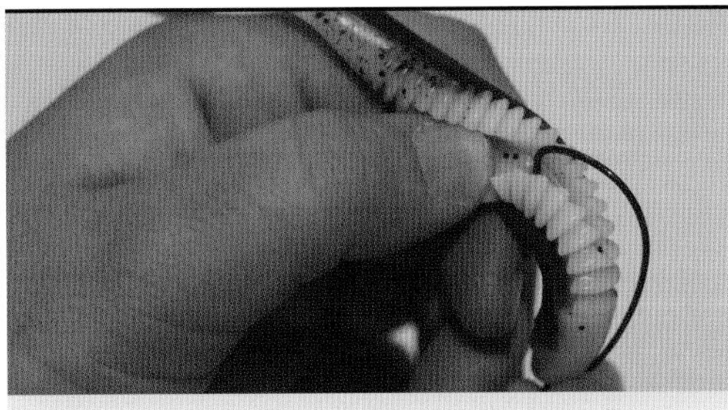

4. Now comes the slightly fiddly part: Pass the hook point through the lure from belly to back. You will need to bend the lure a little – and this is where your earlier 'measurement' comes in.

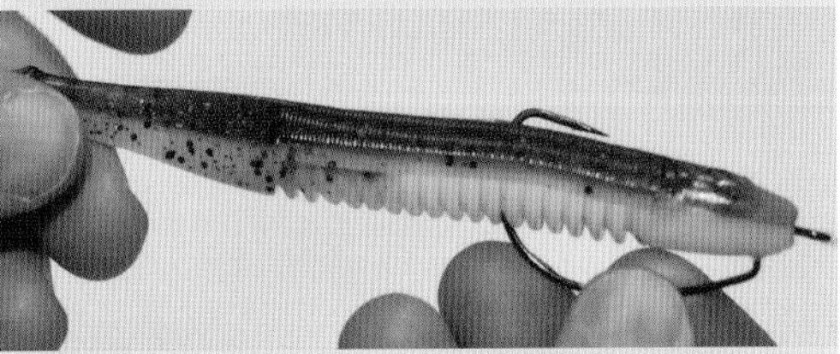

5. The lure is ready to fish. We've pushed the hook point up a little here so it's visible, but you'd usually have it near-enough flush to the back of the lure. This allows you to fish around weed and snags, but the lure should crumple easily if a fish bites, exposing the hook point.

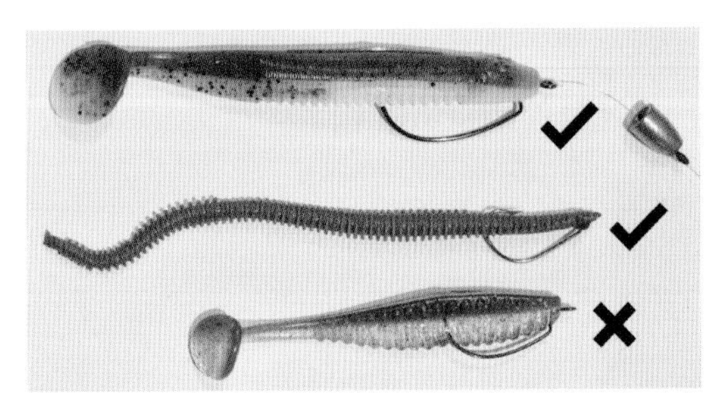

Your chosen lure needs to be able to collapse easily when grabbed by a fish and some models are better at this than others. *Top:* A purpose-made lure with slits for weedless rigging and (*middle) a* scented ragworm are perfect for the job, easily collapsing on the take. However, the final lure (*bottom*) doesn't have enough 'give' in the body to expose the hook on the strike. This is quite a common error: you'll still get bites, but your hook up rate will be poor!

Weedless Jig Heads

A weedless jig head is the easiest way to rig up your own soft lures so that they are snag proof. These jig heads are not always the easiest to find – and sizes can be limited. Another good alternative is a wired, ball type weight, which allows you to match a wider range of weights and hooks.

Texas Rig

This is a brilliant way of rigging your own lures in weedless fashion, with the exact choice of weight and hook you prefer. This is done with a suitable weedless hook, a drilled bullet or cone weight, along with a float stop or split shot to keep the weight snug against the hook. That said, if you use two float stops, one either side of the sinker, you can also instantly make a Carolina Rig; add a floating lure into the mix and you've got a Mojo Rig.

Trace

Fluorocarbon traces are a good, abrasion -resistant choice for snaggy spots, attached to a strong braided main line. As a general rule, make sure your final trace is no greater than 75% as strong as your main line (so you might use a 15lb trace with 20lb mainline, for instance). Trace length will depend on the scenario. For example, with mini species you may only need a foot or so. If rocks and other obstructions are really up and down, and the fish run hard, however, a longer final fluorocarbon trace would be sensible, to avoid your thinner braid rubbing against the ugly stuff.

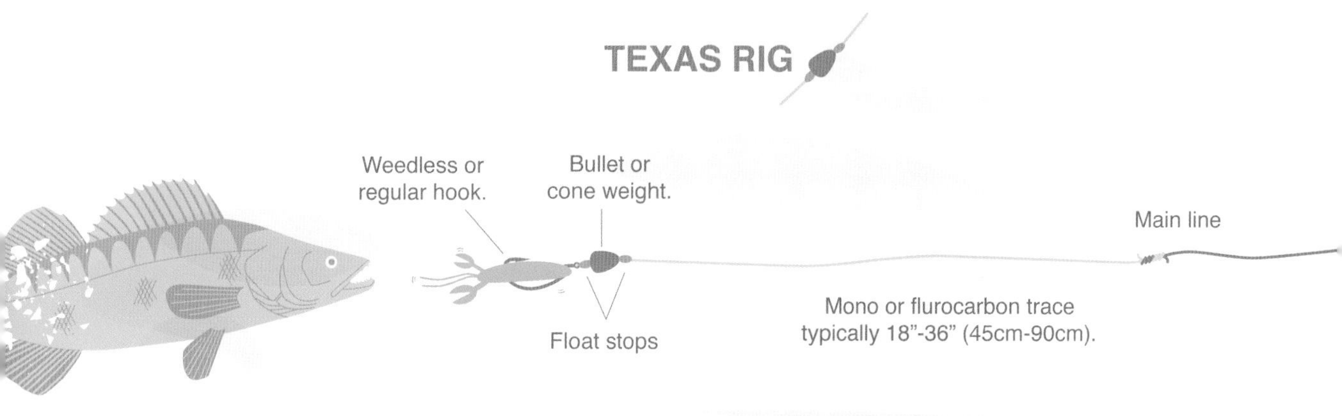

Weight

You can use various drilled weights to thread onto the line. The simplest is a basic drilled bullet or cone design. As for how heavy this should be, aim for weight that lets your lure behave as naturally as possible in the conditions and no heavier. A smaller weight might do for a calm, sunny day, for instance, giving a slow, natural fall. A rocky, rolling sea might demand half an ounce or more, on the other hand.

Hook

Dedicated weedless designs are very much the way to go. The size of your target species and the lures used will dictate size. The hook gape needs to be roughly twice that of the solid part of the lure (minus the slits) to allow enough room to hook a fish effectively.

Add a buffer ...and a rattle!

TIP

Another twist we like to include is to use a glass bead in between the weight and the hook. This was originally thought of to protect the knot above the hook; but for our fishing it is more commonly used for the clacking sound it makes as the lure, bead and hook contact each other!

How to fish it

Snag fishing is not for the faint hearted, but weedless set-ups minimise losses and encourage us to be braver with our casts! Even so, this style of fishing is often a case of no risk, no reward. It's tempting to rush our lures in to avoid weed and rocks, but very often predators will be tight to cover, so while you should experiment, slower retrieves are often most effective.

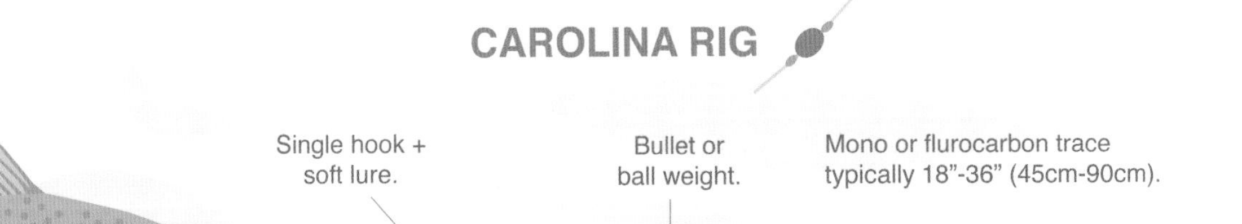

Carolina Rigging

The Carolina Rig (or simply 'Caro' rig to lure fanatics) is a variation of the Texas Rig, using a swivel or float stop to keep the weight at a distance from the lure. This allows our artificial to fall at a slower rate to the weight. It's especially useful when we'd like to use a larger weight with a small hook; for example, when we might need a longer cast on clean sand, yet keep hook sizes small for a mixture of small species. It's also a good choice for calmer conditions and canny fish that might not like a big weight, slap bang next to the lure.

Line: A length of up to 2-3ft (60-90cm), somewhat lighter than our main line is preferred. In clear, calmer water, modern fluorocarbons are excellent because they are very subtle in appearance underwater.

Pimp your Soft Lures!

Lure anglers are forever experimenting and fine-tuning their lures. Competition anglers in particular are notorious for this. There is a definite method to their madness and little edges and tricks are closely guarded. After all, fish can be moody and unpredictable, so even a small change can make a big difference. Here are some simple ideas worth exploring.

Screw-in spinner blades and rattles

An edge that can really make a difference on tough days, **screw-in spinner blades** are a great way to add extra vibration to your lures. Small, willow shaped blades with a screw thread are most common, which will attach to virtually any soft plastic bait. This is a great dodge for coloured water, or fish that need 'waking up' a little. Other similar variations of this trick include metal frames, which clip onto your jig head or soft bait, transforming it into a spinnerbait.

Screw-in, or push-in, blades and rattles are a great way to 'pimp' lures when the fish are moody.

Rattles are another good addition for slow days and pressured waters. While some lures come already rigged this way, you can also buy small inserts to customise other soft baits. Most predators are highly sensitive to vibration, so it stands to reason that rattles can provide a little 'edge' to make your lure easier to locate; or do predators find a rattle 'irritating' and decide to lash out? There are definitely days when predators like zander and pike seem to react to a rattle. Wrasse, too, seem to like an internal rattle for that matter – and we've even heard of bass anglers removing added rattles to avoid constant bites from wrasse!

Flavouring Lures is another obvious way to increase attraction. This can be done by storing them in a container of flavouring or by adding a scented gel. Does it make a big difference? Opinion is divided, to put it mildly. Some think flavoured lures are easier to detect in coloured water, or that predators are inclined to hang onto them for longer on the take. Our suspicion is that flavours probably make a small, incremental difference. So perhaps for competition anglers where tiny improvements count, you might try it. For simple pleasure fishing, though, you might regard this as too much fuss.

Soft lures can be stored in a container of potent additive to really soak in the flavour.

Softening lures is another neat trick used by competition anglers to give them more flex and movement in the water. The easiest way is to immerse them in very hot (but not quite boiling) water. This makes for a super soft lure that will wiggle if you so much as breathe on it. Especially handy at great depths where, unbeknown to many anglers, increased water pressure dampens the action of most lures. The only disadvantage to softening your lures is some reduction in durability.

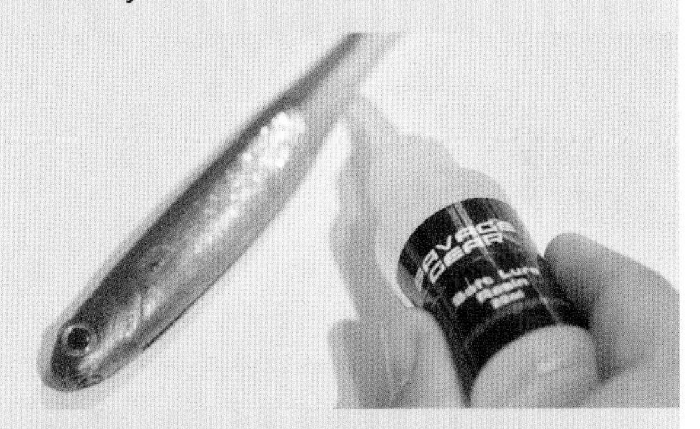

Resins & running repairs
Soft plastic lures are prone to ripping and tearing, especially when your quarry has fierce teeth. One way to boost their longevity is with a bottle of special resin. Several companies now produce liquids that can be applied directly to damaged areas – or even used to reunite two severed halves!

Weight: A simple ball sinker, cone or drilled bullet. Fixed between two float stops, this can be moved as you please. Depending on the scenario, you might want a longer or shorter distance between lure and weight, from as little as 4in (10cm) to 2ft (60cm).

Hook: Your choice here is wide open, depending on the target species and lure style.

Mojo Rigging

The Mojo Rig is a variation of the Carolina Rig but utilises a buoyant artificial. The lure works directly against the weight to produce a lively and very different presentation. It's highly effective over weed beds, as the weight will sit in the salad while the lure floats clear above, provided that the final trace is long enough. That said, it also works over a clear bottom and has proved to be dynamite on venues such as large reservoirs, even for tricky fish that have seen all the usual lures and presentations (just don't tell everyone!) One final tip here is to use lighter fluorocarbon when fishing the Mojo, to allow buoyant lures to rise up more freely.

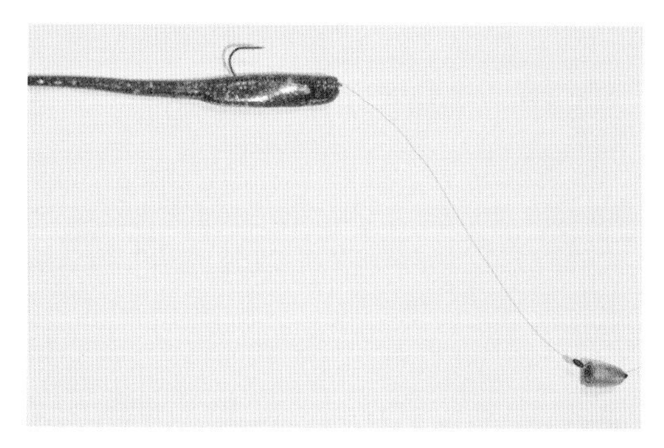

The subtle art of keeping in touch... TIP

While most lure anglers quickly become enthusiastic when it comes to lifting, reeling or twitching their lures to entice fish, it's what happens on the pause or way down that is often equally important! So often, anglers impart action on the 'up' so to speak, but then just let lures fall with little thought. Controlling that fall, by keeping light tension on the lure, for example, or using pauses and 'breaks' thoughtfully is so often key to earning and detecting more takes. The simple lesson here is to stay in touch with your lures at all times, paying special attention to the pauses and movement you are creating 'on the way down'.

Weighted Hook Presentations

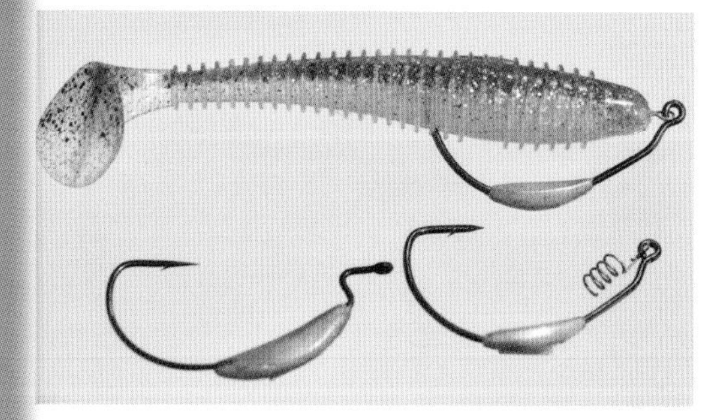

There are a number of hooks with weight on the shank which achieve lovely alternative lure presentations. By shifting the weight to the middle of the lure, you get quite a different action to that of the typical nose-weighted jig. The result is a more lifelike, horizontal fall position, which is ideal for slower sinking and subtler presentations. They are especially good for keen-eyed predators in clear, weedy water, for example. Various means of attachment are used: sometimes the lure is hooked weedless-style, while others use a corkscrew-style attachment through the nose of your artificial. Some hooks have sliding weights which allow you to adjust the position of the weight to change the action of your lures, from a loose rolling action to a tighter wobble.

Caro Floats and Tuned Weights

Usually the preserve of more fanatical light game anglers, 'caros' provide the option of fishing small lures at greater range by adding extra weight. They can be used in conjunction with jig heads and pin tail lures alike; in fact, they can also be used with flies or even Caro Kabura lures (special artificials which are as yet uncommon in the UK).

Caro floats sit on the surface and their main purpose is to get the lure out at distance, with the lure fishing almost vertically underneath. They often glow in the dark and hence win particular favour with anglers night fishing for scad and other species.

Besides the usual models, caro-tuned weights are also available to provide a sinking presentation. Sold in different weights and densities, the main difference between models is the fall rate and angle in the water. You can find those that sink as shallow as a 15 degree angle for rocky or shallow ground, or those that fall faster and more steeply to reach fish over features like deeper reefs or channels. An intermediate is perhaps the best all-rounder.

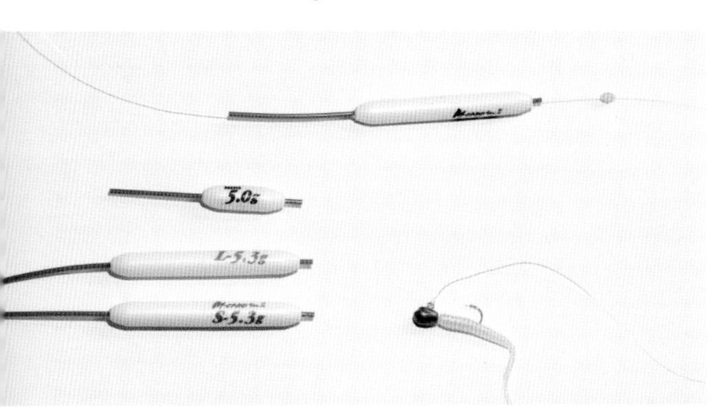

Above: A selection of Caro floats and weights: L will fall away from the angler, M is slow sink, S comes towards the angler. They work especially well with pintail lures.

Line & tackle: It is important to use a fluorocarbon leader above the mainline with a caro, to prevent wear to your braid. The easiest way to fasten the float in place is with float stops, leaving an appropriate gap between this and your lure depending on how deep you want to fish. Line strength will depend on the size of fish and distance required.

Slow Jigging with Metals

While many of us began sea fishing by lobbing out spoons and pumping them in with a vengeance, a new breed of metal lures have introduced subtler presentations and tactics for predatory fish, especially the pelagics. The Japanese, in particular, have developed a whole array of sophisticated lure techniques, but it is 'slow jigging' that is perhaps the most widely used and best suited to UK and European waters.

As the name suggests, 'slow jigs' are designed to fall more gradually, fluttering in the water under gentle tension. This keeps them in the 'zone' for longer, rather than bombing through the depths – ideal for our cooler seas and slower swimming fish. They can be cast and retrieved, but are also highly effective when presented vertically underneath a boat or structure. It's a more patient approach than the standard cast and crank, but well worth experimenting with.

Recommended tackle is rather different for the new breed of 'slow-pitch jigs'. These lures are designed to do the work for you, unlike traditional pirks or vertical metals. The most suitable rods have a parabolic 'through' action with a fine taper; in other words they are less springy and more forgiving to impart subtle control to your lures. The reel is equally important, and should be chosen to match the weight of the rod: a small, strong fixed spool can be used from shore, while multipliers are favoured by boat.

From the shore, it's a case of fishing more patiently than you would a traditional metal. After casting out, try keeping the lure under gentle tension at first, to swing down in an attractive arc. Next, when you've counted to a certain depth, make an upward sweep of the rod tip to animate the lure (it's this movement that gets the lure working correctly, from shore or boat alike, and is sometimes called a 'pitch').

After each sweep or 'pitch', the rod tip is gradually lowered so that the angler keeps in touch with the lure as it flutters down enticingly. There is no constant cranking of the reel; the angler just gradually takes up the slack after each sweep or 'pitch' of the rod. The idea is to encourage an almost vertical fall of the lure, while feeling closely for takes. You can vary the speed, but again, it's a much subtler art than traditional cast and crank spoon or spinner fishing.

From a boat, slow jigging with metals is also highly effective and comes into its own in slack water, when the tide stops running. British anglers are increasingly using this technique to keep bites coming in this quiet period. Due to the strength of tides around the UK it is only possible to fish the method effectively when the tide starts to slacken off or on lesser tides. By boat, these metals tend to be fished vertically, and fine PE braids are a must to maintain 'feel' and encourage the subtle action of the slow falling metal. Whilst anglers tend to fish between 100-250g in weight with 'slow-jigs' from charter boats, on your own inshore craft or on lure-only charters it is possible to use lighter jigs. Whilst this is not purist 'slow-jigging' you can have some exciting light game, bass, and species hunting on lighter jigs using your traditional spinning gear. It's a more versatile method than you'd imagine too, and our skipper friends have reported over 30 different species to slow jigging!

The Ned Rig

The Ned Rig has taken perch fishing in Europe by storm. While specialist lure anglers have been familiar with presentations that resemble a burrowing or bottom-hugging prey item for a while with rigs such as the Neko, the Ned has really gained a cult following. US bass fanatics first popularised the method, after finding that cut-down Senkos (a special floating lure) earned bites from fussy fish.

Best fished on a slack line, the Ned Rig often produces best when the retrieve is slowed right down; try dragging your lure a short distance along the bottom (literally just a lazy, long sweep, from a few inches to a few feet), pausing for anything up to 15 seconds, then reeling in

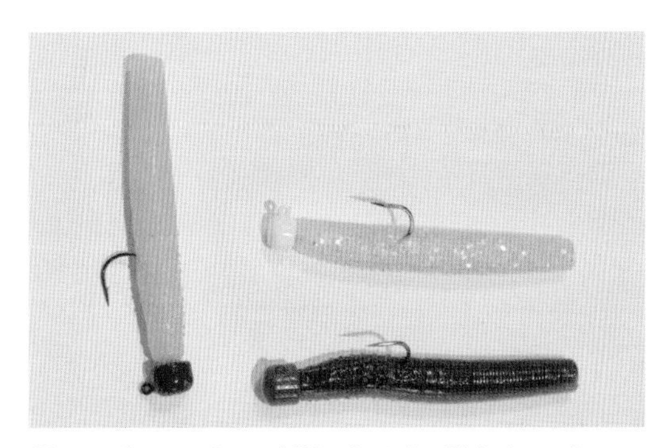

Above: Lures rigged 'Ned' style. This is quite a specialised presentation, so do look out for purpose-made tackle. Your jig head should stand upright at a 90 degree angle from the bottom (like the pink lure above!) Lures can also vary a great deal in terms of buoyancy; the best of them, such as those made by Z-Man, are super-buoyant and incredibly elastic, achieving that effortless 'popped up' presentation and wiggle.

the slack and repeating. It can produce fish from seemingly dead swims, and we've witnessed this in tournaments where virtually all of the fish resulted from dragging a buoyant lure slowly across the bottom close to structure. Whether it's just something different, or the 'burrowing' action suggests a prey item that is about to get away, it seems to drive perch nuts.

Cheburashka Rigs

The Cheburashka rig, or 'Cheb' as it is often known to UK anglers, is a bottom searching rig that uses a special slotted ball weight to accommodate the hook. 'Cheb' weights have detachable internal wires that allow you to attach a hook of any size or design. This versatility is a big advantage. For example, you can easily match a heavy weight with a tiny hook, or a weedless design. Equally, you can also then change your hook pattern in seconds without having to break your tackle down.

Interestingly, the name comes from the lead and wire formation looking like a fictional character from Russia, called Cheburashka, with big protruding ears rather like the metal loops on the ball weights used! The Cheburashka rig is a winner in many scenarios, particularly tricky or extreme situations where, for example, a relatively tiny size 12 hook might be needed alongside a 7 to 10g weight, or for when that little bit more articulation is required for Ned rigging.

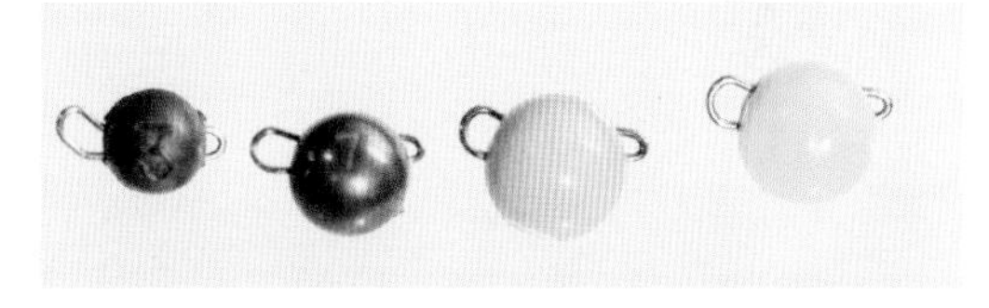

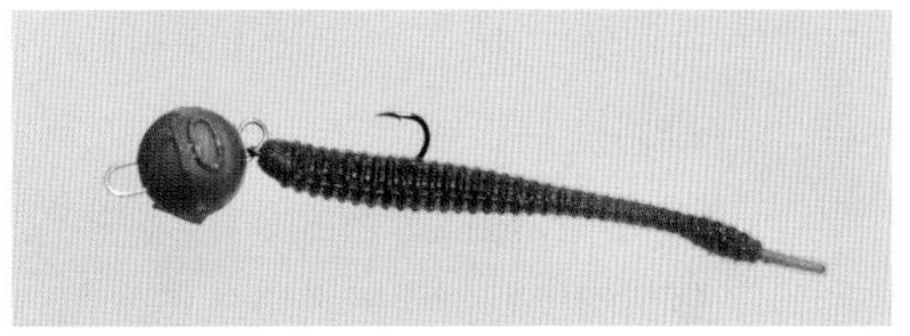

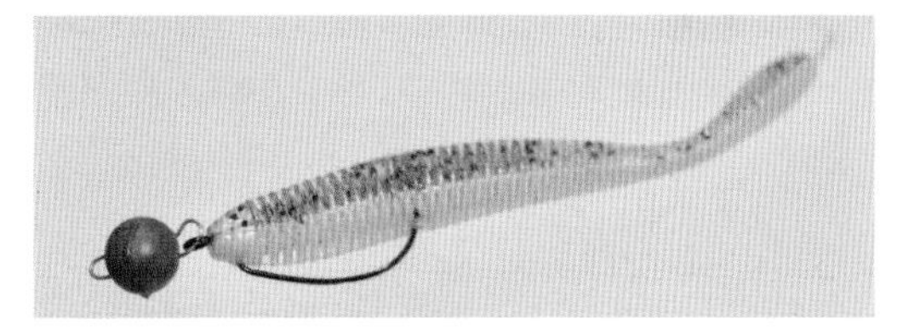

Above: The Cheburashka or 'Cheb' rig involves a ball sinker with detachable wire 'ears' (top). It's great for using a heavy weight with a small hook (middle) and can also be used for weedless rigging (bottom).

Jika Rigs

The Jika is an interesting rig as it offers a presentation quite similar to drop shotting but is used for fishing tighter to the bottom. It allows you to fish your lure perpendicular, at 90 degrees, with the

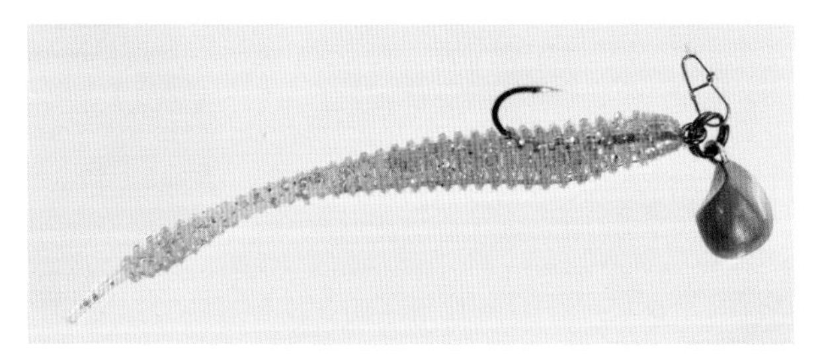

artificial attached directly to the weight. This is useful when more mass is required yet you need to retain a smaller hook size. It's especially effective for wrasse over clean ground and is often preferred by less experienced anglers, who often find it gives added sensitivity and produces easier to hit bites than other weedless rigs, due to the very direct nature of the weight to hook. Dedicated hooks are available for the Jika rig; look for those with a welded ring attached to the hook eye, and attach this to your weight using a split ring.

Wacky Rigging

Wacky by name, wacky by nature, this method involves going against convention and hooking a soft plastic lure in the side or another unusual position. It might look ungainly, but this can offer a completely different behaviour as the lure sinks, and a presentation totally different to the norm. It can often be a trigger on slow days, or for fish that are shy or subjected to regular angling. Wacky rigging can be used with a variety of lures, but the best are arguably those such as a Senko or similar worm-type stickbaits, for a really lively, erratic action.

You can also fish Wacky lures in a weedless fashion by using Wacky hooks with a wire or nylon weed guard. Since the lure is only lightly hooked with this method, losses can be slightly higher with aggressive takes or general wear and tear. To prevent this, they can be fitted with specially designed O rings which help to retain the lure.

Neko Rigging

The Neko rig is a variation of the Wacky rig where one end of the lure is weighted with either a nail weight (or even a real, cut down nail!) or a corkscrew weight. The Neko is the forefather of the Ned rig and offers quite a similar presentation, mimicking a feeding fish or burrowing worm. It can be highly effective for inactive or wily fish and is useful for fishing amongst weed when using a Wacky weedless hook, or in situations where there are fish feeding hard on the bottom.

There are all manner of rigging materials to assist with this rig to lengthen the life of your soft plastic but they can be hard to find so we have shown the rig here without.

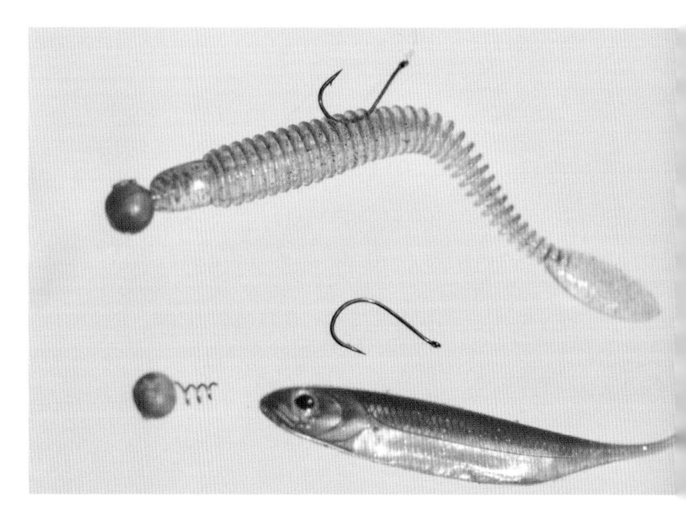

Stingers

TIP

In simple terms, a 'stinger' is an additional hook added to any soft plastic lure, using a short length of leader material. The reason we use them is to hook those fish that strike short, or nip the tail of our lures on a regular basis, thus avoiding our main hook! We should point out that stingers are not essential, and are only an important option for certain species of fish and, as such, should only be used when absolutely necessary. Species like pike, for example, tend to take lures whole; hence extra hooks tend to be a nuisance and only tend to be used on whacking great lures such as hulking great swimbaits.

With zander, however, it is a case of judging whether it is necessary. Sometimes, they will easily grab your lures whole and a single hook is fine. On other occasions, however, when the fish are dormant and not feeding hard, they can tend to 'bite short'. In these circumstances, a stinger hook can convert more takes into fish. Much of this is down to experience; the more you fish for zander the more you will identify these patterns and know when and where to add a stinger. When your lures keep coming back minus a tail, for example, or with teeth marks towards the rear, you might consider one!

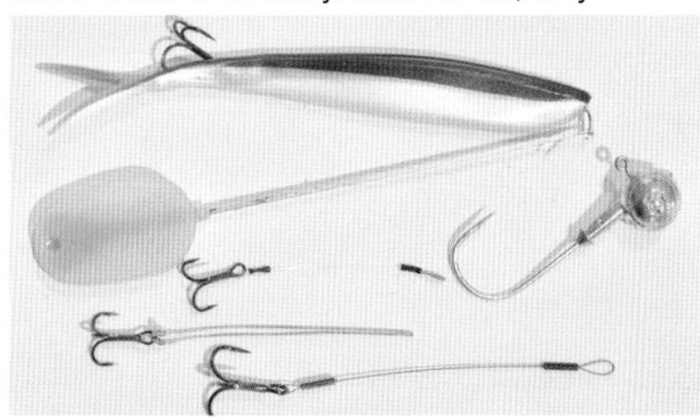

Types of stinger and how to attach them
Most stingers can be bought ready-made, but it's quite simple to make your own. They can be threaded through the body, or indeed left exposed on the outside of the lure. All tend to be attached to your main jig head or hook via a link of line or wire. To make your connection more secure, you could try using a piece of heat shrink tube over the shank of the hook, or plasti-dip for a more snug, safe connection. Fluorocarbon tends to be preferred as a link material for zander, but if pike are present we would strongly advise wire to avoid bite-offs.

Rigging Flies with Lure Tackle

A little-used twist, combining artificial flies with lure fishing tackle, can be a deadly trick to employ. One or two JR Hartley types might get sniffy, but the fish certainly have no objections! And why the heck not? If you think about it, flies fulfil a very similar function to lures; they are artificial creations intended to mimic prey or provoke a response. Made of fur, feathers or synthetic fibres, however, flies have a subtle, deadly action all of their own – and will sometimes outfish conventional lures.

The possibilities are endless, but there are several simple ways we've used flies to great success. You can fish smaller 'nymphs' like shrimp imitations on a split shot rig. Or you could use large dry flies such as grasshoppers and beetles with a controller float. The most useful flies of all for predatory fish are baitfish patterns or 'streamers', however. These are especially deadly when your quarry is attacking fry or needs a slower, subtler presentation.

You can present these lures in several ways. Simply adding split shot to the trace to provide casting weight is easiest of all. Flies can also be rigged drop shot

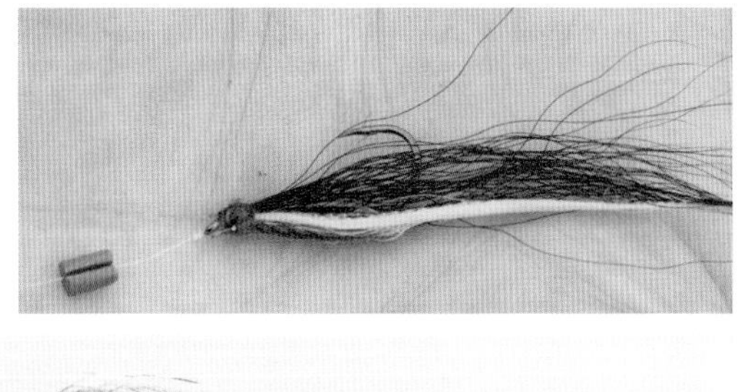

Above: Really simple ways to cast flies with your lure tackle: just add split shot or a Cheb weight!

style, however; or any larger baitfish imitation can be used in conjunction with a Cheburashka weight. Give it a try; only a tiny minority of lure anglers are doing this at present, but it can be dynamite!

Tipped Metals

An excellent dodge for those who fish with spinners or modern metals is to combine the flash and vibration of these lures with a fake worm – or even a real one. This is an especially good way to catch the frustrating thin-lipped mullet, which will often follow a spinner, before a section of plastic or genuine worm seals the deal! Nor are they alone; many other fish have an annoying habit of following bladed lures but refusing to take them! Trout often do this – but more 'occasional' predators such as grayling and rudd can also be tripped up by the combination of spinner plus worm. Try a plastic worm or grub or, better still, a section of scented fake worm. The best way to mount the trailing delicacy is by replacing the treble hook on your lure with a single, and 'threading' the worm up the shank securely.

You could also try a real worm, of course; some may call it cheating, but with fussy fish like mullet it can be a true game changer! Why it should work so well is open to debate, but what seems certain is that most fish have a competitive instinct; could the metal lure simulate a smaller rival? When a predator sees a lesser fish 'struggling' with a meal it can't quite handle, it will often decide to go and commit robbery!

This method also works well with smaller modern metals, too. With a tiny spoon-style lure, you can easily twitch your way along the bottom. Flatfish, gurnards and other species will quickly be drawn to the commotion, before snaffling a trailing scrap of scented soft plastic. Again, do feel free to swap hooks and scale down for the smaller species.

The tipped spinner is a game changing trick for frustrating, hard-fighting mullet.

WATERCRAFT & TYPICAL VENUES

The waters we cast into are so many and so diverse, it would take several lifetimes to explore all of those even in one small country. Nevertheless, while fishing is infinitely varied, certain patterns and lessons in watercraft repeat time and again. This section of the book is by no means exhaustive then, but aims to familiarise you with some of the most common venue types and scenarios.

At the outset, we should state that understanding lures and methods is only half the battle. So much of successful angling is about understanding your quarry and its world. This might mean finding out where predators rest and hunt; the hours they keep and their preferred prey, not to mention keeping a close eye on the weather, tides and other conditions.

Watercraft on the coast

From quiet harbours to estuaries and the open shoreline, the angler will find an endless variety of locations and fish species. So where on earth do we start? With such a lot of water, it helps if we break things down into smaller parts and concepts we can understand. One good tip here is to focus our efforts on just one or two marks

TIP

Keep a diary!

Wherever you fish, it's a great idea to record your sessions for future reference. In retrospect, the difficult days can be as useful as our successes! Human memory can be unreliable at the best of times. So, whether you use old fashioned pen and paper or a digital app, a diary can help you repeat former glories and avoid past mistakes.

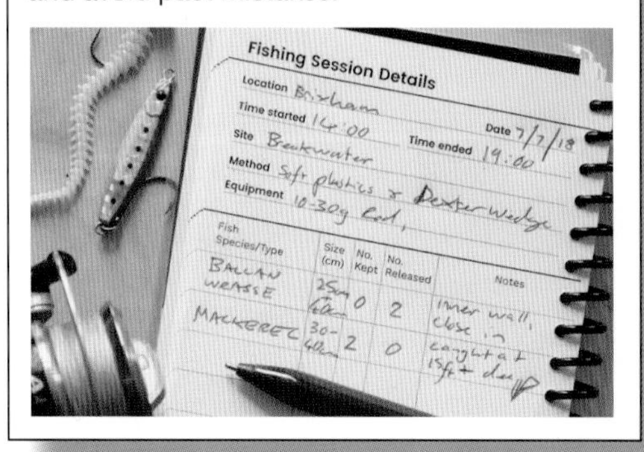

and species at a time. This way, we can work things out piece by piece.

In this section we'll be dealing not only with common settings and locations, but the effects of tides, weather and times of day. Ultimately, it is up to you to do the detective work. Modern technology has definite advantages and can provide some short-cuts, of course, from online groups to Google Earth. However, you'll never find all the answers by looking at a screen or messaging digital contacts. Besides, the process of trial, error and discovery is all part of the rich experience of fishing. So, while we hope to bring some experience and learning to the table here, it pays to be curious, active and open-minded. Indeed, the most satisfying fishing wins of all are those that you figure out for yourself.

Typical shore marks

For the typically light tackle favoured by LRF anglers, sheltered spots are often our 'bread and butter' fishing. It's not that you can't use smaller lures from beaches or rock marks, but with stronger winds and tides this can quickly become difficult. Man-made structures and spots like harbours, piers and seawalls offer shelter and fish-holding features at most states of tide, providing consistent sport and great variety.

It is a great advantage to be close or even directly above the fish you want to catch with smaller lures. The less line between rod tip and lure, the more control and the better the 'feel' you have.

Above: Andy is joined by Maurice Minchinton at Plymouth's Barbican area, which offers several kilometres of fishable walls, steps and structures.

Harbours and marinas

Although man-made, a typical harbour can hold as many kinds of fish as a natural reef. Nooks, crannies, slipways and steps will host blennies and rockfish. Walls and boulders will have pollack and wrasse. Sandy, more open ground will have flatfish and yet more variety. And besides the true 'locals' we also have fish that are not residents but seasonal 'guests' from summer mackerel to winter whiting.

The first place to explore is generally around any existing structure, making short casts or working right under the rod tip. The fact that you'll find consistently calm, sheltered water, even when the open coast is rough, is a huge advantage for the lightest lure fishing. With typical spots between one and four metres deep, you can easily use just a gram or two of weight and feel carefully for bites. The golden rule here is control; use just enough weight so that you can cast comfortably and feel what is happening through the rod tip.

As for the fish-holding features of harbours and boatyards, these are endless. Smaller sea fish love to loiter around jetties, pillars and posts. Boats and floating docks can hold fish too, with smaller creatures often found clinging to the underside; but do cast with care and be mindful of the owners. Whilst harbour fish are confident enough to come into built-up areas, be wary of casting a shadow across the water and stay light footed around harbour walls to keep disruption to a minimum.

A TYPICAL SALTWATER

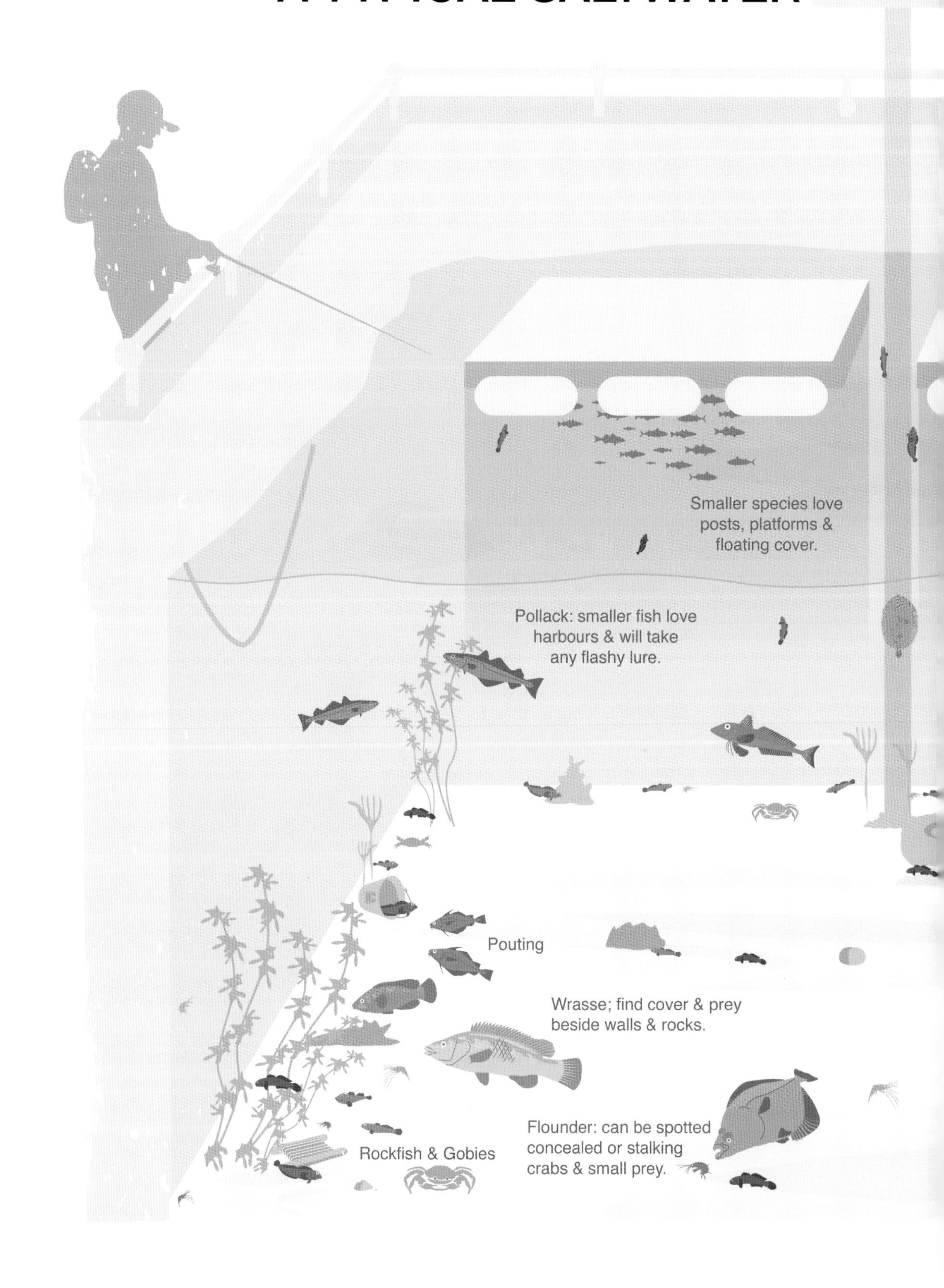

HARBOUR OR MARINA

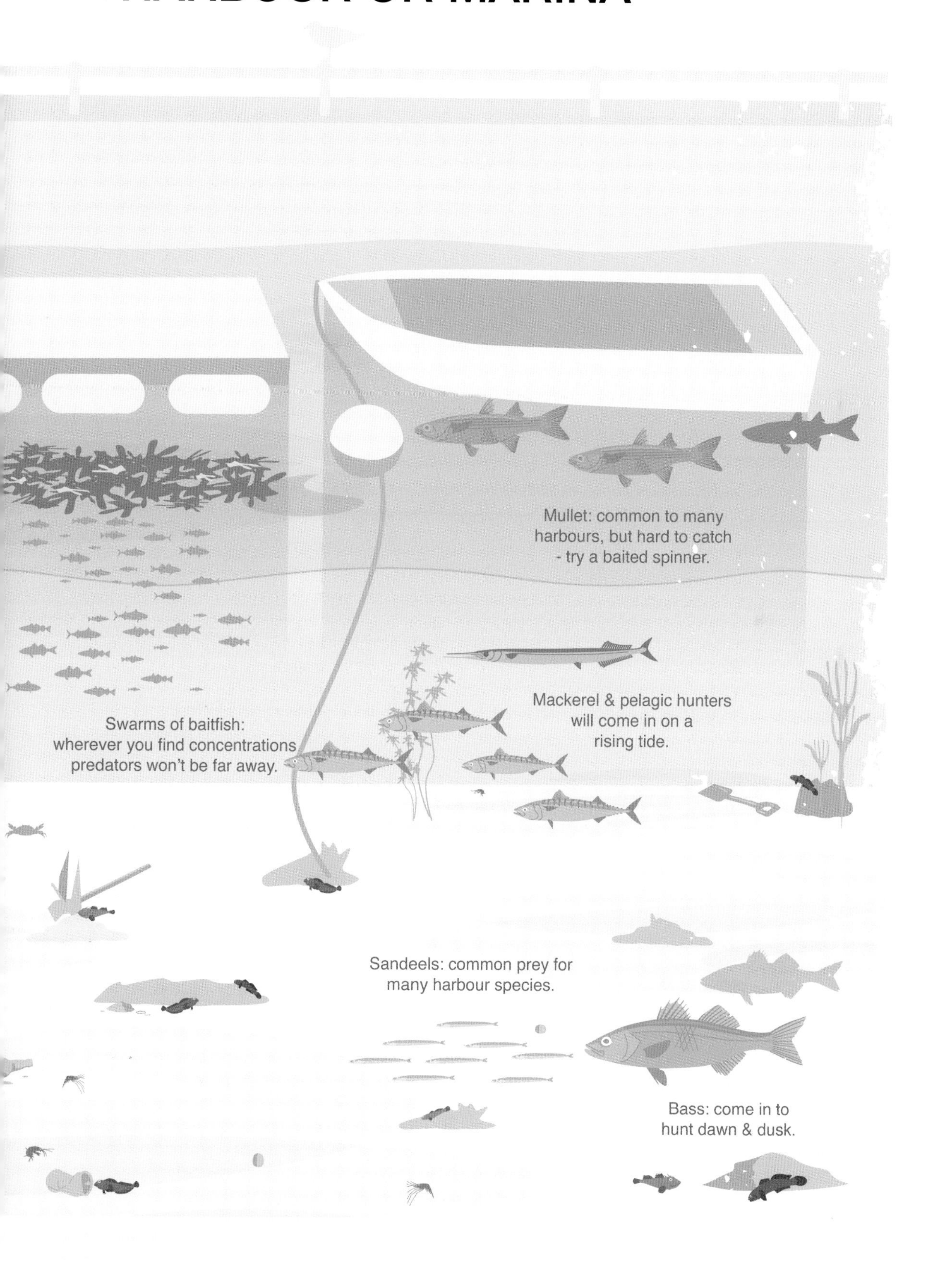

TIP

Manmade attractions

Never be afraid to be nosey when it comes to sussing out good spots. Locations near takeaways, or better still working docks, where fishermen bring in the catch and dump remains, are always worth investigating. Prey and predators alike will often be drawn in wherever there is free food – and smaller fish are often at their most vulnerable when feeding.

Piers and breakwaters

These common structures give the shore angler easy access to deep water. This is a big advantage, as it means you can get comparatively small lures into productive areas with no need for longer casts. The best place to fish, though, often depends on when you arrive. At low tide, the farthest end of the structure might be the only place there is deeper water. Any pocket of cover will still often produce, however, whether it is two feet or ten feet deep. Even when the bigger predators are away, you will tend to find a typical local rogues' gallery of rockfish.

As the tide floods in, larger fish will also venture closer; or perhaps come out of their hiding places from beneath the pier! Casting tight in you'll find wrasse and pollack year round but especially in the summer. Throwing metals and small plugs is great fun for mackerel, school bass and other species such as garfish; and working faster sinking soft plastics and metals across the bottom might find gurnard, flatfish and others, depending on location. Meanwhile, in the depths of winter, pout, whiting and codling are possible suspects.

The only barrier to fantastic light tackle sport will be occasional rougher conditions or high winds, in which case you may need to gear up a bit heavier or find those more sheltered corners.

Above: Andy and Paulina Garnett enjoy a cast at Swanage Pier, working lures right by the 'legs'. One of Andy's favourite venues, its mixture of kelp beds and clean sand beyond produce flatfish, gurnard and wrasse by day, while bass will arrive on an evening flood tide.

TIP

Net gains

While most man-made structures and rock marks hold lots of smaller fish, there is always the chance of a surprise. Don't lose the fish of a lifetime by not having a decent net! A dropnet will work, but better still is an extending net handle that will go to at least 4m. If you hook a better bass or wrasse, you are going to need it!

A rock platform on a rising tide; be sure to pack sturdy footwear and plan an escape route.

Rock marks

Rock marks are dynamic locations that place particular demands on tackle, offering an appealing blend of risk and reward. You're not going to manage an ultralight outfit in a gale or the top of a spring tide, so stronger tackle and snag-proof rigs are often necessary. However, the payback for this testing environment can more than outweigh the challenges. Indeed, find yourself a wild, rocky beach or rugged headland to fish from and you'll often find some mean fish to tangle with, quite often larger ones. Local netsmen tend to give the really hairy spots a wide berth!

Not that all rock marks are a case of swirling tides and crashing waves, of course. From sheltered coves to former quarries, there are also marks that regularly offer more sheltered water alongside access to greater depths at close range.

Many shore anglers will immediately think of rock fishing for the heavyweights, such as big bass, wrasse and pollack, which can be found in these spots most of the year. For this reason, typical gear for this kind of fishing could probably be described as medium rock fishing or even HRF (Heavy Rock Fishing). But these species only tell half the story. As the water warms each year, you will start to find a wider range of species moving tighter to the shore. Mullet arrive in late spring, showing themselves as they move in on the tide to graze on algae and, in some areas, maggots hatching from rotting seaweed. Into the summer, fry and sandeels mass closer to the shore, with pelagic fish like mackerel and garfish in hot pursuit.

There are many ways to approach such a setting, depending on what you wish to catch. On a rising tide into evening, or on neap tides, bass, pollack and wrasse come to the fore and it is especially prudent to step up your tackle and to fish higher in the water as the larger predators will now be most active. First light is another good time to find larger species actively feeding. Wrasse come out of their hiding places and can be caught higher up in the water before the sun comes up.

The places fish lurk and those where we'd prefer to cast lures are often two different things!

Andy's Tip: Work with the elements, not against them

'One huge lesson you will learn in the sea, or indeed any large or flowing water, is the huge effect that currents, waves and wind have on your fishing. At sea, the swell can move your lure as much as two meters in a second. Natural forces can seriously impact a day's fishing, but rather than trying to work with the elements many anglers try to fight them.

As a young beginner at sea fishing, I used to do the same and would automatically tackle up heavier to try and keep control, but this would often cause the lure to sink too fast or behave unnaturally. Think about it: are small fish or other prey items anchored in place? Over time I learned to use the swell and the wind to control the lure's descent to achieve a much better presentation. In fact, you'll often find that currents naturally take your lure to where the fish are waiting. I also found that using my body as a shield from the wind I could retain the feel of the lure as I worked it back.

To some extent the same is true on rivers and freshwater settings, and it is often to their disadvantage that anglers actively avoid windy spots or more vigorous currents. As a general rule therefore, try working with the elements rather than fighting against them where you can. Where you want a slower presentation amongst a bit of a tumult, however, drop shot presentations can be very useful.

Neil Sutherland enjoys a 'bracing' day in Devon; there were still fish to be caught on this occasion in spite of, or perhaps because of, the elements sweeping our lures along.

If you intend to spend a whole day on the rocks, however, you'll find a much wider range of species at different stages of the tide. Hence, bringing lighter tackle and a variety of smaller lures is wise. During slack water at low tide, larger resident fish will generally switch off unless you can cast into deeper water, so at this point it is worth trying the lightest of approaches for corkwing and other small wrasse, not to mention various rockfish and blennies. Pelagic fish such as garfish and mackerel will also feed through most of the day, if you can reach them with longer casts. It's very much a case of experimentation and by bringing different rigs and smaller lures into play, more species can be caught.

Beaches

While anglers tend to be magnetically drawn to prominent features such as rocks and piers, there is also interesting fishing to be had from open beaches. Large, sandy expanses of water are popular with swimmers and human traffic, admittedly, but can also be worth exploring with light lure tackle. For those with limited chances to fish, it's easy to pack a simple outfit and enjoy an hour or two of fishing, even on a family day out.

While the typical sea angler might hit the horizon with a big lead, the lure enthusiast will wade and fish closer in, searching for fish that hold closer to the shoreline. The species present will depend on the beach; those with steeply shelving shingle and greater depths close in will see predators like bass and mackerel come within just a short cast, especially early and late in the day. Watch the water for any signs of chasing predators or distressed fish!

Rockpools represent a delightful 'lucky dip' for the species hunter.

As for the shallower, more open beaches, we would recommend going for a recce with polarising glasses. Across sandy plains, bars and gullies you are likely to find small flatfish and other open seabed dwellers. Indeed, flounder will sit just feet from the tide line and many is the time we've spooked them while wading. A patiently jigged worm or crab lookalike is likely to get their attention. Customers like sand gobies and weavers are also common, if you scale right down with your lures.

Wading and donning polarising glasses or even a snorkel mask will reveal a lot to the visiting angler, even if you don't know the venue. A pair of old sandals or summer shoes is also recommended, not only for greater comfort but avoiding weavers – which you might catch while you're there! Keep an eye out for patches of weed and clusters of rock as the water deepens; look for darker areas within a short cast to find wrasse, blennies and other surprises.

Every beach is different and you never know what you'll find. It's possible to catch gurnard, dragonet and even small turbot from the shore with a little luck and some detective work. Try bouncing small metals or soft plastics across sandy bays to seek out these fish.

Rockpools

When discussing sea fishing venues specifically with the most delicate LRF tackle in mind, we should also consider rockpools. Find any suitably rocky location on a retreating tide and these areas can yield a real pick 'n' mix of species. Suitable spots can be as small as a bathroom sink or as big as a small swimming pool. Virtually any sizeable volume of water will hold myriad sea creatures, including fish, and it's immense fun watching them emerge and seize a lure.

Tackling rockpools takes more stealth than other sea spots. Indeed, when you are so close to the fish and the waters are shallow, you simply must approach cautiously! It makes sense to kneel and

Left: A snorkel mask is a cheap, exciting way to find features and fish. It's brilliant fun and if you team up with a friend you can even watch how the less shy species react to lures!

avoid sudden movements or shadowing the fish. Indeed, a lot of rockpools can appear empty at first, beyond sea squirts, anemones and fronds of weed. But once you drop a small lure into the mix, you can watch the residents come out of hiding. You will often find several natives in one spot, in fact, and they will size each other up besides your lure! It can be like watching a tiny amphitheatre with its own little gladiators.

There are a wide variety of fish you might find in rockpools, especially if you can find the deeper, more substantial spots near the tide line; if you are swift, spring tides can be even more profitable, although you need to plan carefully and fish with great caution once the tide races back in. Various blennies and gobies are a given, but you may also find mini beasts such as topknot, rockling, sea scorpions and smaller wrasse. For maximum sport, the lightest tackle you own makes sense. Tiny jig heads will work, but the split shot rig armed with a tiny barbless hook is the best and kindest way to catch and release the various miniature predators.

Estuaries

Offering sheltered waters and rich mixed habitat within a short cast, estuaries are also prime spots for light lure fishing. From sand flats and channels, through to tide rips and the 'mouth' where fresh and saltwater meet, there is a lot to explore, not to mention various species to target. There will be school bass running in any tide rips and channeled currents. There will also likely be flounder and various mini species in the sandy and muddy ground.

Above: Tides can ebb or flood many square miles of ground on larger estuaries; it pays to be mobile and have an escape plan!

An estuary can often represent a range of spots within a short distance, and the species you find will depend on habitat. Waders can be useful, although you will need to take care because mud can be treacherous and tides can rise rapidly. Every estuary has tide rips and distinct channels cut by the tide. Many will also have a harbour or marina, along with other natural vantage points where passing fish will be funnelled, as well as a definite 'mouth' where freshwater and sea meet.

Perhaps the best starting point to tackling any estuary is to take a walk at low tide and map out any features. Because the waters rise and fall quite dramatically, you might find huge areas of sand and mud that are only covered at high tide. Fish like bass are real creatures of habit with their patrol routes and the areas they like to feed. Quite often you could set your watch by them at a particular location, so although you might need to suss out the best spots and tides, any success you have is likely to be repeated if you identify a pattern.

While there are hundreds of lures that might work, you should also try to get an idea of the prey on offer for resident fish. Crabs, shrimps and marine worms are common in just about any estuary. You'll often find teeming sandeels and sand smelt too, along with young, vulnerable fish of various species, from baby bass to young mullet. So, while you might not need a perfect replica, a rough idea of the size and colour of typical prey might help.

As the tide gains water, especially if

Above: Gulls often betray concentrations of small fry, accompanied by larger predators. On this occasion, the area was teeming with school bass willing to hit anything small and shiny.

this coincides with dawn, dusk or dull weather, you might find bass on the hunt. They love tide rips, where they can overpower sandeels and juvenile fish, but fishing currents correctly can take some adjustment. One general observations is that the quicker the current, the less erratic you will want to retrieve. The current will often bring food to the bass and they will nip in and pick off hapless smaller fish, rather than chasing for many metres. Hence a slower style, letting your lure travel with the current, perhaps with the odd twitch thrown in, will often yield more fish than cranking in at speed.

During slower, brighter periods of the day, however, expect bass to be dormant and sat below weed. It's prudent to work your lure close to cover at these times, or to fish a weedless lure through vegetation to trigger resting fish. When not hunting, they will often choose to sit it out and rest behind rocks and other structured, and you may need to work soft plastics close to the bottom, bouncing them downstream with the current.

The bottom of low tide can also be an interesting time to fish, but this largely depends on the estuary. Some larger venues will still have expansive slacks and depressions worth searching for flatfish and mini species; others will leave barely enough water to fish! Of course, a walk at low tide can be a useful exercise regardless, as it will tell you a lot about the features that you may miss on a flood tide.

Keep a low profile

TIP

In both fresh and saltwater, it's sensible to be stealthy and treat your quarry as a wild animal. A lot of anglers mistakenly assume that sea fish don't spook easily. If the water is clear, stay back a little and keep down; kneeling and making smaller movements is better than standing bolt upright. If you are fishing close to a wall or beneath rocks, make your first cast close in, keeping a couple of footsteps back.

In truth, we probably spook more fish than we ever know; and the findings of our underwater photographer Jack Perks in the making of this book were startling. Sound travels very well underwater; in fact, he could easily make out heavy footfalls and even our topics of conversation, when recording several yards beneath us!

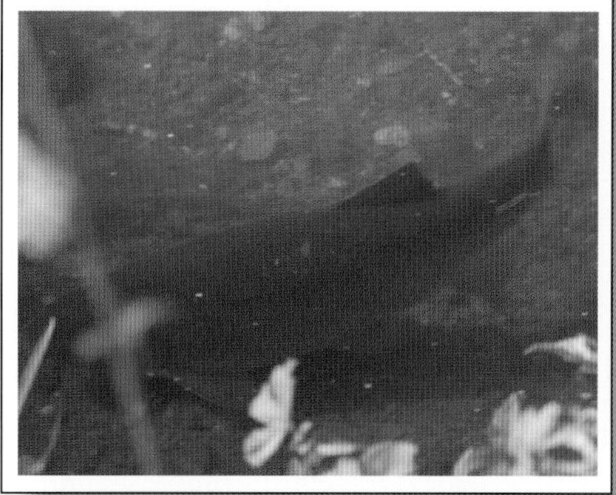

Times and Tides

It goes without saying that the matter of *when* you fish can be just as important as where. Because while some of the smaller fish like blennies and rockfish will be present at all states of tide if there is enough water to hide them, the bigger fish will often come and go.

Perhaps the biggest misconception of all in sea fishing is the idea that it is always best to fish on a rising tide. As the water comes in and covers new feeding grounds, this can indeed be an excellent time to fish, but only represents half of a complex picture. For certain species and marks, a low or dropping tide can be far better. For every fish that loves a big racing tide, another prefers slacker, lower waters. Similarly, there are definitely spots that fish better at low tide than high – and you will certainly get a wider range of species by trying different phases of the day. Of course, whether the fishing is good or indifferent, it is always worth a recce at low tide on a venue because many of the hidden features will be revealed for later reference!

Another point to note well is that not all tides are equal. Spring tides are a prime example of this, marking a bigger influx of water and prime feeding conditions, albeit not always easy for lure fishing. Feeding behaviours change a lot; with the extra depth and swirling currents, smaller fish get buffeted about and the bigger predators grow bolder. You might suddenly see a large bass or even a sea trout! But equally, species like flounder and even mullet tend to get more aggressive. A very high tide can also flush out creatures usually confined to the shore. Sea slaters and maggots are dislodged from seaweed and fish will come right into the rocks to get at them.

Try something different

We are all guilty of getting stuck in a rut at times. We get it into our heads that only certain lures, times or spots produce. So often, a simple variation or experiment brings a completely different result. If you always stick to the same script you'll only get some of the answers. Our solution? Dare to try something new or a bit bonkers every time you go fishing.

Lure colours v depth and water clarity

Do lure colours matter? And if so, what should inform our choices? This whole issue divides anglers, but is worth some consideration. Generally, when we're lure fishing we are appealing to predators hunting by sight; hence the clarity of the water and the depth we are fishing can make a big difference. For example, when faced with murky water and difficult conditions, we might use bigger lures in loud colours that produce extra vibration to help a predator find the target. In a low, clear river for trout or chub, however, the opposite criteria might apply, with small, natural-coloured lures the best option to avoid scaring wary fish.

Looking at things a little more scientifically, different colours in the spectrum stop showing up as strongly at different ranges. This applies horizontally, as well as regarding depths from surface to the bottom. Reds and oranges can cut through muddy water well, for example, but are the first colours to disappear from a distance of around 20ft (6m). Yellow and green fade next; while black and blue

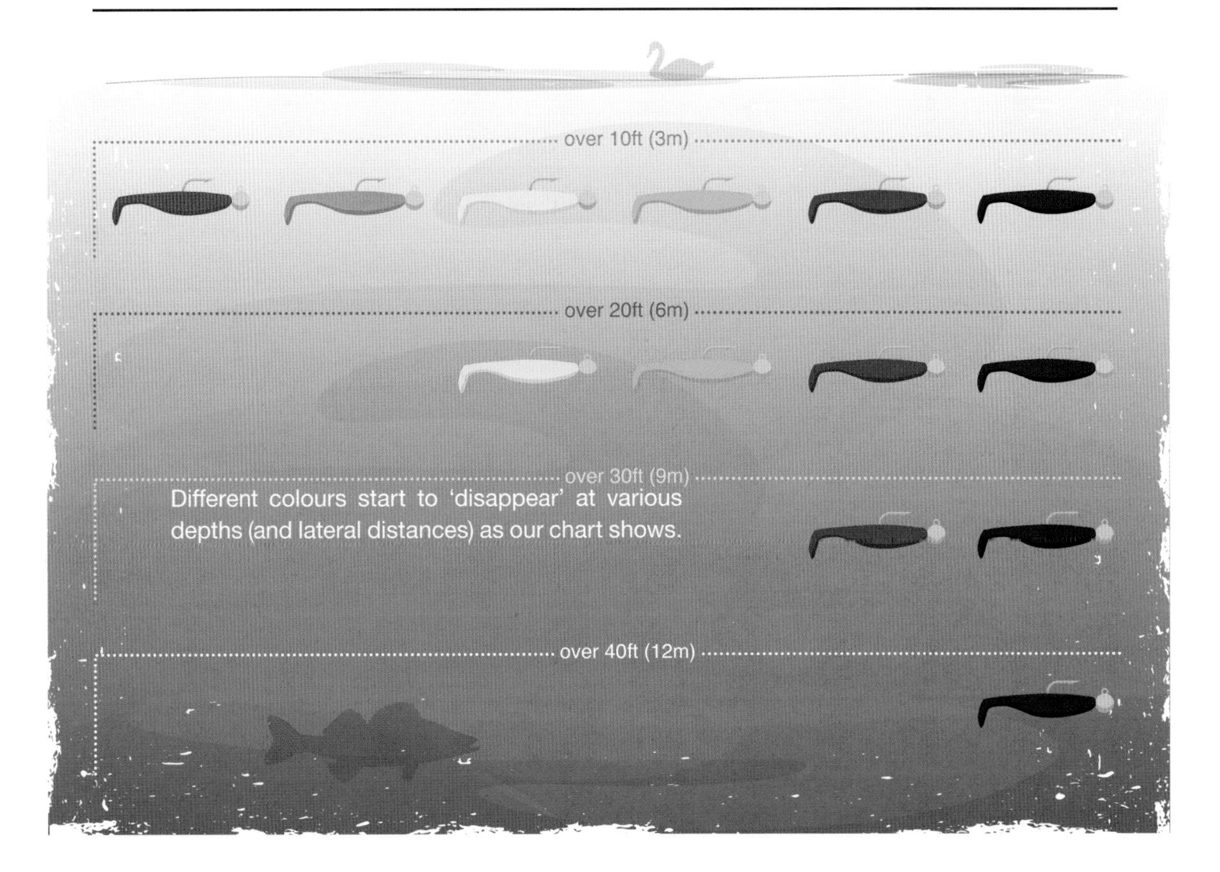

are the colours that show up best at greater depths. It's rarely the most fashionable lure colour, but it could be that black is arguably the best all round colour for clear or murky water alike, and yet one of the least used.

Killer contrasts: Shade lines and colour changes

With many predators thriving on the element of surprise, it follows that they tend to love areas of contrast. This could mean a spot where exposed, well-lit water meets a shadowy feature such as an overhanging tree, wall or jetty.

It could also mean the zone where water clarity changes from bright and clear to murky and stirred up; for example where the wind or tide has stirred up the bottom and a visible colour change can be witnessed. You might refer to many of these as 'change zones' where light and shade or coloured and clearer water meet. All of these offer a big advantage to predators, assisting them in launching attacks on their prey without being detected.

Of course, light is also a factor and many predators don't like bright weather. Perhaps they mind less if the water is deep, or rocking waves mask their presence. By and large, though, hunters feel both more vulnerable and less able to hunt effectively in bright, open water. Fish don't have eyelids either, so they are likely to go deeper when conditions are bright and still. This can vary quite a lot by species, however.

Common Freshwater Settings

While inland fresh waters are perhaps less daunting than the sea in terms of sheer changeability and volume of water, there are still a heck of a lot of different venues to explore. We deal with species specific clues in our dedicated chapters later in the book, but felt we should explain some basic venue types and starting points here.

As with sea fishing, a good tip is to focus on just one venue at a time and, if you are inexperienced, it is much easier to locate and tempt fish on smaller waters. Narrower rivers and canals are especially good places to start, as they tend to have obvious hot spots and don't require huge casts. Those where predators tend to be smaller but more numerous are especially good fun and easy to get to grips with.

Canals & Drains

The recent popularity of lure fishing has triggered a return to the canals for a whole generation of anglers. These are excellent, low cost waters to put a bend in your rod, but with miles of water to search they tend to reward mobile, active anglers. In fact, with most traditional canals being so narrow, they are perfect for light jigging and dropshot. A lot of fish come from right by the bank, or around features, where vertical jigging works well. You can expect perch, and often surprises like zander, pike and even chub on many canals. We'll discuss specific tactics later in the book.

Fish are not always predictable or glued to cover, but special attention should be given to any features, such as lock pounds, junctions, trees cover, bends, bridges and wider sections. Where permitted, marinas and the sides of boats can be excellent too. Of course, other stretches of canal can appear more featureless and you might need to use trial and error to locate fish.

Below: A classic British narrowboat canal. We wouldn't strictly recommend antique cane for drop shot tactics, but Garrett Fallon combined old and new kit to good effect on this occasion, tempting several perch and even a two-pound chub!

If in doubt, bridges are excellent places to find perch on many canals. Indeed, if you had to bet your house on catching a fish, you could do worse than start right under a bridge with a tiny 1-2" soft plastic lure!

One of the most important features of all will be hidden to the naked eye too; namely the 'shelf' or 'drop off' at each side of the canal, where shallow water slopes into the deeper main 'track'. Don't always expect these features and typical depths to be even, however. Where the canal narrows, or turns a bend, or approaches locks, for example, the deepest water won't always be where you expect it.

Fish like perch and zander hug these areas, using them to patrol and ambush smaller fish. In fact, one favourite trick is to suss out where the water drops away to the central depths and then gently explore this with drop shot tactics, so that the lure works right under the rod tip at the bottom of the shelf. You can even do a spot of 'trolling', slowly walking along with your lure rod to cover many yards! This also works right by the bank if the water is coloured, presumably because the fish then feel safe even in inches of water.

Just like on any venue, predators will switch on and off the feed depending on the time and circumstances. Besides the classic early and late in the day feeding windows, another productive time on canals can be just after a lock has drained or a boat has passed through. This could be for several reasons. Perhaps the smaller fish are dispersed and disturbed, making them easier to pick off? Or perhaps food is displaced by the commotion which, along with clouding of the water, suddenly brings both prey and predators onto the feed? On busier canals, such as the famous Grand Union, it is definitely a period to exploit.

Besides our network of canals, drains should also be mentioned on the subject of man-made waters of this type. From the Somerset Levels to the Fens of the East Midlands, there are many miles of often little-fished water available to those with a healthy sense of wanderlust. There are some fine pike, zander and perch caught in these waters, too.

Apart from obvious features such as pumping stations and bridges, however, there are often few distinctive features, so a cast and move approach tends to be the best policy. Indeed, fish are often tightly bunched in particular areas and you will often find a contrast between hundreds of yards of rather barren water, interspersed

Similar to canals in some ways, drains and fens are also great to search with lures. Above: John Deprieelle jigs a tiny Somerset channel, where he found perch in just 16" (40cm) of water!

with large shoals of prey – and the predators that feed on them. Not armchair fishing, but the rewards are there for those who are prepared to put the miles in.

Rivers

Travelling from man-made to natural freshwater settings, a meandering river presents another very different environment to explore. From tiny upland trout streams, to large tidal rivers, all have their share of species to catch. The fish of these venues tend to be fit and strong, as they live their lives contending with currents. The characteristic spots and features that hold fish can also be fairly obvious, whether you find places of obvious sanctuary for predators and prey, or any change or break in the current. We'll cover tactics for individual species in the relevant chapters, but suffice to say that running water is a very different challenge. Fish have more definite lairs and hotspots on most rivers though, if you can get to them.

Slower rivers and tidal venues

The widest range of fish of all occur in the slower, middle to lower reaches of rivers, where steady pace and varied habitat bring a long list of possible species. Chub, perch, zander and pike are usual suspects to encounter on lures, but near to the sea you might well find mullet, flounder and bass, too.

Quite a lot of the time, fishing will be tricky to access and mud, not to mention the surroundings can be off-putting. Murky water tends to deter a lot of anglers, but should not be a reason to be completely put off. The same is true for less tidal, but equally large and imposing-looking waters.

Start by exploring any obvious features and working close to the bank; if there is reasonable depth close in, even the most barren or industrial-looking river can have a healthy head of fish. In fact, those venues that look really grim are often quite underfished. Like a delicious dish that looks messy at first glance, a lot of folks are simply not prepared to take the first bite – so be brave!

Mid-sized and typical coarse rivers

Moving upstream to steadier, more appetising waters, most rivers will have a mix of species present as we reach the main watercourse. Most will have a range of obvious and less obvious features to look for. Find deeper sections such as pools and bends, or interruptions to the flow such as fallen trees and overhanging cover, and you are likely to find a mixture of species present.

Perhaps the absolute classic feature has to be the weir pool, however. In these spots you will find just about all the resident species hanging around, occupying different sections of water. Perch, for example, like to sit in deeper holes and slacks off the main flow. Curiously, there is often an unseen slack right underneath the weir itself, beneath the turbulence we can see on the surface. Pike tend to sit further back in the pool mopping up fish that have been washed downstream; but will also hang in any suitable slack, such as a flood gate or concrete pillar, just off the flow. Steadier, well-oxygenated flows will hold fish such as chub and trout.

Of course, the real art to 'reading' a river is also to suss out the less obvious signs and features. One absolute classic is the 'crease' where faster and slower water meet; another is any 'pocket' of slack or slower water created by a boulder or obstruction. Even in the midst of fast or turbulent water, such pockets can provide attackers with a suitable vantage point. After all, as fit as river predators are, their goal is to waste as little energy as possible, so any sheltered spot that lets them sit just off the flow and able to spring out and hit any passing food is perfect. This is true of just about all predator fish.

Anglers, however, and especially newcomers or those more accustomed to stillwaters, tend to avoid flows and turbulence because it's harder to control their lures. Be brave and don't be put off and you will often be pleasantly surprised!

Smaller streams and upland rivers

For intimate, visually exciting fishing, smaller rivers are perhaps the most suitable places of all to try the lightest styles of lure fishing. Even some of the very tiniest, with just a couple of feet of depth, can be productive for species like trout and chub. Not only can you get away with short, easier casts on such waters, most have quite obvious features and you can actually spot many of the fish, removing the guesswork.

A classic, tumbling moorland river. The closer to source, the clearer and faster a river gets.

Some tiny rivers are narrow and fast flowing from source to sea; in larger watercourses you will find such waters upstream, the closer to the source you get. As we enter these clean, stony, quick waters, species like trout become more prominent, while fish like pike and perch give way to chub and grayling. Shorter rods and waders come especially highly recommended for such smaller, bushier surroundings, where access might be tricky. Depending on the availability of food and how acidic the water is, you may only find smaller fish. It's all relative, however, and even modest

5 Bends

Twists and turns on any river break up the current and create varied habitat. The current will often scour out the bottom and make deeper areas, for example, or create eddies and slacks, much loved by fish such as pike and chub.

Confluence 4

Areas where watercourses meet or there is an inflow are always likely spots to find fish.

Boulders and "Pockets" 3

These features form natural shelter and ambush points for predators. Even in the midst of rushing, white water,an obstacle might form a "pocket" of still water for a waiting predator, such as a big chub or trout. Migratory salmon and sea trout also use these places to rest on their journey upstream.

7 Holes and Undercut banks

Features on rivers are not always obvious, which is where counting lures down and testing depths comes in. Typical holes can be deceptively deep and might continue for several feet underneath the bank itself. Predators love these at most times, but especially in flood conditions. Ideal to search with dropshot and vertical jig presentation.

2 Run

Steady, oxygen-rich passages of water are perfect habitat for species such as trout, grayling and dace. Fish will usually sit facing into the current. Ideal for a small flashing metal or crankbait.

1 Riffle

Shallow, stony areas are larders of prey such as insect life, minnows and bullheads. These can look rather bare, but sometimes larger fish like trout will come and raid shallow areas during low light and darkness, especially if there is cover and deeper water nearby.

Rivercraft and typical features

6 Overhanging cover

Trees and bushes, whether living or fallen, create ideal cover for prey and predators alike. These are spots to reward accurate casts perhaps take a risk or two- weedless and snag resistant lures are an advantage for obvious reasons!

Flood arms & slacks 

Whether man-made or caused by nature, large slack areas are hotspots for predators such as pike and zander. This is especially true when the river is in flood and prey fish stack up out of the main flow.

8 Manmade structures

Bridges, walls, boats and pontoons can be magnets for fish of all sizes. If the river has avian predators such as cormorants, small fish are liable to be especially ght to cover. Typical man-made spots tend to be obvious and easily accessible, however, so clued-up fish might require a different approach, whether it's an unusual presentation or unsociable hours.

10 Weirpool

Weirs are among the richest areas of all on a river. Rather deceptively, you will find deep, scoured out water right under the rushing "sill" of the weir- don't be surprised to find large fish sheltering under white water. Otherwise, look for eddies, slacks and any obvious features. Pike and perch will hug the edges and natural vantage points, given any depth and cover,while the deeper, steadier flowing water will appeal to the likes of chub and zander.

predators are a delight on the most delicate tackle. If you are fishing a tiny river, a one-pound trout might be regarded as every bit as special as a fish several times that size on a huge reservoir.

Another highly noticeable feature of smaller rivers is that the fish tend to require more stealth. If the water is clear and shallow, the fish will be less forgiving of clumsy casts or anglers who stand bolt upright against the sky. Fortune here definitely favours the sneaky! With most river fish tending to position themselves with their noses facing the current, you would also be well-advised to approach the fish from below, casting 'upstream' to them, from behind: more notes on this in our chapter on trout (*see p153*).

Lakes and reservoirs

Often harder to 'read' than rivers, lakes are an altogether different beast to tackle for the lure angler. The most manageable of them are smaller lakes with obvious features to go at. Where coarse and fly fishing clubs and fisheries allow lures, these can offer great sport (our diagram on *p159* shows a typical day ticket trout lake).

Larger lakes can be more daunting, but yield some of the biggest predators of all. Perhaps the biggest challenge is finding the fish, so any clues should be used to the full. Maps and depth charts are one way, as is a simple walk around the circumference – you may well get an indication of where anglers tend to catch fish and spots popular with coarse anglers will tend to have shoal fish drawn in with their bait.

Also watch the lie of the land; generally the bottom will follow a similar shape, so a flat, gradually sloping shore will tend to mean a gently sloping bottom, while

The ever-changing river: from winter floods to summer flows

As we move into the colder months and the rains arrive, river levels can alter substantially. Underwater features can change significantly after major flooding, too, taking the fish to places you wouldn't usually find them (some favourite slacks might have been dry land a few weeks earlier!)

Many anglers are put off by high, muddy looking depths, but in actual fact these conditions can serve to concentrate the fish; if you can only find them! Fish like pike and perch can often be found moving out of the main flow into the less turbulent backwaters during flood conditions. Locations such as flood arms and backwaters are still worth trying as prey and predators may well be found in high concentrations. Although chocolate-coloured water is tough, fish will acclimatise too and can still pick out lures, although you might like to try bigger and brighter offerings with a strong kick.

It is so often a matter of confidence, but flood conditions sometimes lead to the biggest fish dropping their guard. Is this because they cannot see the angler or lines as easily as usual, or they are simply ravenous and ready to lash out after days of difficult hunting? Who knows?

Otherwise, if you are still struggling to catch in dirty water, another excellent time is post-flood as the river levels drop and visibility improves. For a hungry predator, this must be rather like somebody switching the light back on in the larder!

Naturally, the rulebook changes drastically again in the warmer months of the year, when lower levels are normal and weed growth is much higher. Indeed, the problem might not be that the fish struggle to see the lure, but that they find it only too easy to see the angler! Equally, in hot, sluggish water, levels of dissolved oxygen drop, making fish seek

It might seem like pure madness, but even a river that has burst its banks can be worth a cast. Not all become a muddy quagmire.

faster flows. This is even true of typically lazy fish like pike and perch in very hot weather (although we would advise you give the pike a break in summer because they are very fragile and easily harmed).

Another common challenge of the summer and early autumn months is excessive weed growth. With lots of light, the stuff can grow like wildfire; and while your instincts might be to avoid the 'salad', predators love it. Not only does weed offer safety, it also provides a huge amount of food, from clusters of fry to various invertebrates. Pike and chub can often sit right in the thick of the weed and sometimes a topwater lure is the only viable method for fishing for them. It might seem like lunacy at first – until a fish bashes through the weed to grab what it thinks is a free meal! Meanwhile, perch can be moody in warmer conditions, if not utterly lethargic in the middle of a warm day. It's often only in the half-light of dawn or dusk that they will leave the cool of the shade and the deeper holes to go and wallop prey in the shallows. Don't be surprised to find them clobbering prey in mere inches of water during these times!

Getting stuck in!

TIP

Nobody likes losing lures, but there is a lot to be said for taking a risk or two when lure fishing. Indeed, if you only fish 'safe' open spots, you simply won't reach a lot of the predators on any water. After all, they will always make the most of cover and tend to love the spots we hate! Get stuck into those tighter corners and you will be fishing water that the majority of anglers leave well alone. About half of the less accessible water in any setting is probably only fished by 10% of anglers, so avoid the temptation to only try the easy bits! Should there be a bigger risk of getting snagged, however, it might be imperative to tackle up heavier because you might need to wrestle a good fish clear or free your lure from a branch. In such areas, fish tend to feel safer and be less line shy in any case.

a steep craggy bankside tends to mean a sharper drop.

Perhaps the trickiest part of lake fishing, though, is that so many of the key features are invisible. True, you might find obvious weed beds or shoals of prey fish topping, or indeed signs of predation early or late in the day, when hunters will move closer to shore. Man-made jetties and boatyards are another obvious place to try, especially during times like autumn when these teem with fry. But the depth changes

Bigger lakes are perhaps the most challenging of all freshwater venues, but they hold some of the largest predators of all species.

predators love are less obvious. Casting and counting down jigs is one way to get an idea; but for large lakes a fish finder and access to a boat could be key.

Another noticeable characteristic with larger lakes is just how much fish will move around to feed. Unlike a river, for example, where a predator might have a particular 'lair', the fish of a big lake might have to travel many metres to find enough food. Temperatures and cover will also play a part between the seasons, of course, as fish gravitate to certain areas and depths where they feel safe and comfortable. And while there are not the same currents or tides as a river or sea mark, a lake is still dynamic and changeable. The wind is a classic example of this, and one of the best pieces of general advice is to put your comfort second to the preferences of the fish and follow all but the most bitter winds, which will carry warm water and food to one side or corner of the lake.

Perhaps more than any other type of freshwater, though, a larger lake can represent a long learning curve and process of trial and error. Rather like a section of sea shore, it can be best to break that daunting, larger picture into smaller chunks and treat each almost as a separate venue.

Further useful information, techniques and tips on finding the fish can be found in our chapters on perch and zander fishing.

Wherever you fish, being highly mobile and having a sharp curiosity will set you in good stead. The more you look, the more you'll find.

NIGHT FISHING

While it's not strictly necessary to fish at god-forsaken hours, some of the best lure fishing of all is during darkness. Humans prefer to be warm and dry, fish have very different preferences. Indeed, some of the rarest and most highly-prized species feed best in the night. So what's stopping you?

Fishing at night is certainly a different ballgame to a sunny afternoon. Casting into darkness adds an extra mystery and heightens the senses, as you become more dependent on touch and instinct. Of course, the risks can be greater, too, so you have to be well prepared and sure of each step. A quality head torch is a must, as are knowing your spot and the tide.

Why should night fishing be so productive? The true magic happens on the river or sea bed: under the cover of darkness, all kinds of invertebrates and smaller creatures leave their hiding places. Predatory fish feel safer from humans too. They get bolder, as it becomes easier to creep up on their prey. Unlike us, it's in the black that they wake up and start to feed, particularly the truly specialised predators.

Like undesirables in a small town, the most weird and unsavoury characters of all will roam the night, singling out their next victims and generally getting up to no good. Unless you fish during their favoured hours, however, you might never know it was all kicking off; and you certainly won't catch predators by sitting

Andy begins the night shift in Weymouth, Dorset. The hour many anglers have already packed up is often the very time they should be tackling up, as he has learned over many late sessions.

Left: Predators of all sizes go on the offensive at night, from the largest to the smallest, in fresh and saltwater settings. You'll never know the potential of your favourite waters until you give it a try!

on the sofa. Don't just take it from us though, it's time you tried night fishing with lures.

Into the black

Although any spot can be tried at night, the street lighting and safer access of urban locations offer some of the very best late fishing. Harbours, piers and marinas on both fresh and saltwater make ideal starting points. Sometimes the lights actually attract fish and other living things, like lamps drawing insects. They certainly draw prawns, fry and other curious creatures such as squid, that will also bring in predators.

On any water, you'll find that some species come out to play and others go missing as the lights go out. Different fish work different 'shifts' when it comes to their feeding and movements. During the daytime certain residents are active and work their patch; as night falls the shift changes and other fish begin to work the same ground. Seldom does anyone catch wrasse at night, for example, but the reverse can be true for other species – and your chances of codling, whiting, scad, larger bass and various flatfish quickly increase at night. In fact, you may well catch species you weren't aware of just by changing your clock!

In freshwater too, there are great possibilities for night fishing. Zander are notably active at night, but perch will feed into darkness too. Even cautious freshwater predators grow bolder in darkness and moonlit night can be the best time of all to catch a predatory brown trout, or that big, wily chub or pike.

Fishing by moonlight: Are fish still wary? Will torchlight put them off?

While predators seem to get bolder in the later hours, this doesn't mean we can approach clumsily. On quiet evenings they can be especially sensitive to disturbance. In fact, they can be hypersensitive to changes in light and will often flee at the first sign of danger. However, it's also true that different species have very different tolerance to disturbance. Artificial light is a classic example. Bass, for example, can be notoriously spooky around moving lights, but it is quite possible to catch flounder and even watch them take by the light of a head torch at times. Whatever the situation, however, it always pays to keep as low a profile as possible – so move and fish as quietly as you can and use artificial light sparingly. If you can learn to trust your senses rather than a torch, you'll find your eyes quite quickly adjust.

Night Tactics from the Shore

There are many different species to target and methods to try at night. Predators of all sizes will be active, including the biggest. Hence, it's good to start optimistically with large lures at first to see if there's a beast or two about. If there are, you're likely to get a hit sooner rather than later. As the session proceeds, you can then explore different methods and scale down if necessary. Typically we might scale right down to a tiny 16 hook and split shot just to see who's home, before working some larger lures dropshot style, or a straight cast and retrieve for more mobile predators.

Harbours and walls are also full of nooks and crannies that are home to a hundred and one different critters. Even on more open venues, fish will come ever closer to the shore under cover of darkness. Some of these, such as rockling, topknot, codling and others, prefer feeding at night. Lastly, it's always good policy to keep company when you're night fishing, and with two of you, you can always try different lures, methods and depths, comparing notes as you go.

Things that go 'thump' in the night!

The fishing rules change after dark, and if you night fish on a regular basis you're sure to be surprised sooner or later. There's no telling what you might find next, whether it's a giant bass or a migrating sea trout. Perhaps the best night capture of all has to be the angler from Fowey, Cornwall who caught a conger eel on a lure! Several locals knew where the fish lived; indeed, it had snapped off several hooks and lures before it was successfully landed. By day it would never be seen; but under cover of darkness, a menacing head would emerge from the wall, ready to seize anything that got too close.

Gilthead bream caught by torchlight.

Tackle and preparation

If most anglers are messy creatures by day, they tend to lose even more bits and pieces at night. Hence preparation is everything if you intend to night fish effectively; not to mention safely. Once you're out by the water, it's no use realising that you forgot your forceps or spare trace materials.

It definitely pays to keep things simple and orderly. Have a permanent home for things like your forceps, phone and camera so they are always quickly and easily located. Cord attachments and lanyards are a smart move, because you can't easily lose things that are attached to your person!

As for tackle, there's a definite case for taking just one rod, unless you need to double up with a tougher outfit for the chance of bass, for example. As for the finer rods, a little more investment makes sense where superior feel is required – although some night owls also like a luminous, contrasting rod tip to help detect bites.

Brightly coloured braids are also easier to untangle in poor light; and if you intend to night fish regularly it's worth learning to tie your basic knots by feel alone. In terms of other preparation, it might also be an idea to take a hot flask and tell your next of kin where you're going.

Contrast zones

TIP

One consistently useful spot to try at night is to work any 'line' where shadow meets light. Whether there are street lights, or just the moon, you will find these areas capital. Most predators use shadows as cover and can be coaxed out to take a well presented lure. Scad particularly like to work these zones and can often be found in huge numbers patrolling the shadows cast by harbour walls and piers.

Lures for night fishing

A lot of night fishing is about confidence in what you're doing. It might seem crazy that any fish can locate a tiny lure in that dark water, but you can rest assured that predators are acutely well-tuned to their environment, even on the dullest night.

Do fish hunt by sight alone after dark, or does vibration also play a part? You can clearly see on the huge, exaggerated eyes of fish such as a pollack or zander that some species possess exceptional vision in limited light. But predators can also sense the vibrations of prey, whether through their lateral line, through tiny pores, or even through elongated feelers or specially adapted fins.

As to what fish actually see, they probably detect a silhouette rather than identifying the usual colours. On a clear night especially, even small prey items will show up against the sky. You could even argue that lure colours are unimportant at night, or even that dark colours or black are most effective. Black lures would certainly create the biggest silhouette, and it's interesting that sea trout anglers tend to find black lures and flies the most effective of all, for example.

Of course, there are also crazily bright and even luminous glowing lures available! Do they work? Although they look tempting to the angler, we've found they tend to be strangely disappointing. Most predators seem well capable of finding much subtler lures in the dark and could even be wary of luminous offerings? For bass, even the less obtrusive baitfish or stickbait type lures with very little rattle or action can score well at night, for example. From this you'd say that few predators demand a neon signpost to find their dinner!

The choice is yours, of course, and other interesting additions to your rigs include tiny spinner blades and glowing jig heads. Just bear in mind that less tends to be more, because the same lures you like to use during the day will also work at night, provided you fish them with the same confidence.

Weird but rewarding: Squid on lures

We're not even sure it counts 100% as fishing, but squid can be great fun to target at night, from late spring onwards. The best lures for them are specially designed not with one or two hooks, but crowns of many little 'teeth', designed to snare an attacking squid. These tend to be luminous and can be easily charged up with a torch or mobile phone. In fact, the squid's love of artificial light can also be a clue to the best locations to find them. Piers and promenades with bright, gaudy lights certainly tend to be productive.

Try casting out and counting down to different depths, before bringing your special squid lure in slowly and steadily. You'll feel the most strange, pulsating takes from squid, often with the lure just stopping or seeming to gain weight. As soon as you feel the attacker wriggle, apply steady pressure to draw it in! Most light lure rods will do the job; fish a light drag and play your quarry slowly so as not to pull the jig from its grasp. If you are lucky you may also encounter cuttlefish during certain times of year, using the same approach.

Using street lighting

Baitfish will often congregate in areas where artificial light permeates the water, whether this comes from street lighting or another source. It is common to find tiny fish sat out of the flow, 'hanging' in these pools of light, particularly on a flooding tide. In this kind of scenario, small pin tail style soft plastics and weightless stick baits are perfect to mimic prey hovering in position.
Bass, pollack and flounder are just three of the species you'll find working these fish and using the illumination to home in on their prey; the same can be true of freshwater predators on urban rivers.

Seasons in night fishing

Part of the draw of night fishing is that you can fit in short sessions after work at all times of year. Even on frosty December evenings, there is sport to be had from the final dusky hour of light, right into the wee small hours, if you're brave enough! In fact, winter brings some of the most interesting sport as various species take refuge around features such as jetties and boatyards, sheltering from the elements and finding food. For the bold and prepared angler, the cod family of species, including whiting and pout, all provide bites in the cold. Jigging with metals, soft pin tails and worms can be an excellent way to keep the hands warm, and these fish fight gamely on the coldest nights.

Fishing right through the year will also tell you which species are residents and which are migratory. The smaller types, like blennies and rockfish may well be present year round, unlike seasonal visitors that come and go.

Freshwater fishing in the dark

In freshwater too, there are great possibilities for night fishing. Zander are notably active at night, but perch will feed into darkness too and a moonlit evening can be the best time of all to catch a big, predatory trout and chub. The latter can be especially aggressive to surface crawler type lures. It's a simple fact that these fish are bolder and feel safer at night.

Below: This zander was caught well off the bottom, beside a jetty. It fell to a lure that was twitched and given long pauses of 10-15 seconds, hanging only a couple of feet beneath the surface!

Perch and zander will feed at night but don't expect them to be in their normal spots. They will both rise up in the water and go in search of their prey, and both will hunt closer to the shore, too. So, rather than jigging the depths as daytime fishing dictates, it often pays to fish lures higher in the water. If you are fishing soft plastics, try as light a jig head as you can manage and fish slowly on the drop. Alternatively, shallow crankbaits are also good for twitching.

It might seem counter-intuitive in darkness, but lures with silver foil or glitter in them seem to work a treat. One of Andy's favourite lure colours at night is white with silver glitter, in either shad or worm profile. Realistic lures with flash foil in them, such as those made by Fish Arrow, can be especially effective for zander and other hunters.

If you can locate baitfish during the day going into dusk you can be certain there will be a predator of some sort in the location waiting for the cover of darkness to mount their attack, and we've had some great sessions fishing for zander and perch at night by employing these methods.

Don't assume you always need to do something different or special in the dark, however, because the methods that you know work during the day, will often work just as well or even better. The Ned Rig and other bottom-searching techniques are very good for both zander and perch at night.

Night fishing for pike

Although our main focus in this book is lighter lure fishing, some of our most memorable pike sessions have been at night too. During autumn, as the water temperature starts to drop, surface crawler lures can be effective and brilliant fun to try. You'll need to be well organised and it is always sensible to get used to the water by daylight and to fish in company. Having your gear organised is key too and night fishing isn't for those with little experience of unhooking pike.

For the brave, however, it is a truly visceral fishing experience, and besides pike you'll find night-time productive for big trout and other species too (*see our chub chapter, p133*). Suffice to say, if you ever you feel your predator fishing is becoming stale or if you have never tried freshwater lure fishing at night, it really gets the blood pumping!

One of Andy's most memorable sessions of all was early November with his friend John Yates, following a hard frost; perhaps not the conditions you'd associate with surface fishing. The low temperatures, combined with John's Movember moustache suggested it might be an amusing session for all the wrong reasons, but they nevertheless decided on an after-work pike session.

Bites came quickly, with soft swimbaits and jerkbaits banking several pike from a fairly shallow pool. The tone was set early as John broke his PB not once but twice and the adrenaline was pumping – and yet there was another twist to come as the pair reached a bridge where the river narrowed and the water gathered pace.

Switching to a weighted soft plastic with a thumping paddle tail that would get down effectively in the faster water, there was a huge, heart-stopping swirl, but nothing there. Next cast, a four-pound jack nailed the lure, but Andy was convinced that this couldn't be the fish responsible for the previous commotion, and kept

casting. Moments later, after just a couple of turns of the reel, the rod slammed round and all hell broke loose. After a short but brutally hard fight, the fish was unhooked and rested in the net for a quick picture before release; the scales had gone missing but it was a formidable fish, not to mention an emphatic lesson in the value of night fishing and an unforgettable experience.

FLOUNDER & FLATFISH

Tackle: A typical drop shot or LRF rod will usually suffice, accompanied by light braid and a fluorocarbon trace. Flounder and other small flatties are not super athletic, but will give a spirited battle on light gear. There is little need to go below around 4lb (2kg) strength traces.

Other considerations: A decent head torch is a must for night fishing. Pack a long handled net for anywhere fish cannot be landed by hand.

Season: Varies by species, but flounder are present most of the year except for the late winter/early spring when fish move out to sea to spawn (varies from Feb to April).

Not so very long ago, talk of catching flatfish on lures might have sounded like utter madness to most anglers. After all, they don't look like they are designed for the chase. How times change though, because these fish are now widely and successfully targeted on artificial baits. Flat customers such as dabs, flounders and other such species might have a rather innocuous appearance about them, but are highly predatory. All it often takes to catch them is an open mind and a slightly different approach. Their way of life might be different to a bass or perch, but

Above: A harbour flounder. This one took a soft lure, slowly bounced across the bottom.

once we suss out their habits and how to target them, they can be a lot of fun on light tackle.

In the course of this chapter, we'll be outlining some startling discoveries and surprising sport to be had with the sea's flatter residents. Much of our experience has been with the humble flounder; but similar tactics will work for a whole host of its near and distant relatives. In fact pretty much all the flatfish are surprisingly active hunters, wherever you find them around the world.

The excellent news for most of us in Britain and coastal Europe today is that the flounder, our main target from the shore, remains common, perhaps in part due to their relatively low commercial value. They live in shallow, sandy or muddy locations and are found in their greatest numbers where freshwater and saltwater meet. Capable of handling fresh as well as salt water, they will travel long distances up into rivers to feed on small worms, fish and crustaceans.

Typical flounder territory: a sandy harbour, packed with ambush points.

These fish are a catch to remind many of us of childhood fishing with the classic worm and sinker. Specialist lure fishing for them might be quite a recent development, but perhaps we should not be so surprised that modern lures work a treat. After all, if we think of traditional fishing methods, the flounder spoon was a classic, with previous generations using flashy spoon in combination with trailing worm to bring out the inquisitive, aggressive side of these well-camouflaged killers.

Over the past few seasons, we've been fascinated by the antics of these fish at close quarters, with or without a rod. Living close to a harbour and having varied, sheltered water to explore at a moment's notice has afforded Andy, in particular, unlimited chances to observe them in their element over many tides and seasons. Time and again, their habits have proved fascinating and unpredictable to say the least.

A Good Hiding

With impeccable camouflage, just spotting flounder is an interesting challenge in itself. After you have sighted a few in the water you will quickly start to identify the shape of the fish, or even the subtle depression in the sand where one has recently been sitting.

These flatties are sneaky, surprisingly active predators. During the day you will often find them in shallow water preoccupied with small crustaceans and rockfish. They tend to blend in with the bottom and, when totally dormant, you may only see a hint of a fin or a pair of eyes, although on a muddy bottom they tend to show up a little better.

A good starting point for location would be to find clean flat sand or mud on open marks. In harbours, work the walls during high water or under boats during the day, and try fishing as the tide drops. During less active times flounder settle and will burrow into sand to disguise themselves. At these times, they'll sit there like a spider on a web, sensing vibrations that mean potential food or threats.

However, they can also be active and aggressive predators. In estuary marks such as Padstow Harbour and the Teign Estuary in Devon, for example, we've seen flounder not only using ambush tactics, but chasing baitfish with the flood tide. They are certainly opportunists when any food source is abundant or easily cornered – and the size and type of prey they will attack can be startling. After the Cornish Lure Festival in 2015, for example, Andy observed two flounder chasing down a shoal of 6-inch mullet! However, their more

Left: A cracking pound-plus flounder for Andy, taken using a drop-shotted soft plastic.

normal diet covers a whole smorgasbord of smaller creatures, depending on location. Crabs, sandeels, rockfish, prawns, blennies and marine worms are all common food items and it follows that quite a range of lures will work. In fact, we've only found them to be super fussy when they are keying in on one particularly abundant, specific type of prey.

Will Pender with a fine Cornish flattie, tempted on a soft plastic worm.

Below: Spot me if you can: can you locate the flatfish in this patch of sand?

Flat and Dangerous

For their size, flounder and other flatties are very discreet fish; much of their way of life is about avoiding being seen by enemies and victims. They'll edge up on their prey, eye it up and wait patiently to strike. In the right conditions, we've watched large fish of over two pounds creep up on gobies, creeping millimetre by millimetre up to them before inhaling their prey in an instant. That innocuous-looking mouth can quickly billow out to a much more formidable size!

The productive lure angler must often be as patient and stealthy as the flounder itself to be successful. Timing can be key and during the day you won't always see them unless they are fixated on their prey.

Knowing that they actively seek out small rock fish and crustaceans can help you identify where flatfish are going to lie and which way they will generally be facing. They can be found out in open sand or mud, but especially love features, slacks and any little irregularity on the bottom that might offer an advantage, even if the water is only a foot deep. Do also keep an eye on any prey besides looking for actual flatties, because if there are crabs or groups of baitfish moving, there will often be flounder in attendance.

One thing that once puzzled us was why flatfish would often sit facing a harbour wall. After all, the traditional textbooks often suggest this is a fish that likes open sand, not structures. But the more you watch them the more it makes sense; so many little prey fish and other creatures use the walls, rocks and debris as cover so why wouldn't the flounder push up against these to try and trap dinner? The same is true of all manner of other nooks, crannies, slipways and structures. Indeed, once the flounder has its prey in a corner or tight spot, there is little chance of escape!

Night Fishing For Flounder

Contrary to what you might expect, one of the best times for sight fishing for flounder is at night. Fishing neap tides with a powerful headlamp, you can find many fish sat virtually with their noses pressed against harbour walls and other features, or creeping about. Under cover of the night, flounder are rather bolder too, since the low light makes it easier for them to move about and launch surprise attacks without being spotted first.

A good way to identify these fish is to use a decent headlamp to illuminate the shadows thrown by harbour walls. Unlike fish such as bass or sea trout that you also find at night, flatfish seem more tolerant of a torch beam too; that said, an adjustable model is handy so you can tone things down a little once a flounder is spotted.

One really eye-opening phenomenon Andy has seen many times during spring tides is flounder hunting in midwater, or even just below the surface, as they chase shoals of baitfish! At first you may wonder if you are seeing things, but they will capitalise on any abundant supply of juvenile fish. Nor does size appear to matter so much, so long as the prey can be swallowed.

On various occasions Andy has witnessed everything from palm sized fish hunting down tiny fry to the real beasts gorging themselves on 3-inch bass. Indeed, during any seasonal period of abundance, these flatties can quickly become fatties.

When should I strike for flatfish?

Bites from flounder and similar species tend to be quite distinct: a heavy thud, followed by another will usually mean the fish has found the hook. It's easy to strike too early, so do delay a second longer, especially if you are missing bites. The larger the lure, the less of a hurry you want to be in to set the hook. With 6-inch worms, for example, you can usually expect three solid knocks as the fish chews down the lure, by which time you can strike into the fish.

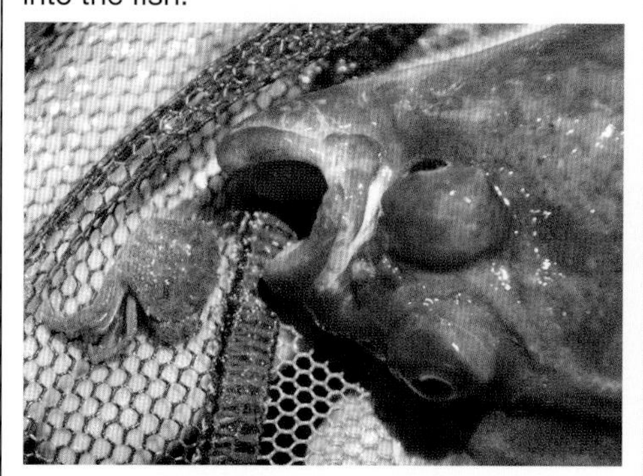

No prizes for guessing what this fish was eating!

Andy's Lures and Tactics for Flounder

I have three main methods for catching flounder and other small flatfish. The first pair will work at most times, while the third tends to be reserved for spring tides. It's fair to say that many lures work, including a large variety of fish, worm, crab and shrimp imitations. Finding out what the fish are looking out for is more important than the latest on-trend lure. Sometimes you will get obvious clues, whether it is the sight of flounder chasing massed fry or even a landed fish coughing up a chewed crab.

1. Drop Shot Rigs for Flounder

My starting setup is often a drop shot rig. This might sound odd given the flounder's love of lying flat on the deck, but it works very well. Typically I start with a size 12 Aberdeen hook set around 10in (25cm) off the bottom in clear water, or 6in (15cm) in muddy water. My lure tends to be a full length soft plastic or artificial flavoured worm (such as an Isome or Gulp!) in a medium size. That said, it can also pay to use a silver or baitfish profile lure in clear water.

One great benefit of the drop shot rig is that it avoids crabs. However, I believe it also renders your lure more easy to spot for the flounder, as they are looking up at the lure. You might be surprised how readily a flattie will often come up off the deck to grab a meal. If you can get the fish to rise off the deck, 99% of the time it will commit itself to bite. Getting the fish to angle themselves up and take a lure off the bottom seems especially vital if you want to consistently catch the bigger ones, and convert interested fish into takers.

As with freshwater fishing, drop shotting is best suited to very close quarters, but you can also make short casts if need be. Whether you are working a harbour wall, or perhaps under or alongside boats, do be sure to cast and retrieve quite slowly. One turn of the handle every 5 seconds is about right, with a pause here and there. In general, try not to go imparting too much action into the lure.

2. Bottom Jigging

My other main method is to use a medium to large worm, shrimp or crab type lure on a light jig head. When it's not too breezy (say under 10mph wind) I try to use a maximum weight of 2g, if possible, and will only tend to increase the weight further if conditions are tricky or I'm tackling more exposed marks on estuaries or the open coast. The thought behind a lighter jig head is to give the lure a greater 'hang time', rather than quickly burying into the bottom. This helps increase the chance of a flounder seeing your artificial and reacting.

That said, there are days when simply dragging a lure across the bottom will also throw up bites. The hitch with this is that crabs can be a menace if you really creep the lure in. At least their presence in numbers tends to suggest there'll be flounders about! One of the most memorable sightings in recent seasons was a huge flattie chasing and devouring quite a large crab that had grabbed a friend's lure, right under our feet!

To Catch a Thief... TIP

As you experiment with LRF fishing, you will undoubtedly locate and catch small crabs, gobies and other creatures that will try to steal a soft lure. These can be a help as well as a nuisance, however. Like all predators, flatfish are naturally inquisitive, and if something else is feeding or distressed, they will quickly take interest. In fact they'll often bully other, smaller fish out of the way if there is something edible to be stolen! The same is often true of your lure. Hence, if you see or feel a goby or crab pulling at your lure, don't be in too much of a hurry to drag it in. Instead, be patient and stay alert. Flounder will often move in to barge these smaller rivals out of the way, hammering your lure!

3. A Cranky Idea?

The final method I sometimes favour perhaps requires a greater leap of faith; but the use of crankbaits and even metals can also be a productive and exciting method on a spring tide. In the era of soft plastic artificials, it's often an overlooked tactic – although a lot of shore anglers will have a tale of the flounder that took their bass plug.

As previously mentioned, on larger tides flounder will move well up in the water column to chase down baitfish as it gets dark. In these conditions, when you have a lot more depth than usual to deal with, it can be difficult to control the usual jigs, especially when fishing off the bottom. At times like this, a small minnow-style hard lure in the 2-3in (5-8cm) bracket is ideal.

Work the lure slowly with subtle taps of the rod tip; sometimes just holding it against the current will get that killer wiggle. Besides the flounder, there is also the added benefit that you may well encounter bass, pollack and other species, as any extra large tide will sweep preyfish around, making them vulnerable to attack.

On one final note, I would strongly recommend that you use a rod with a firm tubular tip when using crankbaits. This is due to the power distribution through the hook points; go too soft and you may not impart enough pressure to set the hooks properly.

Plaice and other flatfish

The tactics that work so well for flounder can also be used successfully for plaice and other flatfish by the versatile lure angler. To some extent, all these species share similar habits. Plaice are not always accessible within a short cast from the shore, but there are ways to target them with LRF tackle.

Above: Wherever you travel in the world, you'll find flatfish lurking around piers, beaches and harbours. Dom had this California halibut on a soft plastic worm and 5lb leader from one of San Francisco's many piers; it pulled like a train!

From piers and breakwaters, any sandy patches can be investigated. From a kayak, you can use the lightest tackle to drift a lure over sand banks and features, although from the shore or wading an estuary you might need a heavier weight. Like flounder, they will chase a jigged or dropshotted worm lure, and you can also experiment with coloured shot, beads and tiny spinning blades.

It's all out there to discover in fact, and if you follow the golden rules of finding the fish and matching your lures to available prey, you have every chance of success. Besides, it's always satisfying to surprise the locals too, when they ask 'what did you catch it on'?

The Topknot: A different breed of flattie!

The common topknot is a highly coveted species amongst LRF anglers. Found all around the coast of the UK, this creature is a peculiar customer with very different habits to its other flat relatives. Rather than sand and mixed ground it is a rock-hugger, as happy on vertical surfaces as it is on the bottom. A genuine challenge, catching one can turn into something of an obsession. Andy takes up the story:

'Since my first foray into LRF, this strange fish fascinated me. Annually, there were always reports and photos of topknots caught during the junior matches on Weymouth Stone Pier. Normally, these were caught more by luck than judgement while youngsters left their float rigs over the side to knock against the wall. Every so often, a surprised angler would find a topknot attached.

I have spent many hundreds of hours looking at likely spots for signs of fish. For ages I'd scoured walls and rock faces but been fruitless – until a strange quirk of fate. One evening, the tides had looked good for a trip on the open coast with friends, but with fierce winds we decided to head back to the more sheltered waters of the harbour.

As we arrived at a favourite spot, I did my usual routine of scouring the walls and structures for any signs of life. Just below the surface, I spotted an unusual shape on a post. In the moonlight, I could just make out what looked like a large blennie's dorsal fin. I was just shouting across to my friend Ade to have a look, when I realised it was a topknot. To say I was excited would be an understatement!

Carefully focussing my headlamp to inspect the alien-like creature clinging to the post, my heart raced. Thinking the small pink Isome worm I had on should be perfect, I set about wiggling the lure as close to the fish as I could get it. A frantic Google search confirmed they are a left-sided fish so I jigged my lure where I thought its mouth would be. It was only after five minutes of frustration that the fish moved and I realised I had been jigging by its backside!

Moments later the fish turned, but rather than have the lure it sucked in one of the split shot from my rig, before spitting it out in an instant. My heart sank; was the chance gone? The next fifteen minutes were a blur of changing lures and frustration as I tried various soft plastics and other lures. Now we knew what we were faced with you would think it would be easier, but the wind picked up and as I kept changing lures and presentations I wondered if it would ever bite.

Crazily, whilst this is all going on Ade had a go for it and accidentally hooked a leopard spotted goby; another rarity which on any other night would have been the star of the show. After a bit of head scratching, I then switched over to a silver and pink metal jig resembling a small prawn. Perhaps it might be more interested in something silver than anything rubber? I got my answer straight away as the fish hit immediately, but in spite of me feeling the weight of the fish as it ran, I simply didn't hook up.

Replaying this in my head and convinced it had gone, I was semi-amazed to see the same fish return five minutes later for round two. Straining for the perfect angle, I lay on the floor this time and dropped the jig between the post and the jetty to trap the line. This was the perfect presentation and straight away the fish rose, sucked in the jig and ran but once again the hooks didn't connect. Arggghhh! To cap off a truly bizarre evening, my friend Christian then had a small conger eel from the same swim!

It was only two weeks later I finally got the chance for a rematch, with virtually the same combination of tide and wind. We were enjoying a fine evening with flounder, bass, pollack and garfish, but my mind kept returning to that topknot. We paused for a pint and some food and, almost jokingly, I told the lads tonight was the night I'd finally catch one. It seemed daft, but I had a funny feeling in my gut that luck would change. Could it really happen in the same spot with the same fish, though?

For two whole weeks I'd been critically tearing my approach apart, determined to make the next chance count. Even so, the pint probably helped to steady the nerves, because no sooner had the light from my head torch cut through the water, I did a double take. 'It's there lads,' I cried out, 'the topknot is there!'

In a matter of minutes I had prepared my rig and managed to keep calm. At least I was fully prepared this time! I'd selected a size 12 Owner Float Rigger hook, 2-inch Berkley Gulp! Fish Fry and a single size 4 shot. As the lads watched, I waited until the jetty moved and allowed me to drop my line between it and the post. I twitched the lure where I guessed the fish's mouth would be; in a split second the jaws flared open. I hit the bite and the fish was on. It fought gamely for its size, at first trying to reattach itself to the post, before moving up and away as it tried to escape.

So tight was the space that I'd suspected we would need to net the fish and perhaps cut the line rather than risk dragging it through the gap. In the end though, we didn't need to and with a huge sigh of relief I admired the catch. I had done it! The fish I had longed for was now safely in my hand.

It was a slightly delirious feeling, the sense of satisfaction felt far bigger than the flat little monster I held, no bigger than a couple of beer mats. The previous week I had won the hotly contested Zandermasters title on Rutland water, but can honestly say that winning the event was less exhilerating than I felt catching that weird, wonderful topknot after so much effort and frustration over several seasons.

Having done quite a lot of homework on these creatures, my friends assured me that the topknot was quite comfortable out of water for extended periods. Often remaining on rocks as the tide ebbs, they can remain in position until the water returns. With this in mind I was happy to spend a couple of minutes studying the fish. It's back had a rough, gritty feel to it but it was strangely soft; it's fins constantly adjusted to grip my hand and, like most flatfish, its colour lightened the longer it was against my fingers. We took a handful of photos and returned the fish carefully, watching it scuttle back to the harbour wall.'

Six great lures for flounder and flatfish

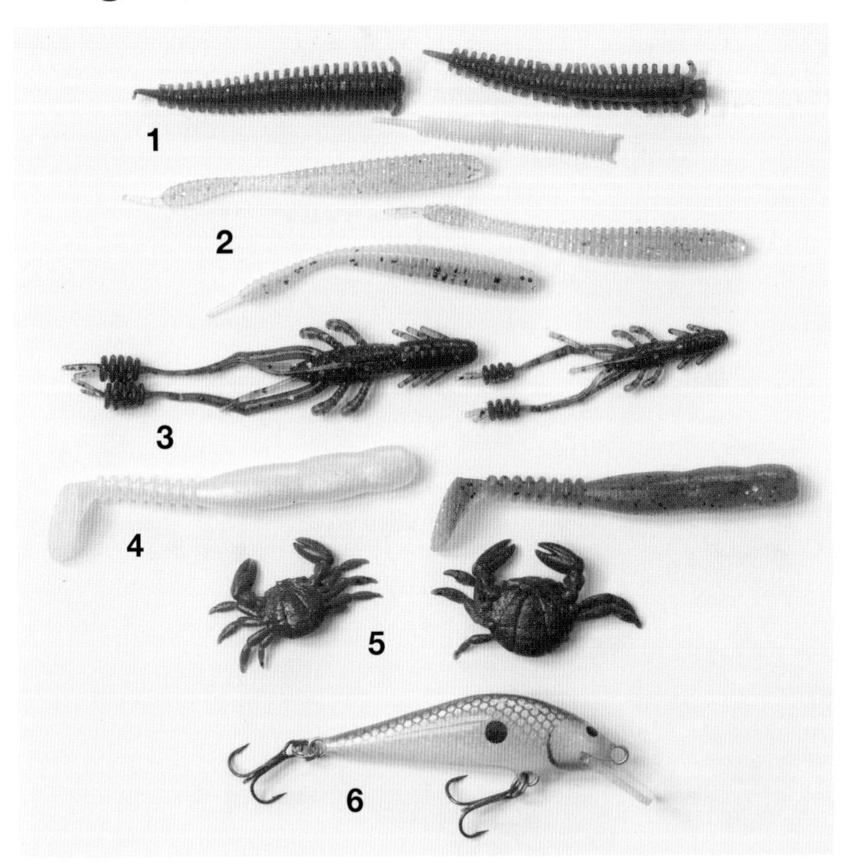

1. Scented soft lures: Flavoured hybrid lures, such as Isome and Gulp! products, are an ideal way to get started when presented on small jig heads or drop shot style.

2. Reins Bubbling Shaker (2.5-inch): Soft plastic worms are another excellent option for flounder, whether you fish drop shot style or jigged.

3. Reins Ring Shrimp: Smaller 'creature' baits cover a wide range of typical flounder prey, usually fished on a jig to hop across the bottom.

4. Reins Rock Vibe: Versatile and deadly for flatties, smaller paddle tail lures of this sort are ideal whether hopped on the bottom, or simply retrieved low and slow. Darker colours are good all-rounders to mimic tiny gobies, while silvery and pale colours are good when the flounder are expecting sand eels.

5.Marukyu Crab: Small crustaceans form a large part of the diet of flounders and other small flatfish. These are best crawled slowly across the bottom, just like a real baby crab.

6. Small crankbait: You might raise a few eyebrows with one of these, but when the flounder are coming well off the bottom to eat small fodder fish, a crankbait can work a treat.

The bend in the rod says it all, as Paul Guest plays a fit fish on light tackle.

BASS

Tackle: Although smaller 'school' bass are excellent fun on the lightest tackle, any serious bass hunter will need to step up to stronger gear to have any hope of landing the bigger specimens. A light to medium lure rod in the 7-28g bracket is ideal. Low diameter braid still permits the use of smaller lures, and with the introduction of high tech lines, such as those from Japan, you can up the breaking strain to 15-20lb whilst still retaining low diameter for distance casting.

Other: Waders are sometimes useful for beach or estuary fishing.

Season: Bass are caught year round these days, especially the smaller, immature fish that you'll find around virtually all estuaries and harbours. Late spring and summer are the most popular times to fish for them, although autumn is when the biggest bass often close in to the shore.

A bold, handsome predator, the silver bass is the most highly prized fish of all for countless sea anglers. With impressive speed and power, along with the potential to grow large, it isn't hard to see why. A huge range of prey items also makes them the perfect quarry for imitative lure fishing in saltwater.

Looking like a giant, silver sea perch, the bass is unmistakable in appearance; from the gaping mouth to the spiky fin on the back, it is a formidable hunter. But these fish have so much more than just a big mouth and fine silver livery in their favour. The locations they are found, from shallow estuaries to offshore

Is there a better looking or harder fighting sea fish than the bass?

reefs, along with their endlessly varied habits and behaviours make the bass a fascinating and challenging quarry. It takes cunning as well as patience to consistently catch bass.

Life of the bass

A slow-growing fish, bass feed on everything and anything they can readily hunt. By the time they reach mere ounces, they will prey avidly on fry, shrimp and sand eels, but as they get bigger, young mullet, joey mackerel and even squid are not out of bounds. That said, their preferences for food can be highly varied by location. In the clear coves and estuaries of Devon and Cornwall, for example, sandeels are among their main fodder. In other areas, however, anything from crabs to prawns or worms can make up the lion's share of their diet. Hence, while they are unlikely to pass up an easy meal of any kind, the successful bass angler is often the one who manages to establish what they are eating and choose the right lure.

Finding the bass is your single most important challenge. It can be a riddle in itself and there are good reasons why the keenest anglers tend to be secretive. It can take trial and error, and many hours if not seasons of work to suss out the best of your local fishing. The good news is that once you have cracked it in any location, they are definite creatures of habit and your success is likely to be repeated. Tackle shops and online resources should yield a few clues, but you are also looking at trial, error, instinct and legwork to catch consistently.

Of course, if you find clearer or shallower water, you might be lucky enough to spot bass. The rest of the time, though, you are looking for obvious features and feeding opportunities. Bass will hunt at various depths and over all kinds of ground, from sandy beaches to deep, craggy rock marks. But wherever you fish for them they seem to like current and areas where their food can be trapped or overpowered. Indeed, vigorous, 'lively' water is especially attractive to the species, which not only seems to make them bolder, but helps to disguise the angler's presence. An incoming tide rip at the mouth of a river, or rocky spots where churning currents flush out small fish and prawns are exactly the kind of places you'll consistently find feeding bass. Even the larger fish will often come crazily close to the shore in these raids, bringing them within easy reach of the lure angler.

Whether it's by luck or judgment though, finding and hooking a bass is only half the battle. Be warned: these fish are as strong and relentless as the tides that formed them. A big one will make the tightest clutch groan, and will snap light line like cotton. They fight with cunning and about as much sense of fair play as a prison football team, too; they will be intimately familiar with every local snag. So, while we definitely recommend light tackle in this book, our advice with bass is to err on the side of caution. Modern braids allow mainlines of 10-15lb plus to be used on even the lightest rods, while quality fluorocarbons provide tough trace material for potentially bruising encounters.

Under the surface: The keen-sighted bass

Although feeding bass can be bold and aggressive, there are also times when they show a great deal of wariness. When you consider that any bass of four pounds plus is probably a decade old, they certainly have time to grow wiser than most sea fish, to say nothing of their keen predatory senses.

During the making of the book, we were surprised by just how tricky bass were to photograph in their element. Even in an aquarium setting, where presumably they see plenty of humans, bass were among the most cautious of all the fish we captured on camera, eyeing dummy lures (minus hook points!) with great suspicion. One smaller bass at Fowey Aquarium gave our soft lure a tentative peck, but that was it; all the others made themselves scarce pretty quickly.

Is it their keen senses, or are bass more 'intelligent' than other sea predators? We cannot say for sure, but wherever you find bass in calm water or at close quarters (and sometimes that means literally inches from the shore) you should move and cast with care and avoid spooking them. If you are fishing by night, keep your head torch off as much as possible.

Talking of keen senses, the eagle-eyed bass angler will be surprised how often bass turn up unexpectedly, even in shallow water. Our friend and bass addict Richard Cake has often seen them sneaking into shoals of mullet, seemingly using these docile creatures as 'cover' before they pick their moment to seize a young or injured fish!

Left: Aggressive, yes, but bass can also be spooky, clever and selective feeders.

Left: Alex Garnett casts into the mouth of the River Teign, a noted bass spot.

Estuaries

Estuaries can be great places to fish for bass. Indeed, the establishment of many brackish locations as 'nursery' areas with no commercial fishing has been very positive for the species. You will find the average bass small, typically, but excellent fun with smaller lures and lighter gear. That said, larger bass are always capable of springing a surprise, with the odds tending to get higher the bigger the tide and closer you get to the sea itself. The trick, as always, is finding the right time and location.

The state of tide and time of day are particularly crucial for locating bass in estuaries, which can travel miles. When the water is rising and growing, fish will be on the move, entering newly flooded areas, even those just inches deep. If you are in any doubt where to look, areas of concentrated flow or natural points and passing places are always likely.

Of all the prey species in estuaries, sand eels and sand smelt are especially common, which is why a slender floating plug or soft plastic in olive and silver will work. That said, there are estuaries where they show a definite preference for crabs and will also feed heavily on marine worms in season; and both these creatures breed and become vulnerable in the spring.

The only way to suss out the best times and locations tends to be through trial and error. Once the water is really up and in full flow on an estuary, for example, the fish will sometimes retreat or sit down out of the flow, behind rocks, sandbars and alongside or often in weed. It is still possible to catch them at this state of tide, but you might need to work lures close to or through the weed, or bounce them along the bottom around rocks and sand bars. While in some areas you will get away with a simple sandeel lure fished with a ball weight (effectively a Carolina Rig), as has been the trick for decades, other spots are much easier with modern weedless soft plastic lures, which are ideal for casting into juicy-looking features without getting snagged up.

So much depends on the mood of the bass and the state of the water, however. As the tide slackens off, for example, it is sometimes possible to find bass feeding near the surface and chasing fish that are retreating with the falling swell. Topwater lures come into their own at this point: try working a surface-waking plug in bold, sweeping glides to see if there are fish around. If the bass are on the attack, a vigorous retrieve can induce explosive takes.

Lastly, bass will move into different areas at night on estuaries; just because a

Left: A typical school bass. Where fish like this abound, we'd recommend fishing slightly lighter and debarbing your hooks.

spot is dead or empty during the day it does not mean that bass will not frequent it at night. If you find artificial light or slacks out of the main flow, take note and return later, for it is often these areas bait fish shoal at night.

How fast should I fish a lure for bass?

The speed at which our lures are retrieved can be absolutely critical for bass and is one of the most common things that anglers get wrong. The correct speed will depend on the scenario and type of lure used. Where bass are chasing fish such as sandeel and joey mackerel in open water, a vigorous retrieve is often the way. These are relentless predators and you would struggle to reel in a lure too fast for a bass to catch.

Surface poppers also can be pulled fast, although you might also try to break up each retrieve with some juicy pauses for effect. If you see fish following but refusing to take, your natural instinct might be to slow down, but if anything, speeding up the lure is more likely to induce a take. Inject too little action or give the bass too long to inspect the target, especially in calm water, and it is likely to lose interest or sense that something isn't right.

Bass are such versatile predators, though, that the rules change according to the setting. For example, one common exception to the above is with bass feeding in a tide rip. In this instance the most productive approach might be simply letting a smaller sandeel or fish lookalike get swept along with the flow, applying the odd single twitch. Create a decent impression of a small or damaged prey item struggling in the tide and sooner or later you are likely to get that sudden whack.

Nor is this the only scenario where the rules change and less is more on your retrieve. Creatures such as prawns, crabs and small rock-dwelling fish don't shoot about at a hundred miles an hour, so nor should your imitations. When fishing tight to walls, rocks and structures for bass that are ambushing prey such as blennies and prawns, much slower, twitch-and-stop retrieves are the way to go. With more time to inspect lures, these cagier feeders tend to need a more careful presentation and more natural lure choice.

The golden rule with bass is therefore to keep moving to different spots, experiment and try to understand what the fish want and consider their modus operandi. What are they feeding on? Is their prey fleeing into open water, or trying to hide in cover? Where are the natural 'kill zones' which stack the odds in the favour of the bass? If you can answer these questions, you're halfway to catching.

This fish was tempted by twitching a soft plastic lure beneath a rock overhang.

A quiet cove with clear water and prey-filled pools and gullies; perfect bass territory.

Rock Marks

Bass love to hunt in rocky, weedy areas. In these locations they not only find a wide variety of prey, but regular opportunities to ambush and corner their victims. A mobile, 'rock-hopping' approach is an adventurous way to target these fish. Perhaps the best way to start is to reach your chosen mark at low tide and study the rockpools for clues. Crabs, prawns and smaller rockfish are common in these locations. Young pollack are another classic, providing a real mouthful for a bass and we've found a suitable olive and pearl plugs excellent, when kicked around rocks in a pitching tide.

Most of your livelier shads, grubs and creature type lures will work if you can locate feeding fish around the rocks. By all means start in the upper layers, to see if the bass will take a small shad or eel around cover. But if the bites are not forthcoming, you can search closer to the rock faces with crankbaits or soft plastics. Darker lures, including various creature baits like prawn or crayfish designs, are all worth a go. Whilst not popular with the ardent bass anglers, those with an open mind often pick up bass around rocky areas on non fish-like lures. These, however, might need to be fished in a slower, stop-start manner, to resemble a prey item momentarily dislodged from the safety of the rock face. Again, some observation and field testing can help.

Snaggy gullies and overhangs are perfect for such tactics, especially later

Fish of dreams: a double figure lure-caught bass for Richard Cake. It took a Fiiish Black Minnow in 120 size.

in a flooding tide, when the deeper rock pools you found earlier are now under twelve feet of water. Focus your attention on one mark for a few trips and you will learn which locations fish best from low to middle and flood tide, not to mention figuring out a reliable escape route!

The best spots are not always blatantly obvious, but once you do find your feeding bass, you can be fairly certain that they will be there again at that state of time and tide. In fact, you could often set your watch by them. They are certainly creatures of habit that learn when and where opportunities are. They can also be highly localised in their presence, which is why serious bass anglers are notorious for being tight lipped and being sticklers for releasing every fish.

Finally, we should also advise you to take some precautions around any rocky, unforgiving mark. Stronger tackle makes obvious sense, stepping up to 20lb (10kg)

Late season fishing and bigger bass

While not common these days, bigger bass can always make an unexpected appearance, even in the very shallowest water. As a rule though, the biggest tend to come from late summer into autumn and winter. At this time they feast harder on sandeel, smelt and juvenile mackerel and mullet, to see them through the slimmer pickings of winter and pack on weight before they spawn out at sea (a fish of 6lb will hold hundreds of thousands of eggs and be over twelve years old, so please release all large bass!) These days bass are caught right into the winter, but it must be said that once the sea temperature drops below 9 degrees, their metabolism slows and they are less inclined to chase lures; for the diehard it might be worth trying slower presentations.

and heavier rods. So does a generous-sized long-handled landing net, which could mean the difference between landing and losing your best-ever bass.

Of course, for the most demanding marks, having sturdy footwear with good grip is just as important – and you must always check the conditions and avoid unnecessary risks. No fish is worth your neck, so please study the conditions and tread carefully!

Open beaches

Open beach marks are a potentially even greater challenge for the shore angler because there can be so much space to cover. Where do you find the bass in acres of what might seem featureless water? The trick is to start by looking for any vantage point or food source. Donning a pair of waders or old sandals and exploring a little to see where your bass and any food sources might turn up is a wise way to start.

On flatter areas of beach, the tide can come up and flood huge areas of water, so any features could be worth exploring. Bass will be funnelled around sandbars, groynes and other structures. Be sure to look out for groups of small fish, shrimps, crabs and shellfish, too. Low tide is the time to look for any depth changes or hidden features on most beaches.

Once the tide races in, the change can be dramatic. Bass and other species can move large distances to patrol and look for food. Perhaps the one thing that tends to go against many anglers is putting our preferences before those of the bass and seeking out calm seas and picnic weather.

It's not that bass won't be around in these periods, but they will be warier and less likely to feed hard. Far better, therefore, to head out when there is more breeze and current to mix things up and help dislodge and disorientate all the smaller creatures that bass predate on. Flatter, more open beaches are not a dead loss then, but more than rock marks,

Bass & Sand Eels

Of all the common prey you find close to shore, sand eels are amongst the most universally important for bass of all sizes. Look out for shoals of these while you are fishing and try to get a rough match in size. You'll find that in the 'early' season, from spring, the bass will avidly feed on smaller 'bootlace' eels that are present. These eels have yet to grow to full size and are best imitated by slender soft plastics. Tiny artificials and even flies are excellent – and can be given added weight with a bubble float or even trailed off the back of a larger lure.

As the season continues and waters warm, the sand eels grow bigger, as well as being joined by their larger relative the greater sand eel, or 'launce'. Bigger bass love these, so be prepared to step up your lure sizes. As always, a 'match the hatch' rule is sensible. However, besides natural silver and olive realistic eels, sometimes a chartreuse or even a pink imitation will work better – especially when the water is teeming with the real thing and you want yours to stand out.

Sand eels are an absolutely critical bass food source.

harbours and other locations, require some local knowledge and a sense of timing.

Schoolie bass will frequent even the calmest beaches through summer, often coming in shallow to feed on sandeel and whitebait. They may not always be feeding, but if you work the bottom with lures such as a Fiiish Black Minnow you will learn if anything is at home.

On beaches where there is a significant swell, and on surf beaches, the bass come in close to the shore, often riding the breakers as they chase whitebait which are being pushed in with the waves. They can often be picked up by casting soft plastic lures into rips or gulleys and letting the tide roll your offering around. It is in these sort of vantage points that powerful, fast-swimming bass are able to come and clean up any dazed and confused small fish that are less able to handle the tide.

Inshore boat fishing

There are many ways to fish for bass from a boat. For some of the big specimens off deep water reefs and wrecks, heavier tackle can be a necessity. For those who seek bass closer inshore, however, the lighter gear can stay, whether you cast from a dedicated fishing boat or take to sea by kayak.

The most fun way has to be to seek out areas of cover and cast into boiling tide races or tight to rock formations. From late spring onwards, it is often possible to find shoals of bass working their prey near the surface. Sometimes you will be fishing blind and simply tackling features and likely-looking spots; on other occasions you might be lucky enough to see fleeing prey fish, or perhaps locate bass via the seabirds who are making the most of the disoriented prey that have been driven to the surface.

A small boat offers major advantages for the bass angler. Left: Chris Guest keeps things steady, while Gary Pearson takes aim on the south Devon coast.

Working different depths makes perfect sense to begin a session, at least until you establish the level they are feeding. For instance, you might start with a topwater or shallow running crankbait, but if a bite isn't forthcoming, try switching to a well weighted soft plastic lure to explore deeper. If you are fishing in company, so much the better because you can try different depths and retrieves until one of you cracks it.

In the absence of any signs of predation taking place, it's always a good idea to keep moving and seeking out likely spots. As from the shore, you may find that the best fishing occurs in less than ideal conditions. On many boat trips, the harder it gets to stand or sit up straight on deck, the better the fishing gets!

As we've discussed, one feature that always seems to produce bass is any area of water with vigorous movement or a run of current. The racing water around headlands can be productive even on flat calm days. A boat allows you to access all kinds of spots you would never hope to reach from the shore, from the bases of cliffs to craggy islands of rock beyond the longest cast from the beach.

One useful tip for horribly snaggy spots and areas such as tide races is to anchor away at a small distance and cast into the feature. Caution is the watchword here. Another sensible way to proceed is to take turns so that one angler works soley to control the drift and position of the boat, while the

This fine bass was tempted right on a colour change zone, where murky, stirred up water met clearer depths, a good distance from the beach. It fought like a tiger! By the time it was subdued, the kayak was virtually grounded on the beach, where the fish was carefully supported in the water to recover before release.

Do I need to fish at night for bass?

While a lot of the old textbooks declare that bigger bass tend to feed best at night, going on our own experience and that of friends, we have rarely found this to be true. The first and last hours of light can be excellent, it's fair to say, but you needn't be a night owl to catch decent bass. On any tide where you find prey items there is a chance of bass, even if it's in the middle of the day – and it is the unpredictability of bass that is part of their charm.

Locations and feeding times do vary, however, and after-dark fishing can be useful on occasion. In the maelstrom of swirling seas around rocks, for example, bass are unlikely to be shy of our presence. On a flat calm, sandy location, however, it could be a very different story. This is especially true in the summer, when popular beaches throng with swimmers and surfers. Bass won't venture too near with all this disturbance, so you may indeed be wise to return at night. Keep your use of head torches to a minimum and don't be afraid to try really close in.

other casts. Successful methods in such scenarios include working a Portland rig as you drift on the race or by using metal jigs presented vertically.

When casting a lure into racing water, you don't always need to impart much action – and often just the odd twitch will replicate a small fish helplessly flailing in the current. It can be thrilling fishing, but you *must* take the utmost care around turbulent water. Tide races and tide-swept rocks can be unforgiving and challenging to navigate and are no place for the inexperienced.

Of course, it is also possible to troll around much of the coastline, but where light lure tackle is involved, drifting and casting is much more fun (in our humble opinion!) For those without access to a bigger boat, kayaks can also be excellent for this style of fishing.

Recommended lures for bass

Provided your presentation is good and you don't spook them, bass will take most lures. However, they can be unpredictable in their tastes and this can vary from mark to mark. Why colour should be so important, for instance, can be a mystery – but colour can be a game changer! Whether it's prey colour, sea colour, conditions or weather that make the difference, these are all part of the puzzle. Of course, the right lure is just part of the mystery that makes bass fishing so enjoyable. You'll notice a lot of bass prey is olive and silver, for example (sandeels, sand smelt, pollack), hence these colours dominate our lure samples below. But that's not to say the same lure in pale white or bright yellow won't work better on a particular mark or occasion!

A range of lures from surface runners to deeper divers might be required at different marks. Shallow diving lures are among the most useful of all though. When bass are actively hunting down prey, lures with a healthy wiggle that dive 30-40cm are especially good. Rattles are also worth trying, especially in coloured seas. Perhaps the greatest strength of the shallow diving lure is that they'll work almost everywhere; reefs, rock marks, estuaries or open sandy spots, at any state of tide, day or night. Just change the speed of retrieve accordingly until you find the perfect speed for the lure to be worked at. At night, try slowing the retrieve down and you should feel the lure moving side to side via feedback from the rod.

Metals and soft plastics also have their place, of course. Metals seem especially good in tide rips and for bass that are in chase mode. When the bass are sitting deeper, or sitting in and around cover, weedless soft plastics can be excellent, from fish-shaped lures to various 'creatures'. Above all, experiment according to the time and venue. Fishing with a friend can also help, so you can try different colours, depths and other variations to establish effective trends.

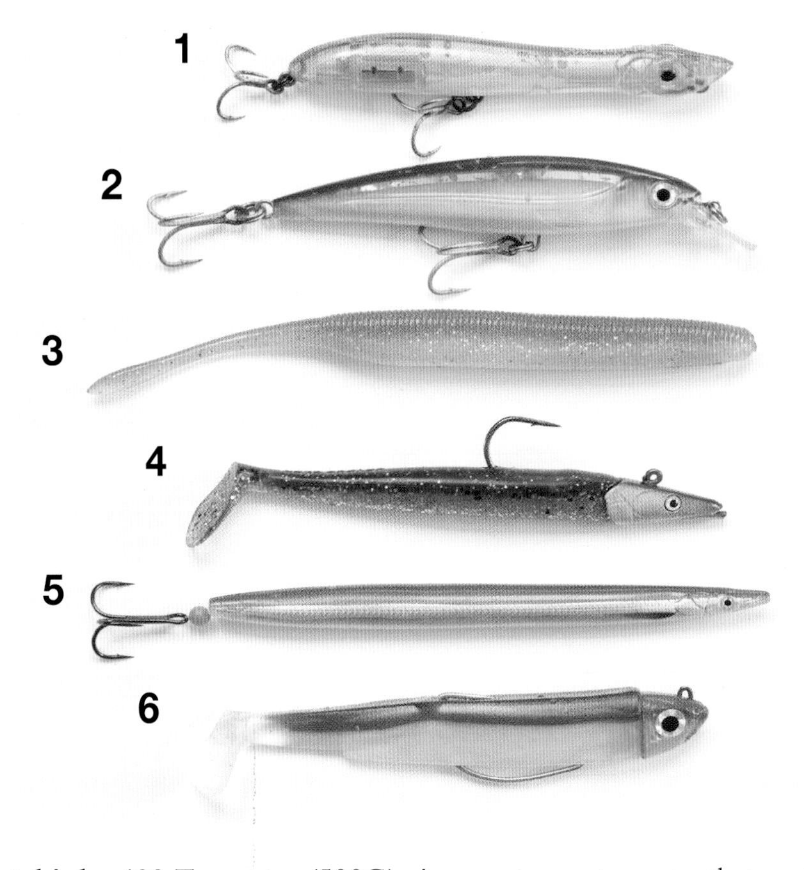

1. **Patchinko 100 Topwater (500G)**: An easy to use topwater that works wonders with a simple 'walking the dog' retrieve.
2. **Rapala XR-10 SW (AYU):** A huge range of floating, diving plugs will work for bass, but this is a favourite.
3. **OSP DoLive Stick (6-inch, olive/pearl):** Ideal for fishing unweighted, to give a slow needlefish presentation. A good choice for night fishing, too. This style of lure will not create much disturbance and is perfect for wary fish.
4. **Savage Gear Sandeel (12cm, blue & silver):** Effective sandeel imitation with a stronger action. Ideal for heavier seas and straight retrieves.
5. **Savage Gear Line-Thru Sandeel:** Perfect choice wherever you find bass feasting on the real thing. Casts miles. Rigged through the body, so lure slides up the line when a fish is hooked.
6. **Fiiish Black Minnow 120:** A phenomenal lure for any snaggy area. Excellent hook-up rate, in spite of weedless design.

ROCKFISH, BLENNIES & MINIATURE MONSTERS

Tackle: The very lightest tackle you can find is ideal for getting maximum sport from smaller species. A rod that will cast just a gram in weight is spot on. Shorter blanks of 6-7ft are also ideal to poke your lures into little nooks and crannies. Typical lines are 3-4lb (1.5-2kg), with low diameter hook lengths of similar strengths. There is always the chance you'll accidentally hook a bigger species when 'scratching', so we rarely go beneath 3lb line.

Seasons: Most small species you'll find around rocks and structures are not migratory. Therefore they can often be caught all year round, which can make them a welcome target when larger fish are absent.

Lures: The smallest soft plastics, especially those that are scented, are ideal. Split shot rigs with tiny fragments of marine worm imitations (such as Isome and Gulp!) are very effective. Dropshot presentations and the tiniest of jigs can also work well.

One of the greatest pleasures of fishing with the smallest lures and lightest tackle is the huge range of cute, curious and downright sinister sea fish to discover. Creatures like sea scorpions and gobies may never put much bend in a broom handle style beachcaster; but with a pencil-thin rod they become a different prospect, providing bold bites and punching well above their weight. It's gloriously intimate sport too, and with the action often right under your feet this is an ideal way to introduce kids and complete beginners to the joys of lure fishing.

The more you delve into rocky places, and the less you worry about mere pounds and ounces, the more variety there is to be found. Of the smaller inhabitants we might describe as 'rockfish' there are literally dozens to find in UK waters alone (*see our reference guide at the end of the book*), while the tactics outlined in these pages will also work for countless others, wherever you find yourself on the planet.

The smaller critters of the sea truly prove that size isn't everything. Species hunting is becoming more and more popular as more anglers discover the huge and unexpected range of fish to be caught. Once you start talking about rarities such as the black faced blenny or fifteen-spined stickleback, it can all get a bit obsessive. But it all starts with simple light tackle, and a very different approach to saltwater lure angling.

Scratching the sea

Catching smaller fish on lures is all about scaling down your approach. Tiny jig heads, with small hooks and weights down to just a gram or less, are ideal where conditions allow, and most of the tiniest worm and grub type soft lures will work. But most of the time the best scratching set-up of all is the split shot rig

Below: Andy searches beneath steps at low tide; such spots may appear too shallow to hold fish, but gobies, sea scorpions and other critters will still be present.

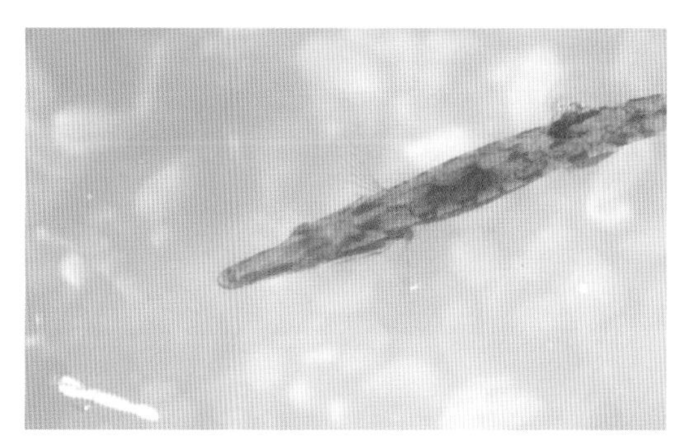

Left: Tiny hooks are often needed to connect with the smallest and most unusual species. Here, a rare fifteen-spined stickleback is drawn to a tiny soft lure on a size 20 hook.

with a broken off section of Isome, Gulp! or any scented soft lure. This comprises of no more than a small hook tied onto your trace followed by one or more split shot pinched a short distance away (*see our chapter on rigs, p41*).

We can't emphasise enough the need to scale down here. The very lightest rods available, designed to cast as little as just half a gram, are the tools to use. As for your hooks, these should be fine and super sharp for this style of fishing. When you're targeting gobies, blennies and other small predators, normal sized hooks for the sea suddenly look very crude. Instead, go for smaller patterns in sizes 10-24; something like a size 14 would be an ideal starting point, but as a general rule, the smaller your hook, the broader the range of mini-species you will encounter.

Many of the hook patterns used for coarse and match fishing are ideal here, and pre-tied hook lengths can also be very convenient and save faff when you're on the move. Blennies and other smaller customers can be greedy things, so we'd recommend barbless hooks too; not only are they easier to remove but they penetrate bony little mouths better.

Talking of bony mouths, watch your fingers with these little fish! It is one of the strange facts of fishing that a 20lb pike will make no effort whatsoever to bite you, and then a tiny rockfish can give you a vicious nip! These rockfish are hardy creatures that are quite comfortable out of water for short periods, provided they are damp. The finest of forceps are useful for removing hooks, but we would also advise taking a coarse angler's disgorger, which is better still for small, toothy jaws and any tricky hook holds.

Typical settings & techniques

This is intimate, close range fishing in essence, and one of the nicest things about it is that you can find various species at all states of tide, all year round. You can try it from rocky open shore marks or piers when conditions allow, while larger rock pools can also be excellent fun at low tide. But with the light tackle required to enjoy fishing for the smaller species, sheltered waters are the best of all settings, as they offer accessible and easy fishing in all but the harshest conditions.

Harbours, marinas and breakwaters all tend to offer calm water and literally hundreds of suitable hiding places for small predators, with a blend of man-made and natural habitat for fish. In fact most of the smaller rock-loving species are just as at home around walls, bridges, quays and piers as they would be on a natural reef.

On the subject of man-made structures, don't always assume blennies and other small predators will be glued to the bottom. They also make their home on walls, posts, moorings and even the underside of boats. Look at the body and fins of many rock-huggers and you will see exactly why; some have almost claw-like

Above: Fishing several feet from the water right amongst the cracks in boulders might look odd, but you'd be amazed what lurks beneath. Dividing a rod so that you have just the top section or two is a good dodge for fishing such small spaces. Quite often the tiniest and unlikeliest places produce fish, from the gaps between boats to the legs of a pier or tiny gaps between rocks. Find a good hiding place and you'll usually find a fish.

fins; others even have little sucker type appendages. Gravity means little when you can hunker down and stick to just about anything, so don't feel you always need to present a lure at maximum depth. Lures presented in mid-water, close to walls, jetties and other features, can tempt fish too.

If you can spot the fish firsthand and observe their antics, this is always fun and instructive. Most rockfish will find holes to retreat into, as they wait for their next victim to pass by and will often dart out at lightning speed to capture their prey. Blennies and gobies, for example, are very much ambush predators and will think nothing of attacking worms or fish almost the same length as themselves if they pass too close.

The angler will sometimes be making short casts, but more often than not will be presenting lures right underneath the rod tip with this style of fishing. They'll guide their artificial into small gaps and the shade of walls and rocks, feeling it to the bottom and working it slowly, with little lifts and taps. Keep lifting the lure off the bottom to entice a bite, but remember when the fish and your artificials are this small, even a tiny movement can be a lot and too much can actually put the fish off or make them miss the target. The movement of the tide will often give your lures a life of their own too, so if there is a bit of swell all that may be needed is to hold steady and let your artificial waft.

While it always pays to keep mobile, this style of fishing can demand some patience. You will often get little plucks and tingles, but must time your strike for when the fish has properly taken the lure. In fact, one of the commonest mistakes is to keep striking at every little tingle. Instead, watch the tip closely, try to feel what's going on and allow the fish to inhale the lure before striking.

A lot of fish will self hook but with blennies you may need the fish to get a tiny lure behind their teeth for the point to get sufficient purchase. Blennies like to grab and run so resist striking instantly, but wait for the point where it feels like the fish is moving off before lifting to set the hook. The sequence is often the same – a tiny tap or three before a more solid bounce and angle of the line changing as the fish decides to move off with dinner. The slight extra 'give' of a solid tipped rod is preferred to allow them to inhale the lure before feeling pressure.

Don't expect all bites to be subtle, because you will often be surprised by a moment of real savagery from the smallest of blennies. Of course, there is always the chance you might bump into something bigger, like a wrasse or even a mullet, so it can be unwise to fish too light – and for this reason you can set your drag fairly loose so as not to smash light hook lengths (which tend to be in the 3lb (1.5kg) bracket and low diameter) on 'bonus' fish.

Rockfish are not always tiny; Jack Perks caught this fantastic, fist-sized sea scorpion from a sea wall at Poole.

As we've mentioned, split shot rigs tend to be king for mini species, but we should also mention that other methods may be called for if you encounter rougher weather, bigger tides or deeper water. In these circumstances, a dropshot setup can be very useful, allowing you to stick with a very small hook and lure but add anything up to half an ounce or more of weight to allow better control.

Of course, there are other occasions when a jig head and soft lure is the way to go. This is particularly true when you need to make a slightly longer cast to get your lure among the fish: if you're casting and wading on a beach, for example, or the cover you want to search is a few yards out – and we often cast this way to reach fish like dragonet and weaver. Nor do all the miniature monsters have small mouths; on the contrary, a sea scorpion or dragonet will easily take a soft plastic mounted on a size 6 hook. If you are casting and retrieving for the smaller ambush predators though, be sure to retrieve patiently with tiny twitches of the

Now you see me: Rockfish like this tompot blenny are suckers for any little vantage point, natural or otherwise.

Beneath the surface: Rivalry, ambush and aggression

Smaller sea predators fish will attack both out of hunting instinct, but also a natural curiosity and territorial aggression. Sessions spent with Jack Perks filming and taking underwater photographs were a fascinating window on their habits. Some fish would attack a lure instantly through hunger, but others would come out and look, back off and repeat. Some of them would bristle up angrily – as if squaring up to the lure – and one or two appeared to bite out of pure aggression, as if repelling an invader on their territory. Sometimes the main reason a lure wasn't taken instantly was a little face-off between neighbours, with the bigger creature barging its rival out of the way before taking the lure. In the case of two similar-sized aggressors, the lure would sometimes become part of a tug of war contest!

With big heads and strong, claw-like fins blennies and other small rock predators are perfectly adapted for their habitat.

rod tip and regular pauses, because these fish will seldom move very far from their lairs on the seabed to grab a lure.

The Life of a Blenny

It is quite common to encounter a whole colony of blennies and other little monsters living in close proximity (well, 'living in harmony' wouldn't be accurate, given their regular battles and rivalries). These are remarkably fierce and well adapted creatures and their body designs alone tell you a lot about their lifestyle. A lot of species have pronounced lower fins, for example, with claw-like appendages or even suckers and other adaptations. These help them grip to different surfaces and some species can even 'crawl' from one rockpool to another over wet rocks!

When species hunting with friends, we often fish the rockpools to find blennies. It's very much big kids' fishing, but the miniature power struggles in each location are a source of endless fascination. Indeed, like gangs of hoodlums in a small town, there tends to be a visible hierarchy in every rockpool.

It is quite magical how an empty-looking puddle little bigger than a kitchen sink can often contain an entire

A black goby; one of so many species that like a mixed bottom.

collection of blennies and rockfish. Try dropping a lure into the mix and they will emerge in dribs and drabs, nipping, jostling and squaring up to each other as a potential food item brings out their competitive side.

Like warriors in a miniature Roman amphitheatre, these bullish little fish will charge out and do battle with each other before hitting your lure and running for their favoured crevice in the pool.

The fact that you never quite know what you will find next is the real magic of tackling up for smaller fish, whether it's in a rockpool or a harbour. It is a matter of perspective and in their own shady world, these creatures are every bit as rare and special as much larger fish.

Some specimens, we should point out, are rare and populations can be fragile – so if you are lucky enough to catch something exotic or seldom seen, do treat it with the utmost care.

Our fish species section at the end of the book details all the usual and not so usual suspects, from the abundant shanny to treasures such as the black-faced blenny.

WRASSE

Tackle: Wrasse are strong fighters and often live in craggy spots that are harsh on tackle. So, while you might get away with dropshot or LRF tackle for the smaller types of wrasse in less demanding spots, (typically 5-8lb lines, rods rated from 3g and upwards), the bigger ballan wrasse demand heavier gear. Try a medium lure rod (7-28g or even a little heavier), with strong braided line (20lb minimum) and a tough fluorocarbon trace of at least 15lb breaking strain. A rod of around 7ft with added 'grunt' seems optimum. If you are struggling, or you find some monster wrasse, go even heavier!

Other considerations: To have any chance of landing a good wrasse, a tough, long-handled landing net is a must. We'd also recommend an unhooking mat and long nosed pliers, not to mention sturdy boots with good grip for typical marks.

Season: Wrasse were once believed to be a summer species. However, these days lure anglers catch them year round. You may, however, find the smaller fish retreat to deeper water during the winter leaving only the bigger fish behind. Be particularly gentle with these fish in the spring, when they might be preparing to spawn.

With beautiful colours and brute strength, wrasse are a quarry with great appeal to many saltwater lure anglers, providing some of the most visceral, exciting sport of the year. A common presence around much of the country, these bold, muscular fish can be found just about anywhere there is abundant rock and weed cover,

Dom fishes a rock ledge in classic wrasse territory on the Dorset coast.

from man-made structures to natural reefs. That said, our home stomping ground of Dorset, Devon and Cornwall probably represents the cream of wrasse fishing in the UK, offering countless rocky marks for the lure angler to explore.

While many of us were brought up catching wrasse on bait as kids, the way in which they will savage modern artificial lures has been nothing short of a revelation in the UK. Unlike other fish such as bass, wrasse also appear to have avoided the worst of commercial exploitation too, with good-sized ballan wrasse in the 2-4lb stamp still relatively common.

Of all the fish you might catch in British and European waters, the wrasse is one of the most beautiful and exotic looking. The smaller subspecies, such as the Corkwing and Goldsinny are delightfully cute and colourful (*see our identification guide at the back of the book*), while the larger ballan wrasse are a fish to test your nerves and tackle to the limit.

Below: Killer clown: Wrasse combine wonderful colours with anger management issues.

As common and catchable as the various wrasses may be, however, the tactics required to catch them can be quite varied and exacting. Indeed, compared to predators that love a chase, such as bass and pollack, the rules for this rock-hugger are quite different. So it would make sense for us to start with a closer look at the habits of wrasse.

Perhaps the real fascination with the species is that lure fishing for them is quite new, and has turned a lot of the old thinking about the species on its head. Once upon a time, the rules seemed set in stone: wrasse are a summer fish caught on a rising tide; they are strictly territorial and only respond to bait. Today, all of these claims have been challenged, if not blown out of the water.

Above: Wrasse are built for maximum manoeuverability in strong currents; note the oversized paddle fins on this four-pound plus specimen.

The Way of the Wrasse

If you approach any harbour wall or rock mark from mid spring to mid winter and peer into the water, you will see small, dark shapes moving between the shadows. Now and again, particularly early or late in the day, you might see something much larger emerge in surprisingly shallow water.

Perhaps one of the most common misconceptions about wrasse is the idea that they are strictly territorial, with each fish hugging the same few yards of rock and weed. In reality, these colourful creatures can patrol over quite a wide area; especially the bigger fish, whose appetites require wider hunting grounds. Nor are they always quite as predictable or rock-hugging as you might think.

One of Andy's most vivid encounters with large wrasse was casting around Mevagissey Harbour in Cornwall. Surrounded by tourists eating ice-creams and sunbathing, he had been peering into the clear, sandy waters and spotted what appeared to be a single rock sitting on the bottom. Suspecting there might be a blenny or two hiding around it, a tiny lure was cast alongside the dark shape. All of a sudden, to his complete surprise, this 'rock' came to life, shifting forward to demolish the lure! After an explosive battle in which the fish careered right around the harbour, an excellent ballan wrasse of 4lb or so came to the net.

Perhaps it just goes to show that although well-camouflaged and at times downright sneaky, the wrasse is a more savage and aggressive fish than its cheerful looks suggest. Quite why it might lash out at a lure is a topic that has been subject to a lot of debate. Is the wrasse really a savage predator, or could it be curiosity or even annoyance at a perceived threat?

While an instinctive aggression probably plays a part, it is certainly true that wrasse are adaptable all-rounders when it comes to food. With its muscular body and oversized fins, it is superbly manoeuverable and purpose-built to hunt in powerful currents and rocky nooks and crannies. The formidable teeth can make short work of shellfish and even well-armoured prey, such as crabs. However, shrimps, marine worms and other creatures are also fair game; and in spite of the textbooks suggesting a fish that likes to use its gnashers to chomp limpets, a large wrasse is equally capable of putting it's laughing gear around smaller fish and other victims whenever the chance arises.

That said, the wrasse is not a fish that chases prey over any great distance. It is a close-quarters predator, designed to harry and trap prey in amongst rocks and kelp where it simply overpowers smaller creatures at the mercy of currents. Perhaps this explains why it took anglers so long to discover that they would take lures; because it's only once in a blue moon one will take the usual crude sea angler's spoon or plug. Indeed, if you want to catch wrasse consistently, you will have to adopt a more patient style of lure fishing, getting right into the rocky, kelpy world they inhabit.

Tactics for 'fun-sized' wrasse

Although there are similarities in fishing for wrasse of all sizes and types, the techniques and tackle for catching a variety of smaller fish are very different to the methods to get the really big bruisers. The good news for the beginner or fair weather fisherman is that the smaller wrasse are often numerous and obliging. They are also fantastically colourful, ranging from marbled greens and reds to ripples of electric blue. Piers, harbours and breakwaters are often just as good as rock marks and you can often catch them incredibly close in. Tide is not always critical either, if there is a little depth to give them cover and security.

Tactics such as dropshotting, split shot rigs and tiny jigs all have their place. Small weedless jigs and Texas Rig set-ups are especially handy. For such close quarters lure fishing, an ultralight rod is great fun, coupled with say 6lb mainline and a slightly weaker final trace.

Various lures will work, but for the smaller fish, small soft plastics, grubs and small sections of scented plastic worm such as Isome or Gulp! are ideal. Where you are purely fishing for bites, a fairly short section of worm, as little as 3-4cm is ideal, owing to their habit of biting the tails of larger lures and avoiding the hook.

Casting close in, or simply aiming beneath your feet, the trick is often to feel the lure to the bottom before applying little lifts and twitches. This is no hurried affair however, and it is often a slower, more patient approach that is best. You will quickly receive little taps and shivers on the rod tip if there are wrasse present. It is best to resist the temptation to strike at these, however, and to wait until you get a bigger hit or the line starts to move away before striking. If you are not connecting, do try pausing longer and letting the bite develop. Even a 3oz wrasse will give a spirited battle on light gear, convincing you that there is something a lot bigger attached.

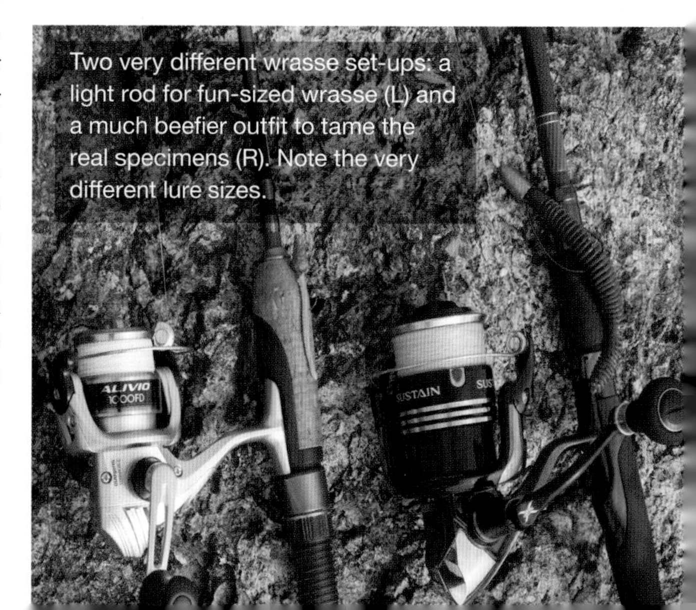

Two very different wrasse set-ups: a light rod for fun-sized wrasse (L) and a much beefier outfit to tame the real specimens (R). Note the very different lure sizes.

'Tapping' into Wrasse

Searching rocky areas for wrasse can be a hairy business, but one great way to bring them on is to literally let your jig head tap on rocks as you retrieve. Try sharp little lifts every so often, before letting the lure drop and strike rocks. This serves like a dinner bell for aggressive wrasse! Another option is to add rattles to your lure.

Wrasse are naturally curious, hence 'tapping' rocks can work. However, when filming and photographing them underwater with Jack Perks we also found they were less quick on the attack than other fish, sometimes studying the lure for a few seconds before either turning away or suddenly lashing out. Perhaps this is why 'hanging' a bait to waft in the tide can work?

Targeting bigger wrasse

Once you have got to grips with catching the smaller fish it is time to step things up a notch and have a crack at the larger ballan wrasse. These can be punishing fish to do battle with, and take a little bravery and stronger tackle! Ballans are supremely strong fish that can grow as big as 10lb; although any of 2lb or more would be regarded as a good catch, while a fish of 5lb plus would be regarded as exceptional. They fight hard and dirty to put it mildly, and with even a 2-pounder, a punishing customer, you will need adequate tackle. A medium lure rod with some backbone is a sensible starting point, with a firm, fast action for dicey locations where there are snags under your feet. 7ft is optimum, although in tricky spots the extra reach of a longer than usual rod can be useful to keep fish out of danger. Pair this with a mid-sized reel (say 3000 or 4000 size) loaded with 20lb braided mainline. This heavier tackle will give you enough power to take on these snag-loving predators and win. If it is a big fish spot, you might want to go heavier still.

Along with the stepped-up tackle, however, you will also need a little guile to find the better fish. For the ballans in particular, we often find a pattern of feeding which, if followed, should lead you to larger, actively feeding specimens.

As mentioned, wrasse are less territorial than the text books suggest and timing your visit to periods when the fish are moving close inshore and hunting can be key. After many years of fishing in the south-west, we've found that larger wrasse will move into water as shallow as 18-inches on a rising tide. The best time of day to encounter these fish is first light and the last two hours of daylight, with a good pair of polarising glasses. Sometimes it's possible to watch large fish moving between rocks and into position in the boulder fields where they like to feed.

Our observations do not suggest these fish are strictly territorial. Some of the areas they can be found are devoid of water on spring tides so it would be impossible for them to make their homes where we catch them. With this in mind, you can fish the same swims with confidence from one session to the next knowing that different fish will patrol the area; whereas if they were strictly territorial we would expect sport to die off or repeat captures to occur.

Effective lures and presentations for bigger wrasse

We've spent hours fishing varied patterns of lures while chasing larger fish. This started with Texas-rigged worms, crawfish and soft plastic fish.

Results varied, but most slender soft plastics in browns, reds, purples and silvers seemed to work pretty reliably. Others have stated that the colour blue triggers a reaction from dormant fish during spawning times, but Andy's feelings and extensive records suggest that blue is no more productive than other colours, and often less productive than browns, olives and reds.

Worm designs work well for numbers of fish, but a lot of the most magical, bone-crunching hits from the real monsters seem to happen when fishing lifelike fish lures such as the Black Minnow (a soft weedless fish lure, with the hook well protected until the fish bites) or other weedless rigged fish imitating shads. Whatever the case, big wrasse seem to have a nasty habit of beating up little fish who get too close!

Presentation can important with these fish-imitating lures. For example, wrasse will often hit those that are 'hung' (i.e. paused and left to flutter in the tide) much more readily than those that are worked quickly and erratically. Indeed, a lure twitched and then paused close to shade of rocks and boulders will often result in a sudden thump even after 10-15 seconds left rolling in the swell.

Perhaps for all true lure anglers it might feel counter-intuitive to just leave their artificial alone for several seconds at a time, but with modern soft lures, the motion of the tide alone will create movement and life.

One way to achieve this 'hang' is to try to balance your lure so that it sinks on a very slow fall; on windy days you may be tempted to increase the weight of the lure to remain in contact but this will work against you as you will begin working the lure in a manner where it is moving too quickly in and out of the 'feeding zone'.

Do not be alarmed; if you are using braided line, the bites will be transmitted to the hand quite noticeably if a wrasse strikes at the lure. You will find that, like the smaller wrasse, larger ballans sometimes just pluck the lure. Do not strike at these smaller taps; try to keep the lure in the zone and you will find a larger bite will usually follow.

Wrestling with wrasse

Hanging on for dear life: Gary Pearson gives no quarter as a large wrasse attempts to snag him. For the real monsters, a kayak or larger boat can be handy to get directly above the fish.

Once you hook a good wrasse you have only the briefest time to take control of the situation. Think of it as a head-to-head fight between you and the fish; whoever gets the initiative first is likely to win! Always fish with a tight drag and be prepared to take control at a moment's notice because most of these fights are won or lost within the first five seconds.

The instinct of wrasse is to dive into crevices between rocks or into thick kelp and weed beds. If this should happen and everything locks up, however, all is not necessarily lost. Releasing pressure on the fish by opening the bail arm of the reel, and then watching your line, will sometimes allow the fish to swim out and battle to recommence.

Wrasse conservation and further thoughts

Another of the old clichés of wrasse fishing is that these fish tend to be brittle and not easy to release. We think this is probably due to their tendency to swallow baits such as ragworms or crab quite quickly, however, and suffering from excess handling and crude unhooking. With lures it is a different matter, as the vast majority of fish will be cleanly and lightly hooked, and that tough mouth very resilient.

Contrary to the old text books then, we have found wrasse to be a hardy, durable species; but this does not mean they will stand up to rough handling or being dropped. Care and respect are essential and we would also advise bringing an unhooking mat, to avoid them being dropped or damaged on sharp rocks. Should you want a picture, you could always retain fish in the landing net for a few seconds while you set up the shot. Barbless hooks are also advised, or crushing the barbs down on your lures. With wrasse, battles tend to be short and brutal. With tight line maintained at all times, you won't lose many more wrasse due to the lack of a barb.

As a final thought on wrasse, it is up to us to protect these beautiful fish as best we can. They make poor, bony eating in any case and populations can be highly localised, so we must handle and release them with care. Lowering them back in a net and allowing a large fish to recover is always better than dropping or throwing them back, especially when a good one has fought hard and might need a few seconds to recover properly.

Wrasse are currently plentiful in our seas, but the same was also once true of other species that are now threatened. Apart from overfishing in popular spots, the main current danger is the risk of exploitation in commercial salmon farms, where they are now being caught and sold as a means of parasite control. The Angling Trust and other organisations are fighting hard to stop wrasse being taken en masse and we would urge you to support all efforts to preserve this great fish, which probably represents the best chance for any lure angler to land a real net-filler from the shore in the current era of commercial overfishing.

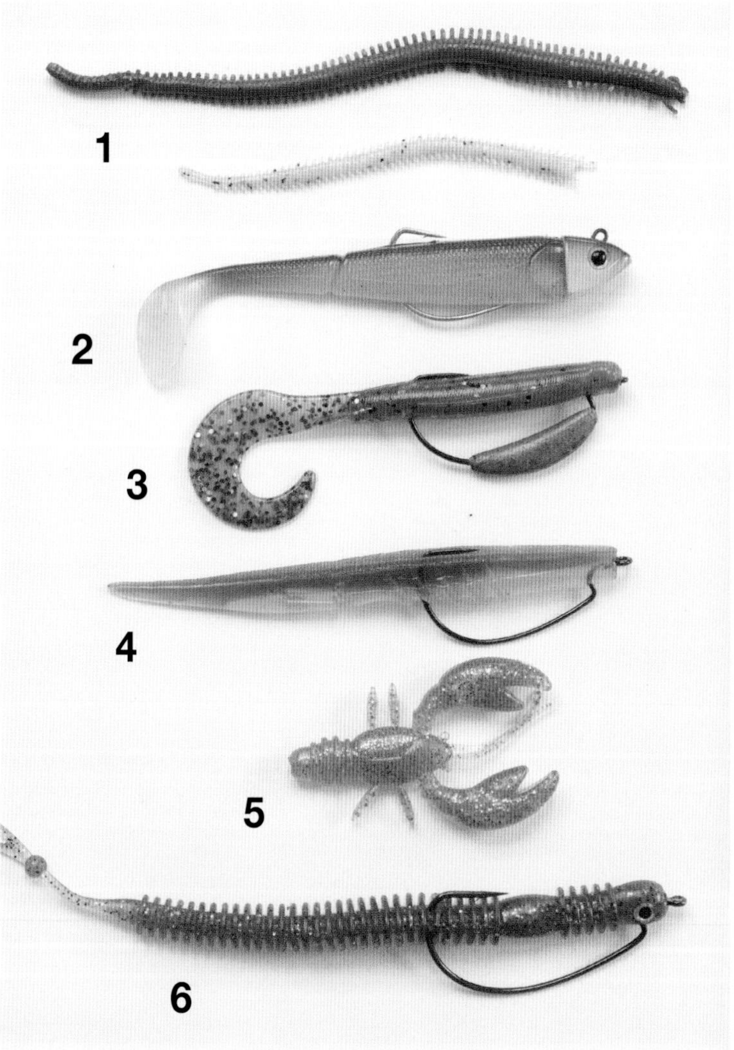

Six great lures for wrasse

1. Berkley Gulp!:
Scented, flavoured hybrid lures are the easiest way of all to get catching. Start small for bites from all the different wrasse, or step up for a big ballan.

2. Effzett Kick Minnow (90mm/12g):
Weedless soft fish are a must for bigger wrasse – and models like this one fish straight out of the pack for great convenience. Take plenty of spares, because wrasse will smash them to pieces!

3. Eco Gear Paramax:
Weedless rigged plastic worms are always worth a cast. This one is shown set up with a weighted hook.

4. Megabass Xlayer:
A soft 'stickbait' with an inbuilt rattle.

5. Major Craft Fighting Claw:
Crayfish imitations will also be taken as crustaceans by bigger wrasse. This model has buoyant claws for added realism.

6. Quantum Dude Worm:
Ribbed worms give off added vibration and are a good choice for wrasse of all sizes. For the real beasts, don't be afraid to offer a worm type lure of 6in (15cm).

'Pelagic fish' are creatures of constant movement. From the smallest to the largest, they are renowned for their power and athleticism.

PELAGIC PREDATORS

Tackle: For most of the smaller pelagic fish that venture inshore, light tackle is a must to get reasonable sport. Typically this would be a light lure rod in the 2-10g bracket, with lines of 5-8lb (2-4kg). For more demanding conditions, or for the travelling angler in search of larger species, a medium to heavy outfit may be required, however.

Season: Varies by species. For the shore angler, summer to autumn is when you are most likely to see migrations of fish like mackerel, scad and herring, which come closer to land as waters warm and shoals of small prey fish gather en masse. However, with annual variations and climate change, you might find pelagic fish in spring and even winter in some locations.

Other kit & considerations: Although the vast majority of our lure fishing is catch and release, pelagic fish like mackerel are perhaps a notable exception – and there is no harm taking a fish or two where they are prolific. Should you wish to keep fish, a trout priest or other implement is a good idea to humanely dispatch them. A freezer bag and bottle of frozen water will keep them super fresh and out of reach of the dreaded seagulls!

While some fish are localised or only make short migrations, other hunters are true long distance travellers, covering miles of coast to reach different feeding grounds.

Wherever you travel around the world, seasonal migrations of pelagic fish are a feature. For our light tackle purposes though, species such as mackerel, scad

and garfish first come to mind, as opposed to the larger predators in this class you might find.

Look at the beautifully streamlined, efficient shape of a fish such as a mackerel and we can instantly see that these are fish made for perpetual motion. Such creatures are athletic and tend to form big shoals, with large eyes and greedy mouths always keen for food. With so much competition and such a fast-paced lifestyle, they tend to bite first and ask questions later, making them very obliging fish for the lure angler. To their own cost, they also tend to be highly prized both on rod and line or on the grill.

Traditionally, methods for pelagic fish have been all about efficiency rather than pleasure. But we would hope that modern lure anglers are more interested in sport and sustainability. The classic broom handle rod and team of feathers might catch scores of fish, for example, but is pretty crudely unsporting, not to mention wasteful compared to lure fishing. The finesse of catching fish individually, on light, well-balanced tackle is an absolute thrill by comparison, and also far better for catch and release fishing.

Below: The dazzling flank of the mackerel, a classic pelagic killer.

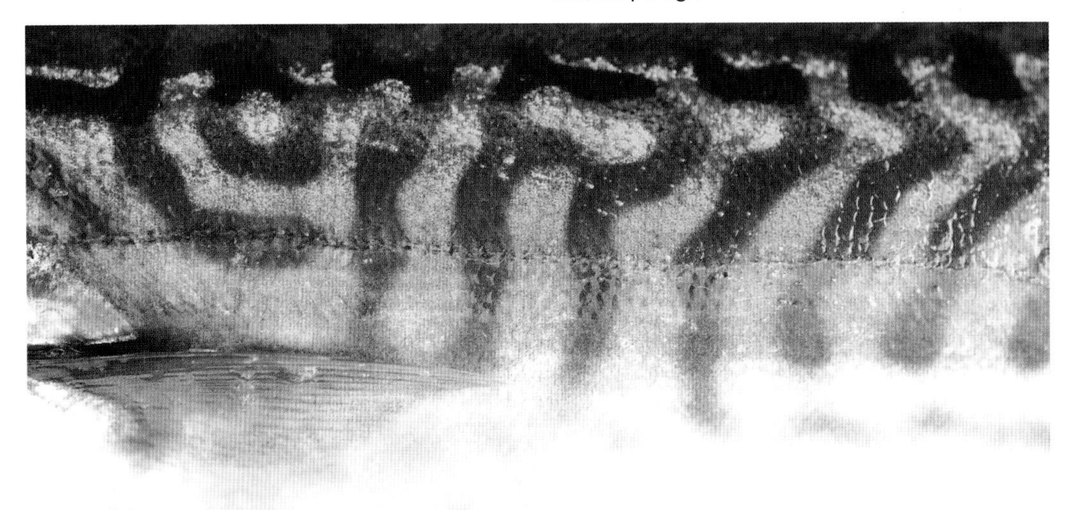

Hope's Nose, Torquay. A typical deep shore mark for mackerel.

Gathering shoals and golden rules

When shoals of bait fish swarm and the great shoals of mackerel, herring and other pelagic predators chase them in, the scenes can be dramatic. These invasions tend to be seasonal, with huge gatherings of fish present one month, gone the next. So, catching shoaling predators can be easy in the extreme, or a complete loss, depending on your timing.

With climate change throwing things further off kilter, the ranges and numbers of pelagic fish are changing. The humble mackerel has been rather sparser and harder to predict in the south of England over the past few years, for example, while the species has expanded massively in northern seas. Scottish waters have been thick with them, for instance, while they are now also caught from Iceland, something unheard of before 2008.

Fast growing and crazily aggressive, they are the ideal beginner's fish. How easy are we talking? Well, old hands down in Devon will tell you stories of mackerel caught on hooks dressed with sections of chocolate wrappers or the foil tops of milk bottles. Clearly it is more a case of where and when you fish rather than what you use.

For simple practicality, though, classic spoons and metals are among the very best options. This is because they are so easy to cast out, cutting through even a stiff breeze, while sinking fast and working at quite high speeds. The key to catching pelagic fish is so often about getting the right distance and depth, so counting down each cast and keeping tabs on how deep your lure is fishing are vital. Mackerel, for example, often tend to keep to depths of 10-18ft (3.5-6m) from deeper shore marks; garfish tend to be shallower, lurking in the top 6-8ft (2-3m).

It sounds like childsplay then, but the depth you fish at can vary depending on the type of line you use, the weight of your lure and also how strongly the wind and tides are. Counting down to different depths meticulously is the way to go, along with retrieving lure at consistent speeds. Perhaps the trickier challenge is remembering exactly the depth you counted to and how fast you were retrieving when you catch a fish, as these details often get lost in the excitement!

Using metals

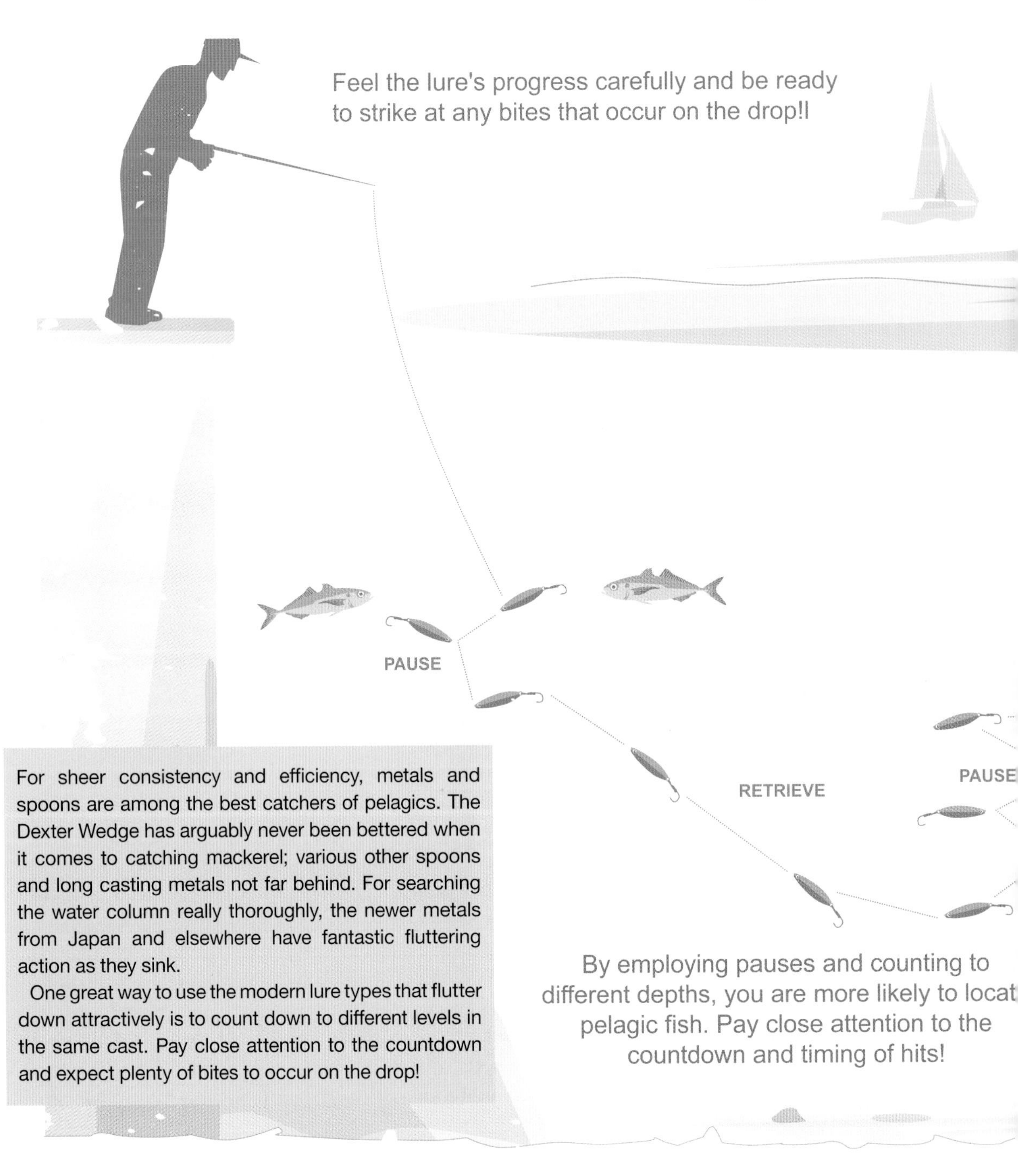

For sheer consistency and efficiency, metals and spoons are among the best catchers of pelagics. The Dexter Wedge has arguably never been bettered when it comes to catching mackerel; various other spoons and long casting metals not far behind. For searching the water column really thoroughly, the newer metals from Japan and elsewhere have fantastic fluttering action as they sink.

One great way to use the modern lure types that flutter down attractively is to count down to different levels in the same cast. Pay close attention to the countdown and expect plenty of bites to occur on the drop!

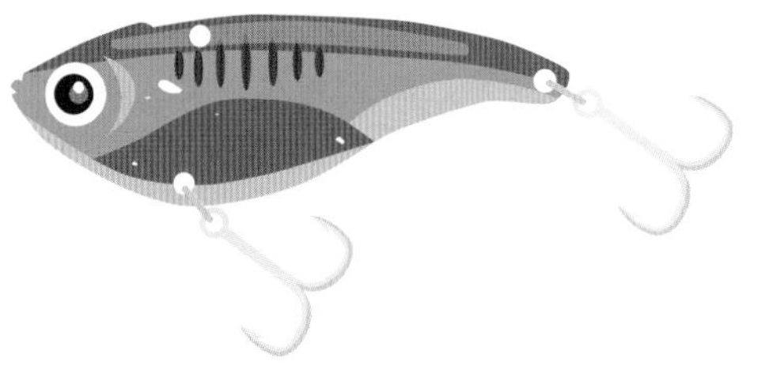

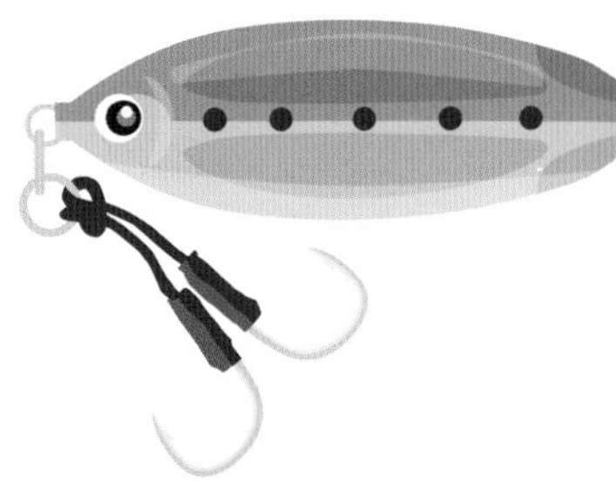

for pelagic predators

Other key considerations include picking the right size of lure. A lot of the popular sea fishing lures are inefficient because they use large, crude hooks; you'll catch a heck of a lot more fish with a quality size 4 to 10 than you will on those meat hooks! Finally, do always take a hook sharpener for this type of fishing. With typically bony-mouthed shoal fish and more wear and tear than most other lure fishing, your hook points will quickly lose their keenness, so take a few seconds to touch them up every trip.

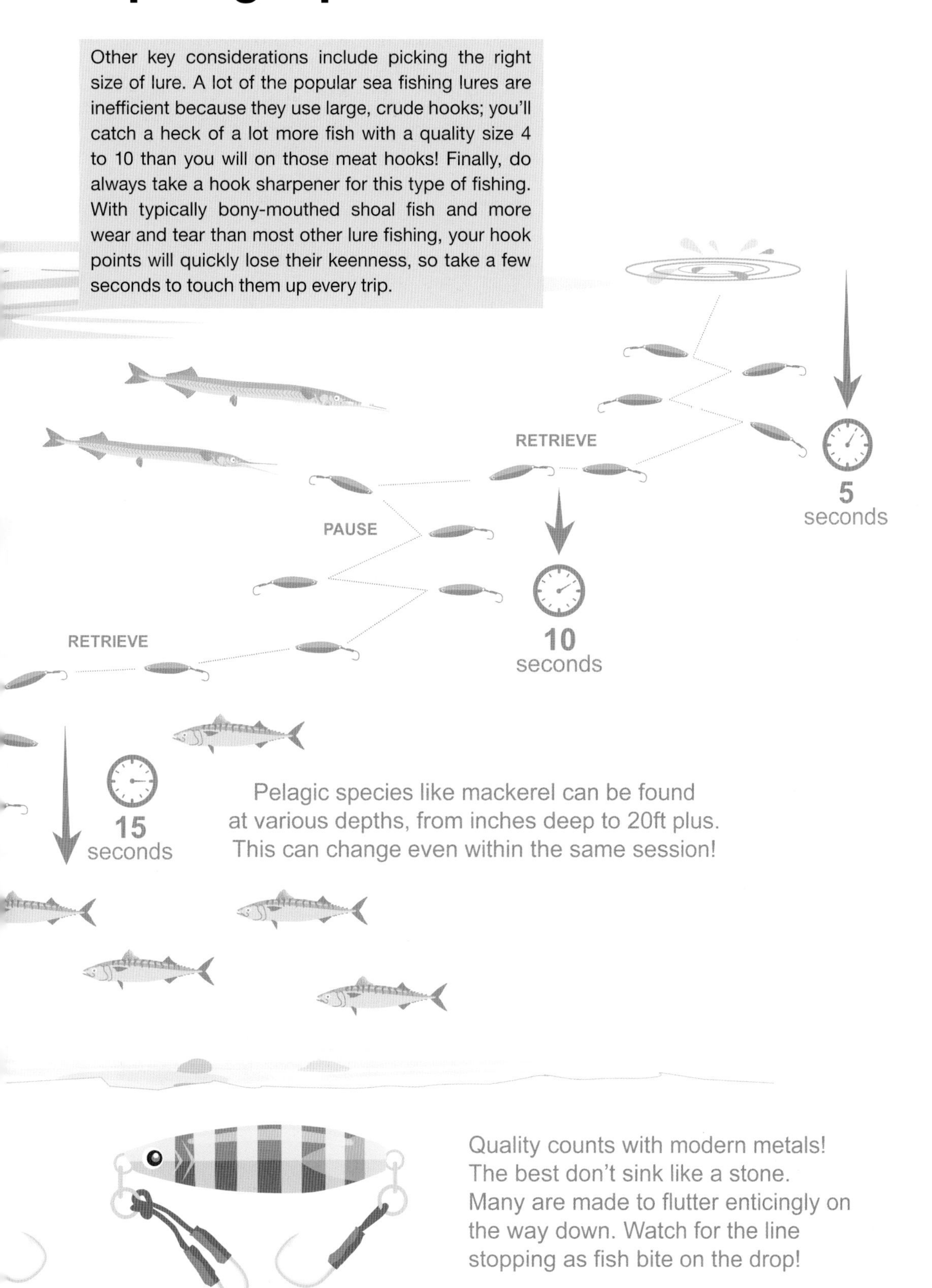

Pelagic species like mackerel can be found at various depths, from inches deep to 20ft plus. This can change even within the same session!

Quality counts with modern metals! The best don't sink like a stone. Many are made to flutter enticingly on the way down. Watch for the line stopping as fish bite on the drop!

Left: The scad, or 'Aji' to our Japanese friends, is a cracking fish to catch on light tackle.

Catching the Scad, or 'Horse Mackerel'

In the development of ultralight lure fishing, the scad is a fish with a special place. Japanese anglers have long prized these small but athletic fish, perfecting the use of tiny artificials and delicate tackle to catch them. In fact, many of the rods and miniature lures that have become so fashionable in the UK and Europe have their origins in Asia, and in scad fishing.

Closer to home, they are a seasonal visitor that can bring fantastic sport in the late summer and autumn. Timing and location are everything. Fish the wrong mark or time and you might never see one. But once you find their whereabouts the shoals can be huge; and like holidaymakers who don't want to return home, they can hang around for weeks on end. Harbours and ports can be especially likely spots – including Brixham Breakwater, Plymouth Barbican and Southampton Docks.

Small, straight soft plastics and metals are best as far as lures are concerned. Finding the right depth is critical, but when they are inshore and active they are quite often in the upper layers of the water. While they will chase and take a steadily moving lure, many fish will hit on the drop, and a twitch-twitch-pause type retrieve tends to be the most productive way to catch numbers of fish. They can be caught in the day, but night fishing is often superior. Slow jigging and the use of caro floats are also excellent presentations (*see page 51*).

Scad are brittle-mouthed fish that are quite easy to lose, so a loose drag and soft-tipped rod is the best combination. In fact, many Japanese 'Aji' lure rods have been developed specifically for the species, with a soft tip section, yet a parabolic with power lower down in the middle of the rod.

Catching Garfish on Lures

Of all the fish we might class as pelagic, the garfish has to be among the weirdest and most challenging we've encountered. A common presence from April until as late as October, they are fish that hunt close to the surface. Many an angler's eyes have widened as one appears from nowhere, snaking after their lure to attack it without getting hooked.

The gar's beak-like mouth is the source of the frustration. This sinister appendage is useful for hunting, but so tough that hooks won't easily penetrate it. Hence the angler has to get the fish to suck in the lure further, so that the hooks can find purchase where the mouth meets the head. Easier said than done!

One particular evening at Weymouth, we found this to our cost with an amazing number of garfish that would chase any lure going, but fail to get hooked time and again. After turning the air blue and doing a bit of head scratching, we finally managed to get even. The answer that

A tiny lure with added stinger; one way to hook a garfish.

Above: 10th time lucky! Dom and Andy endured literally dozens of bites on a balmy evening, before Andy managed this garfish.

day was to scale the lures right down. We tried soft plastics, but the only models that worked seemed to be tiny lures just over an inch (2-3cm), with a small, razor sharp jig head and perhaps even a second 'stinger' trailing off the back, such as a size 12-16 fine gauge hook. That said, we've heard of lure and fly anglers swapping hooks for velcro patches with some success!

With conventional lures, though, these fish can take some adjustment. Your natural reaction when you feel nips on the lure is to keep reeling. However, we tend to find that pausing the retrieve and giving the fish some slack gives the best chance of a hook-up. This is because of the way the fish takes a prey item; the gar will hit with its beak to stun and trap a small fish first, before moving it back into its mouth. If you strike too quickly, your lure will simply bounce off the beak, whereas our aim should be to give the fish enough time and slack to try and swallow the lure.

Since early frustrating experiments, however, we have also found that small metal lures will work to greater effect. Again, it is necessary to pause momentarily when you see or feel a hit, to try and persuade your garfish to swallow your artificial.

Most of the time, garfish will chase in the top few inches of water – so it's best to fish with a high rod tip. When a fish bites, however, dropping the tip whilst feeling the lure down can help provide enough slack for the gar to inhale your lure properly. The same can be said for when fishing with soft plastics: giving the fish a bit of slack will greatly improve hook-up rates.

For those curious to catch more garfish, further experiments could also be rewarding, and we've heard of enterprising anglers using small lures coated in wool and similar materials in place of a hook, to snag the teeth of the garfish!

Can I catch and release pelagic species?

Pelagic fish are designed for speed and long distance travel, but they tend not to be very hardy. This is partly why feathering and methods that use multiple, often crude, hooks are wasteful – because unwanted fish are going to be hard to release. As well as the fine, easily dislodged scales, it is also thought that the grease from human hands can react with the skin of fish like mackerel, causing damage. Although new research has disputed this idea, pelagic fish do seem quite brittle, so handle them as little as possible. Small hooks with crushed barbs are kindest; and with long forceps you can often flip your catch back into the sea without even touching it. If you do hook a fish badly and have to handle it, you are best to eat it or use it as bait.

Six great lures for mackerel, scad and other common pelagic fish

1. Dexter Wedge: As old as your grandad, but still a long-casting and highly effective spoon. Lots of companies have copied the original cheaply, but the version made by Dexter in Wales is still the best.

2. Eco Gear Metal Vibe: These distinctive little lures will vibrate aggressively on a simple, straight retrieve, but will also wobble enticingly on the drop.

3. Major Craft Jigpara: Simple straight metal jigs are ideal for most pelagics. Fish a simple, steady retrieve, but pause at intervals to let it flutter and flash as it falls.

4. Fish Arrow Flash-J SW, 1-inch: Fished on a darting jig head, this style of lure makes an excellent impression of a small, distressed prey fish.

5. Major Craft Aji-Pin: Japanese style pin tails work a treat, fished on small jig heads. The thin, supple body gives tons of movement, but collapses quickly on the take and is easy for any small predator to engulf.

6. Eco Gear Strawtail: So universally effective, we couldn't not include it. Works on a standard retrieve, on the drop or with a darting jighead.

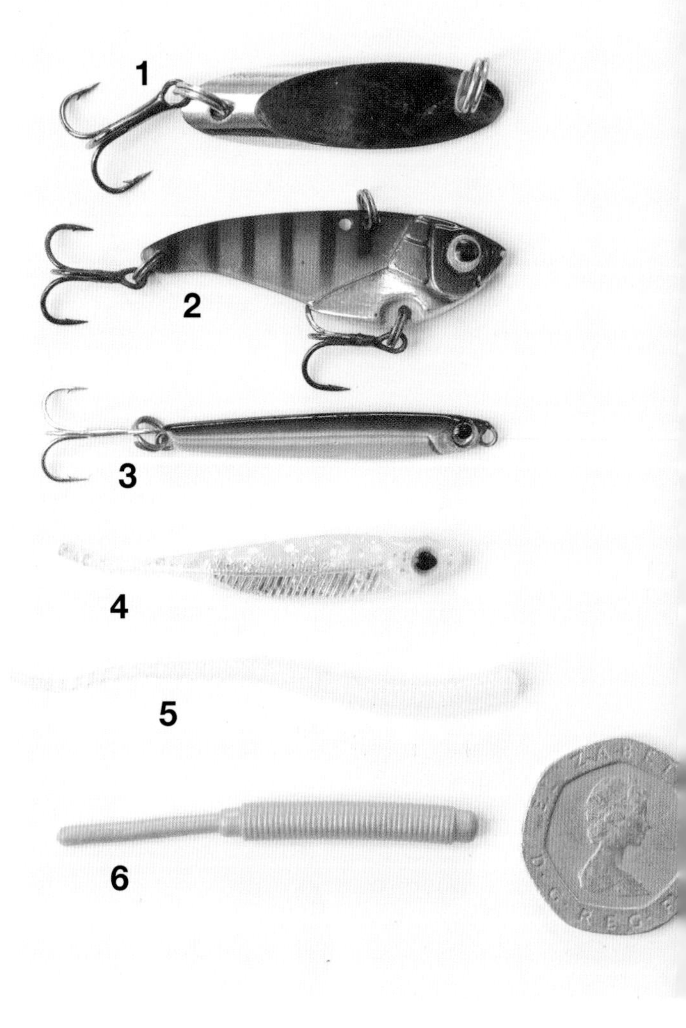

CHUB

Tackle: Chub are not the hardest fighters, but as they grow quite large and live in snaggy places, feather-light tackle is seldom an option. Typical lure gear would consist of a light rod of say 2-10g, with quality 6-12lb (3-6kg) braid. A lighter fluorocarbon trace of 5-8lb (2.5-4kg) then makes an ideal low-visibility link to lead up to your lure.

Seasons: Chub are catchable year round. A closed season applies on rivers, but canals and other waters may still be fishable. Summer fishing, when the fish will bask at the surface and can be stalked, is perhaps the most exciting time of all.

With keen eyes and a bold mouth, the chub is a worthy quarry for any lure angler. With something of a Jekyll-and-Hyde character, they can be quick to seize anything that looks like an easy meal or threat. But if their greediness is legendary, so is their instinctive wariness. They are long-lived, cunning fish that can infuriate the angler, which explains why the capture of a big old specimen is rightly regarded as a great achievement on a lure.

To catch them consistently requires stealth and watercraft. It often needs a willingness to gamble, too, because many of the best chub haunts are far from accessible. Just getting a cast in can be a challenge and these are fish to test your skills at cramped, close quarters fishing.

Chub certainly grow to a good size and you'll often have little trouble spotting them. Adult fish on even modest rivers can reach 2-3lb, but in the right water they'll comfortably double that. Some canals and stillwaters also have populations of chub, which goes to show just how adaptable these predators are. They have sharp eyesight and various hunting strategies, from seizing fallen insects to terrorising shoals of minnows. Lure fishing might not be the most ruthlessly efficient way of catching them, but it is definitely the most exciting method – especially where sight fishing is possible. Lure-caught fish tend to be a bigger than average stamp, too, compared to those caught on worms and maggots.

Chub swims and tactics

Approach is everything when you fish for chub. If you scare them, it's game over before you even cast. One or two golden rules that can help us, however. The first is that they tend to swim facing upstream, with their noses into the current, much like trout. You can use this knowledge to your advantage by approaching from behind them, or downstream, where possible.

Chub are notoriously wary of anglers. Standing bolt upright or 'skylining' is never a good idea and the successful chub

A shoal of chub averaging 3-4lbs shelter beneath a tree. A small surface lure is likely to draw a curious or aggressive response.

angler might have to stoop, climb or crawl along the bank slowly and unobtrusively. Quite often the very best spots are those most anglers avoid. Anglers willing to take a risk will often catch chub where others fail. Shorter rods and underarm casts can help too.

As to where to reliably find chub, the same kind of spots tend to produce, time and again. Look for any significant cover or vantage point such as an undercut bank or a depression in the riverbed. Overhangs, bushes and floating rafts of weed and debris are classic.

Their behaviour can also help us catch chub. For one thing, in spite of having regular patrol routes, they are fiercely territorial and opportunistic – and this natural, reactive aggression can often be their undoing. Chub are quick to come out and inspect anything that comes into their particular 'patch'. Fish sheltering under an overhang or weed raft will take almost anything edible-looking that falls in, from a grasshopper to a mayfly or even a mouse. We suspect that hunger is not always the prime motivation and they are also quick to repel unwanted guests and rivals, which could explain their aggressive reaction to small, vigorous plugs and fish-type lures.

Andy recalls a time where he was waiting for dusk in order to fish a marina for zander. Arriving at the canal early he decided to watch the water and fish around under some moored boats outside a pub. On the far bank was a common elder tree, berries were falling off into the water and he knew this is usually a good time to catch fish feeding on the berries. He scaled down to a tiny 1-inch dark lure and, first cast, a good fish thumped it on the drop. It turned out to be an excellent canal specimen, much to the surprise of the onlookers from the pub.

Notes on lures

A whole selection of artificial lures and baits are worth a try for chub, with typical sizes in the 2-3in (5-8cm) bracket. Spinners can be an excellent starting point. Casting ahead of the fish and working it back across the current is often effective, but they will often follow only to turn away. There are ways to get them to take though. Sometimes throwing a break or pause into the retrieve works, or an extra twitch or burst of speed as the chub gives chase. One key time to get a response is just as the lure tracks downstream of you, and the spinner suddenly turns and catches the push of the current. If you're really tearing your hair out though, tipping a spinner with a section of soft plastic lure or a real worm tail can also convince those more cautious chub.

You can also try and mimic natural food around cover, of course, with a small floating lure. Some specialists make their own micro baits to resemble crickets, bugs and all kinds of things a chub might find falling into the river. Smaller surface poppers are excellent fun to try: cast low and accurately around branches and cover. Try twitching and pausing, and if a chub closes in on the lure, give it an extra pull.

Small floating plugs and surface poppers can also be excellent, not so much to mimic food but to provoke the aggressive side of chub. Models with a tight, vigorous wiggle are especially good to try; don't worry about being too subtle or lifelike, but aim to wind them up with a strong vibration right through their home turf. Where weed and trailing branches are too treacherous, weedless lures are obviously worth a try too.

This chub accepted a spinner. After following for several yards, it decided to hit only as the lure turned and picked up speed in the current.

Alternative tactics: Weighted flies, jigging and dropshot

Although chub are a fish you can catch on most lures, they are also definitely a species to tear up the rulebook. Sometimes they'll turn their noses up at a realistic, perfectly presented lure, but hit something totally unnatural very next cast. In short, the chub angler should never be afraid to experiment with lures and mix up the realistic with the outlandish.

Another excellent trick is to use ultralight tackle with a weighted fly. This is best achieved with a rod that will cast just a gram or so upwards. A couple of split shot will send any suitable fly to your target fish. Woolly Buggers, Humungous or any small streamers and trout lures are worth a try. In the height of summer you might try a large floating 'terrestrial' fly such as a cricket, beetle or daddy-longlegs given casting weight by a small controller float. Is it strictly lure fishing? Yes, absolutely, you are imitating a living thing.

Winter fishing can also be productive with the weighted fly, but this time of year is also when soft jigs come into play. All the usual grubs, minnows and even crayfish have their day, and with a suitable weight these can be crept into deep holes and difficult spaces. Chub are also happy resting up by bridge pilings and walkways, or any snaggy spot whether that's a wild river or an urban canal.

Detecting and hitting the takes with subsurface chub is very different to those you can watch striking at the surface. The takes can be quite sudden and hard, but the bigger fish are more than capable of just gently sucking in a small lure, so be ready to strike at any second should you feel even the most minimal yet suspicious touch on the line.

Once you hook a decent one, you'll find that chub are not insanely powerful, but will put a serious bend in an ultralight rod. They fight dirty too and will know every snag in the river, using their wits as well as their muscle. They can be easily lost, should you play them too brutally or with overly heavy tackle – and the ideal rod has a smooth action and a bit of power in reserve. They have rubbery lips, but hooks can easily come adrift if you're not careful.

A word of warning: a big chub will quite often come in disarmingly easily, before saving the biggest fireworks once you sink the net. Be ready and have your clutch set a little light if anything!

Night fishing for chub

Lure fishing for any species at night is exciting stuff, but even more so when it comes to chub as this is the best time of all to get them to smash big lures at the surface, and you could well net a really big fish. Balmy evenings are especially atmospheric, when the sky sparkles with a thousand stars and the smell of summer fills the air. There is nothing better than an evening walking the river bank in search of chub.

Under cover of darkness, even the biggest are less wary of human presence and it is no coincidence that a disproportionately high number of fish over 5lb are tamed after dark. They have a real mean streak at night, with a preference for surface-waking artificials. Depending on the type of river you are fishing, you might opt for a weedless

Above: A weighted fly cast on ultralight lure gear fooled this fine chub. You can watch how Dom tackled up and landed it in a special YouTube film.

lure such as a frog, mouse or the all-time classic Heddon Moss Boss JR. Or alternatively, choose a 'crawler' style lure such as an Arbogast Jitterbug or similar. Such lures will scuttle across the surface giving an audible popping or glugging noise which will quickly alert any nearby chub. Perhaps the only disadvantage is the likelihood of drawing pike too; but thankfully after dark you can get away with a lot heavier tackle and even a wire trace without unduly worrying the chub.

If there is one style of chub fishing that highlights just how aggressive and unpredictable these fish are, it's night fishing with big lures. In fact, besides those monsters caught by design, the authors know of some huge fish caught accidentally by anglers casting really large pike poppers that most of us would consider way too big for chub! So, if you want to catch the real brutes in your local river on a lure, you know exactly the time to try your luck.

Six great lures to try for chub

1. Eco Gear CK50: Small, vigorous floating plugs can work well for chub, which often hit them from sheer aggression.

2. Duo Shinmushi Topwater Lure: Skittering surface lures resembling larger insects and other accidentals are immense fun to try for chub.

3. French-bladed spinner: Classic spinners are always worth a try and will work vigorously in the current.

4. Fish Arrow Minnow, 3-inch: Chub eat lots of smaller fish such as minnows and bleak, which explains the appeal of realistic soft lures.

5. Reins Bubbling Shaker: For deeper-lying fish, soft worms and grubs are ideal on jig or drop shot tackle.

6. Eco Gear Bug Ants: On many rivers, crayfish are an important prey item. 'Creature' style lures mimic the young ones that chub find especially hard to resist.

Common as muck, yet loved by anglers everywhere: the humble perch.

PERCH

Tackle: Perch are good fighters for their size, but don't grow huge. A rod of 1-5g is delightful for smaller waters and more intimate fishing. For practical purposes such as longer casts or deeper waters, however, a rod in the 3-15g bracket might be needed. Reel lines and leaders range from 4lb to 8lb typically, unless the risk of encountering other predators demands heavier gear; regular pike attacks might necessitate a wire trace as 'insurance'.

Season: Perch are caught throughout the year, although in some areas a closed season applies. When the fish are spawning (typically from late March to early May, depending on location) they should be left alone.

Found in all kinds of freshwater throughout Britain and Europe, perch could be called a true everyman's fish. Aggressive and distinctively marked, they are a beautiful, instantly recognisable species. The big mouth immediately denotes an out-and-out predator, while the bold, muscular profile marks it out as a manoeuvrable all-rounder rather like a bass. Their curiosity and greed are legendary and they are often the first fish an angler ever catches. But while the small fish can be bold and stupidly greedy, the larger fish are a different prospect. Perhaps this is why the perch is a creature that captivates both expert and complete beginner alike.

The wide diet and bold, curious nature of the perch makes it just as willing to grab a jig or spinner as a worm. They are a prolific and versatile species. All but the very largest form groups from just a handful to several dozen. When young they will eat anything they can find, from various insect larvae to freshwater shrimp. Once they reach a respectable half pound or so, however, they tend to dine almost exclusively on fish, whether that means minnow, roach or gudgeon.

Of course, not all perch or perch waters were created equal. There are those packed with prey that never quite live up to their potential; perhaps the perch compete too readily with each other to get large? There are waters that look too cramped or poor quality which produce big surprises. It just goes to show that perch are among the most versatile of our freshwater predators, thriving in locations from grotty little canals to giant reservoirs.

So what constitutes good quality perch fishing? These days, social media and tackle companies are awash with huge fish, but most of the time a 1lb perch is a good one, while a fish of 2lb (1kg) or greater is a 'specimen'. A 3 or even 4-pounder is a season if not lifetime's best for most. Even with a current spike in popularity as pronounced as that dorsal fin, there are more large perch out there than you might think. Part of the reason they're not caught is because anglers don't fish single-mindedly for them. Sure, you will catch the odd monster by accident when pike fishing, but to really get the most from perch, you have to focus your approach.

Perch locations and feeding habits

Location is everything when catching perch. For anglers of all experience levels, they can be reassuringly predictable in their whereabouts. These are fish of the textbook bridge, bend or even the side of a boat – at least on smaller waters such as rivers and canals. In terms of both where and when they will feed, there are also definite patterns that are not rocket science. From the outset, however, we should probably distinguish between two types of places that we will find our perch. There are definite areas where perch choose to take shelter and other areas they like to hunt; and these are often quite separate.

In the first type of area, perch love structure, features and hiding places of all kinds. Whether it's an overhanging bush or bend in the river; a deep hole or the side of a bank, any natural or man-made

Left: Perch thrive in urban locations and love man-made as much as natural cover.

vantage point is likely to shelter perch. Cover is good; so is any change in depth (often called a 'drop-off'). Find an area where suitable cover meets a change in depth in the same spot and it's an even safer bet there will be perch present – probably a big perch or two. Perch tend to have a hierarchy, with the biggest fish hogging the best lairs.

Perch sometimes ambush prey from favoured hiding places, but a lot of the time they will simply be using cover to stay safe and keep a low profile. All but the biggest perch must be wary of bigger predators such as pike, otters and cormorants, after all. Areas of cover form their 'living room', to put it one way. So, while you might sometimes catch them here, you may also have to find out when and where they break cover to go on the feed.

Locating prey shoals can be very helpful, therefore. If you can witness attacks actually taking place and small fish fleeing, this is even better. Anywhere there are concentrations of immature preyfish you will often find perch moving early and late in the day,

Feeding spells and locations

One of our favourite locations for perch is on a rural canal, where a large fallen tree holds a shoal of four or five big perch. Just a few yards away the water is wider and more open, with hordes of tiny fish. During short periods of low light, typically at dawn or dusk, these local bullies come out of their den and work as a team, terrorising shoals of small roach and bleak. A lively, darting soft jig or spinner works well. It can be quite important to 'match the hatch' loosely though and come up with a decent rough copy of their current meal choice (usually a silvery lure of 2-3in imitates a baby roach nicely).

Arrive just half an hour either side of these magic spells and tempting these fish becomes a very different task. They can still be spotted in clear conditions, lurking around their favourite sunken tree. Sometimes they can be tempted, but now the successful lure is likely to be a jig or dropshot-rigged lure, teased close to cover. And when they seem reluctant to move, a bright or provocative lure can work far better than carbon-copy realism. Perhaps this is because we are trying to provoke a fish that is not on the offensive?

As for sunny, still weather and the middle of the day, catching one is another matter. They can be like stone statues until the light begins to disappear and the little fish get jittery. There are always exceptions to rules though – sometimes a lazy fish will still react to a lure, just as a period of leaden skies or a lively breeze will sometimes put the perch back in feeding mode. Wherever you fish then, the key is to try to identify reliable spots and feeding patterns and think of the preferences of the perch first rather than your own when it comes to antisocial hours or awkward spots!

Left: This perch took at the social hour of 1pm. Such catches tend to be the exception rather than the rule, however. On this occasion, overcast conditions and coloured water helped.

harrying and striking at them. It can look chaotic when victims jump clear, but perch are adept pack hunters too, effectively herding and trapping their food like an organised gang.

Much of the time, however, such feeding spells are fairly short and the angler cannot afford to rely on finding fish that are actively hunting.

Instead, we are trying to tempt perch that are sitting on the bottom, hiding, or merely holding position in between meals. These perch might not be willing to lash out at anything edible-looking; but they can still be tempted if we get it right, whether we fool them into thinking an easy meal has come along, or simply trigger their curiosity or territorial aggression.

On the subject of perch and their hunting preferences, it is often said that they are pack hunters and to a large extent this is true. Find one fish and you will almost always find more of similar size, whether it's a huge gang of juveniles or just a small handful of brutes.

Perch lack the frightening burst of pace or grasping teeth of pike, zander and others, and are adaptable all-rounders rather than specialist sprinters or ambushers. Perhaps this versatility explains why they are able to colonise even quite poor habitats where other fish struggle? Their mob tactics are a big advantage in the hunting stakes, and they are masters at penning and cornering fish with their shoal mates. Perhaps this is why spots like sluice gates and the corners of harbours are ideal, as are sudden depth changes and hidden structures even in very deep water.

Perch bites can vary quite a bit and also tell us more about their hunting habits. One noticeable trait is that they will not always devour prey in one go; they often nip and damage prey to render them helpless, before making a final, decisive attack. This could explain why you will quite often feel a nip or two when perch fishing, before a bigger touch. Don't be disheartened if you feel a brief snap and then nothing when retrieving your lure, therefore; keep that lure moving and be ready to set the hook as soon as you feel that more decisive hit. Some anglers have even used stinger hooks for perch, but this isn't generally necessary. Instead, keep reeling and stick to sensible sized lures.

Thoughts on perch lures and tackle

Many of us were raised on spinning for perch, but there are so many different lures and presentations to catch them these days. Size is perhaps our first consideration. Tiny fry imitations, plastic worms and even flies can be worked on the lightest of lure rods – and this can be awesome sport, even for the fish of mere grams (affectionately called 'wasps' by today's lure anglers) while those of a 1lb or more really pull back. Perch certainly punch well on the very lightest rods and size isn't always everything.

If you want to be more selective though, it makes sense to scale up and also try lures of 3-4 inches (7-10cm) to single out bigger perch and tackle distances and depths where the tiniest of lures are impractical. It also depends on the size of fish you find and what they are eating. Suffice to say you can set out your stall to catch any perch that swims, or tackle up to try and pick out a big one, but a 2-inch lure will catch most perch that swim.

Every touch counts

The competition scene in the UK has seen anglers honing their skills to deliver huge numbers of perch on tiny soft lures, using jig and dropshot presentations. Winkling them out by the dozen is a real skill and it has become common to see anglers fishing hooks as small as a 16 and cutting down lures of just an inch (2.5cm) or less to maximise their chances with small perch.

Experts such as Adam Kirby and Steve Collett have really taken this to the next level and, interestingly, anglers who have come to perch fishing from an LRF background tend to do very well with their customary fine tackle and small hooks. Indeed, many adverts, magazine articles and beginners' sets feature hook sizes of 2-6, but you will catch far greater numbers of perch using micro jigheads and lures, or finesse drop shot tactics with tiny hooks in sizes 8-16.

It can be a skilful business too, involving a delicate touch, total concentration and razor sharp reflexes.

TIP

Andy's Tip
Try bigger lures for bigger perch!

Andy has spent many hours fishing lakes for perch that have large concentrations of silver fish. While pleasure fishing one particular venue with a simple waggler he noted just how big some of the roach and rudd were, and these fish would even accept tiny 1-2in lures when a change of tackle was made! Surely a bigger lure had to be the answer?

The lake in question was quite overgrown in places, so the artificial chosen was a 5in (12.5cm) weedless paddle tail, which would avoid the worst of the weed but collapse around the hook nicely on impact, should a big perch inhale it.

Andy went on to take two real brutes of over 3lb among plenty of other solid fish, all caught by working the lure slowly on the drop, allowing it to flutter down onto the weed then bumping it up like a fleeing fish. Since that day, a similar presentation has worked for plenty of other big fish on similar waters.

Do such bigger lures create a bigger pulse or even annoy the better fish? Or do the largest perch single out larger victims individually rather than hoovering up numbers of tiny prey?

Who knows; what is certain is that offering them a bigger target works often enough to make it worth a try in just about any scenario when you are struggling to find that monster among the runts.

Left: Would you like to go large, Sir? Ok, go on then...

Such tactics do produce the occasional bigger perch too, but if you want to tangle with the real brutes more regularly, you may have to substitute quantity for quality and scale things up. When it comes to targeting specimen perch in the 2lb plus bracket it is certainly prudent to fish with lures of 2.5in (6cm) and above.

Specific lure types and presentations

Among all the lure types used to catch perch, soft plastics are the most versatile. There are many presentations, but the basic jig head plus soft plastic lure combination will suffice for a great deal of close range fishing, and is perhaps the easiest method of all for the beginner. Rigged with enough weight to keep our artificial deep and allow us that critical 'feel' through a light rod, the simplest of soft plastics can be jigged around under the bank, or cast around the contours of walls, boats, bushes and other features. A stop-start retrieve is often the way to begin, keeping things lively but not overly fast in order to explore each spot fairly thoroughly. You want to try and keep the lure low and slow for perch if there is any great depth, because reeling faster will cause the lure to ride up off the bottom. Another great trick is to let the lure sink right to the deck and lie 'dead' for anything up to ten seconds, before suddenly lifting it off the bottom.

Naturally, drop shot presentations are also ideal where you can find distinct features to work around and are also excellent when the perch are more static and not willing to chase. Don't always expect takes to come instantly; even if your lure touches down just where you want it, it can take a really patient retrieve and several seconds of inspection for a perch to inhale a soft plastic lure. Drop shot is often the very best method for catching these more cautious fish, or those

Never neglect close in for perch, especially if there is any depth, cover or colour to the water. Keep back and try right under the rod tip at first; there was a lovely fish sat right beneath this concrete platform in just 24' (60cm) of water.

glued to a particular spot or depth (*see p41* for further drop shotting pointers).

Once you have mastered the basics, there are countless ways you can catch perch on soft plastics. They can be rigged to fish at just about all speeds and depths, while the variety of modern artificials and presentations is vast: drop shot, Neko, Jika, Ned, Mojo, Texas, Carolina… you name it, it will work for perch in the appropriate setting.

Old school perch lures

Of course, way before the huge selection of soft plastics arrived on the scene, traditional styles of lure caught plenty of perch, so we should also mention their value to the modern angler. Starting with bladed lures, most waters will still respond to the time-honoured spinner or small spoon at some stage. If the perch are busy striking at shoals of flashing fry, a spinner can be excellent. That pulse and flash create more disturbance and catch more light than the fanciest soft plastic too, making them well worth carrying on any day, even if it's just to throw the occasional curveball at fish that won't cooperate. We'd definitely advise a modern, in-line version, to save a headache with twisting your reel line.

On the subject of bladed lures, smaller spinnerbaits can also be very effective and are much underrated. They have excellent weed resistance and a real pulse and presence if you fish dirty waters, and nor will they twist your line like the old Mepps or Ondex would.

For the record, you can even spice up your favourite soft plastic presentations with a spinning blade at times. Although not as widely available as other jig head types, a bladed version is always worth carrying in your perch box. On several occasions during competitions, for example, we have seen bladed jigs score when the fish had seen every trick in the book.

Last but not least, we should also briefly cover crankbaits, micro jerkbaits and one or two other ideas. Cranks certainly have their day in smaller sizes, especially those that create a nice tight wiggle and will get down to the necessary depth fairly quickly. We suspect that, realistic or otherwise, these lures also have a high annoyance factor and will occasionally trigger perch that are behaving with their customary bully-boy aggression, rather than a natural feeding response. Who knows? It works often enough to give it a try.

In addition to all the usual and not so usual perch lures, one final trick even many of the die-hard perch anglers could learn is to use flies on their usual tackle. Whether dropshot rigged or given casting weight with split shot or some other means, streamer-style flies have a lovely fluid motion and sparkle all of their own, quite unlike any solid or soft plastic lure. For pressured fish that have grown wary of OTT lures, that greater subtlety can be key.

As you have gathered here, variety is certainly not lacking when it comes to ways to tempt perch without a worm or live bait in sight! The key is not in knowing a gazillion fancy lures and rigs, but in figuring out when to use the different presentations in your armoury. Rather than a scattergun approach, these skills are best learned gradually, figuring out one new method or water at a time.

DIFFERENT VENUES FOR PERCH

Canals and smaller waters

If we had to recommend a suitable venue to start fishing for perch with lures, the easiest would be canals and other smaller waters such as ponds and drains. Not only do these fisheries have quite obvious features, but present the angler with consistent, manageable depths to search.

We'd include clear drains and other waters in this, and if you can find a setting where perch can be observed following and taking lures this is even better, because it will teach you a great deal about how they behave around your artificials. Find a clear little lake or a canal with minimal boat traffic and you may have found the most visually entertaining place of all to test lures and presentations.

Small waters are just the place to try the very lightest tackle and take your fishing back to basics. One of the first lessons to learn is just how close in to the bank the fish often are – and to work your lures patiently, right under the rod tip whenever you find extra depth or some cover close in.

On clearer, weedy waters, you may well reap the rewards by fishing little spinners and any small, flashy soft plastics. Small floating plugs and suspending lures might be essential where you find a lot of weed growth. For the newcomer, or the angler used to more featureless waters, weed can be off-putting. But perch love such territory, and there is a lot to be said for getting stuck in and exploring the spots others avoid – even if it means using weedless lures or painfully slow-sinking presentations.

On more busy, urban or industrial waters, perch can be just as prolific, but

A typical narrowboat canal. These venues demand an active approach, hopping between spots.

fishing around structures may well be your main plan of attack. Whether you target bridges, pontoons, walls, marinas – and right down the edges of the boats there – the rule is very much 'explore everything'. The presentation par-excellence for so many of these features is drop shot, with quite a patient approach; having observed shoals of perch under the water, they do not always bite instantly and can require extra patience.

Without massive variations in depth or width on a lot of canals and drains, the best spots can often be similar year round, too. The big difference is in the typical presentations and the speed you work your lures. In general terms, summer fish are livelier and more willing to chase, attacking higher in the water and even sometimes suspending in the upper layers in warm temperatures. In the winter, you will need to work lures more slowly and patiently and really nail down those times and locations fish are willing to feed. After a bitterly cold and clear night, for example, that usually productive first hour of light can be a dead loss, while the fish unexpectedly wake up and start to hit in the late afternoon, with just a little extra light and heat.

There isn't the space here to describe these tiny waters in vast detail, but Dom's previous work *Canal Fishing: A Practical Guide* has further guidance, with details of many UK canals, along with hotspots, lures and advice for these waters.

Reservoirs and Large Stillwaters

Large reservoirs can be daunting places to locate perch, to put it mildly. From the wild lakes of northern Europe to the giant reservoirs of England's Midlands, they have huge potential for larger perch but can take some careful sussing out. Bank fishing can often be quite limited too, necessitating the use of a boat.

Above: Bigger lakes and reservoirs hold some stunning perch; but access via boat or kayak can be a must to get at them.

If you are approaching any large lake, a fishfinder is a big help. If you have no technology though, all is not lost. Many fisheries have maps and even contour charts if you do your homework. Otherwise you can count lures such as jigs as they go down, and common places to start are next to towers and structures, steep sloping shores and under bridges and jetties. The latter features often contain concentrations of fry and smaller fish that the larger perch will be feeding on. If in any doubt, some fish spotting early or late in the day will often reveal dimpling prey or hunting birds.

Failing the location of substantial prey shoals, a sensible way to proceed is to start by boat fishing at different distances from the shore, working your way at different distances from the bank. Using the various methods we've discussed, explore weed lines first, starting close to shore and following the weed out as far as you can. See if you can locate where the weed finally thins out and the bottom drops away; this will often follow a particular line to your right and left, so start off by exploring this drop-off for perch sat up against the weed.

Another method is to move further away and cast toward the shore with crankbaits or spinnerbaits. Starting with a spinnerbait, cast it toward the bank and allow it to fall and touch the weed then burn it back to the boat using a quick retrieve. Crankbaits can be used in the same way. Selecting lures to dive to the required depth and run just above the weed beds can produce some spectacular results with perch darting out of the weed to nail the lure.

You will find reservoir perch stacked up in shoals up to 30ft deep, often resembling a christmas tree on your finder. This may be against structure but they can often be found moving from place to place. Without electronics we'd suggest moving and covering as much water as you can. During the hours the sun is high in the sky, the perch will often be deeper or close to structure. Tailor your approach to suit the depth of water, and bear in mind that low and slow is often the way to go during the day, as the shoal builds. Besides the usual presentations, using the Ned Rig or simply dragging your lure like a feeding fish or a burrowing worm on a Neko Rig can be deadly if the bottom is relatively clean. Alternatively, if you find a shoal stacked high then there is a chance that the larger perch are sat on top or the edge of the shoal, so underweight your lure or use a lure that falls slowly to pick off fish from the edges of the shoal.

The best time of all to catch perch is often as the light falls. Under cover of duller conditions, this is the time to find perch moving into the margins to attack shoals of small fish, especially around boat yards, bays and steeply shelving areas. During the late or early shift, these hunters will often come up higher in the water to feed too, so faster retrieves and a 'match the hatch' approach should work well.

Running water perch are among the fittest and best looking you'll encounter.

Rivers

For many traditionally-minded anglers, rivers offer the real cream of lure fishing for perch. Running waters come in many sizes, but all have similarities where the habits of perch are concerned. They are often totally underfished for smaller predators too, since so many visitors only have eyes for pike. There are some huge perch in British rivers such as the Wye, Trent and Severn, for example, where the best swims seldom see a perch angler. If you can wade, push or boat your way to those bits others can't reach, even better.

By their very nature, river perch must contend with flow to hunt and lead an active lifestyle. This makes them among the fittest and most beautiful of their kind. It also makes the rivers a great target in the dog days of winter when they are likely to be more active than their stillwater cousins.

Flooded rivers and perch

Floods might seem a tricky time for perch, but sport can be surprisingly good if you can only get a lure close to them. Raging currents and rising waters cause them to seek out and hug areas of definite shelter. Any deep holes are likely spots, as are man-made features such as flood arms and defences or any backwaters out of the main flow. You might assume that in such conditions perch just keep their heads down and sulk; but when you consider that their smaller, weaker prey are even more at the mercy of the rising waters, all is not lost! Indeed, slacks that are choc-a-bloc with prey can be well worth a try with a lure, even where waters look dull and unappealing to our eyes. Find a deep, sheltered spot packed with prey and you might just catch a really fat perch! This is the time to try a vigorous paddle-tailed soft plastic in a bright colour, or choose a brightly coloured UV lure on a drop shot and fish it slowly through these areas.

A good perch swirls in murky water; it took a bright yellow lure, presented beside an outflow where the bottom was scoured a little deeper.

River perch are as predictable as any others and just as big fans of sanctuary and depth changes. These fish are sometimes unfairly described as 'lazy', but in truth, like any predators, they simply live as efficiently as possible and refuse to waste energy. Hence any break in the flow could hold perch: sunken trees, weirs, wider sections, bends and pools are all worth trying. Equally important though, are the depth changes and features you cannot see so easily at first glance, so it always pays to spend time learning the riverbed and making notes where there are holes and structures. All of these riverbed features will be used by perch to hide behind during certain times of day or when the river is in drought or flood.

Certain depressions in the river bed are always likely spots to try. These areas are affectionately referred to as 'perch holes' by matchmen on some rivers, who love drawing such hotspots.

Once these holes are known it is down to your skills to work out what the fish want. Often there is no great secret once you've located the fish, and you can often enjoy many trips back to the same holes to find them there. The knack is to get them to bite, as often they won't be in attack mode. As the rivers build up speed during periods of heavy rain, perch will often tuck down even more tightly into these holes or move into others that offer shelter. On rivers with marinas, flood arms or other large artificial structures, expect the fish to move in.

It is helpful to think of perch as having different modes or moods on any given day. They can seem difficult or absent during periods of inactivity; whereas it is not uncommon for perch to show themselves during the most vigorous feeding periods, often coming into shallower water to feed on unsuspecting baitfish.

Overall then, they have very definite times and areas for rest and attack; but while we can identify tendencies in most waters these are not set in stone! A classic example of this was once when Andy was out fishing on a tough day. He and his friend Martin were surprised not to contact any perch in their usual deep holes on the riverbed. Thinking of giving it up as a bad deal, Martin then took one 'last' cast into mid river expecting nothing. Just as the lure came back to the margins, a large perch hit it. Nor was it the last. So much for the deep holes, these fish were hunting in barely a foot of water late in the day!

It was Andy's turn next with a spectacular perch of 3lb 6oz, taken inches from the bank where you wouldn't have dreamt of fishing in any 'normal' day on the river. Of course, the other moral of the story is one of sheer persistence – predators can switch on or off the feed in an instant and you never know what the next cast will bring!

Tricks and tips for difficult perch

When you consider just how suicidally greedy they can be, there are also days when you might scratch your head at just how difficult the humble perch proves. There can be a few different reasons for this and, frustrating as it might seem, figuring out these conundrums is a lot of the real satisfaction and craft of lure fishing!

Conditions can play a huge part, especially if temperatures are plummeting or fluctuating. Could anglers also be playing a part though? Given that perch can

Flies offer something completely different for tricky perch and can be fished on dropshot tackle, or simply with split shot to add weight.

be surprisingly long-lived (15 to 20 years for a big one), it's certainly possible that they might 'learn' to avoid certain lures or heed the warning signs we put out. Doing something different is certainly worth a try wherever you find pressured fish. Here are five ways to beat the stalemate:

1. Try a suspending lure: Suspending lures can often get a result for clued-up perch. For one thing, they break the monotony of the standard up-and-down jigging presentations, giving cagey predators more time to react as they 'hang' within easy reach.

2. Add a blade: An extra kick can help counter dirty water or agitate dormant fish. Look for jig heads that incorporate extra vibration.

3. Trail a fake or real worm: If perch follow but won't bite, try trailing a piece of scented rubber, or even half a real worm from the hook of a metal lure (it's ok, we won't shoot you).

4. Ned rigging: Working a buoyant lure along the bottom offers something completely different, making this an ideal plan B on many perch waters. Worked with longish pauses, up to ten or more seconds, it can be dynamite!

5. Try flies: More fluid and subtle than any lure, hard or soft, why not try a fly? Either use dropshot style, or simply add a couple of split shot as casting weight. Turrall now produce special 'Drop Shot Minnow Flies' designed by Dom for the purpose.

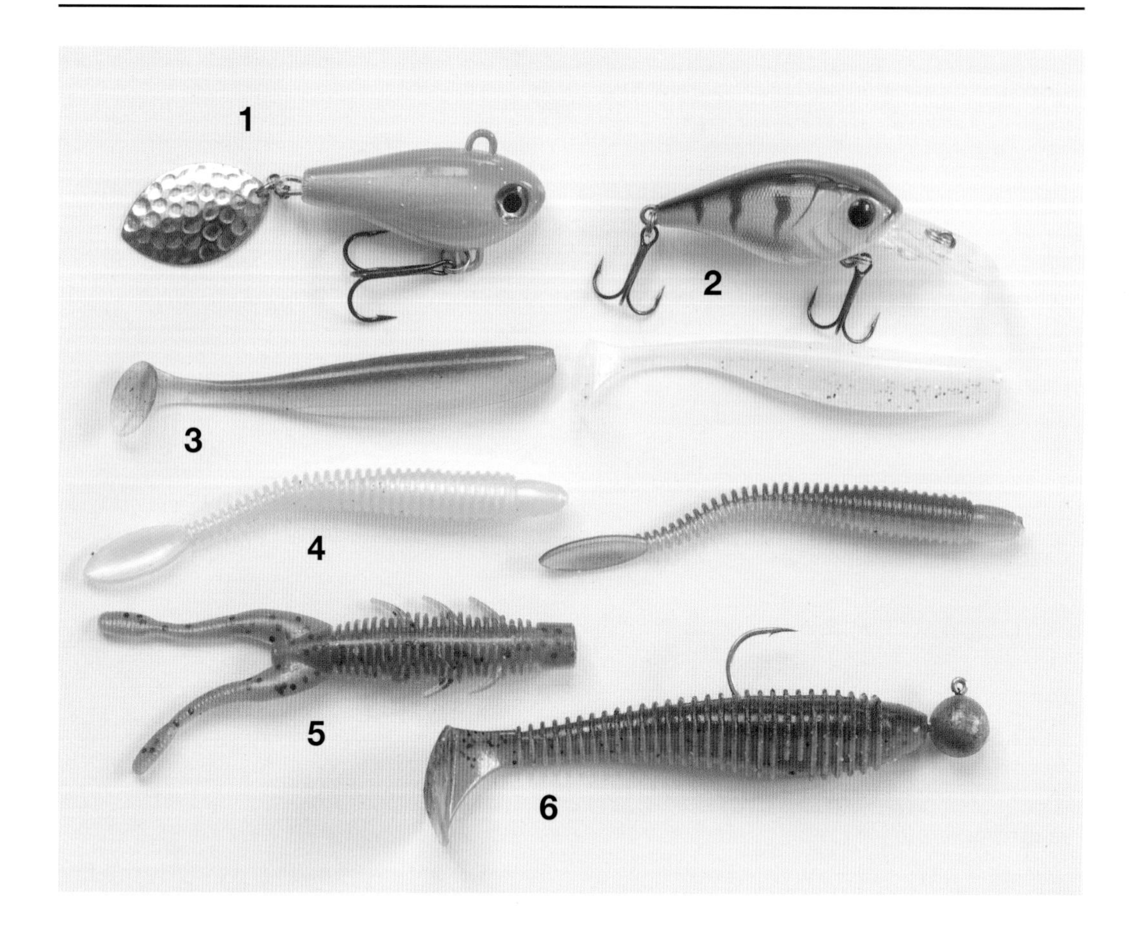

Six great lures for perch

1. SpinMad: There's still something magical about a vigorous, bladed spinning lure. Modern, in-line spinners like these avoid line twist.

2. Major Craft Zoner 40S: Small, vigorous crankbaits continue to catch perch, whether you try them in natural or outlandish colours.

3. Keitech Easy Shiner: Small shads such as this can be fished on a small jig head or drop shot tackle and will catch every size of perch.

4. Lunker City Ribster: Ribbed plastic worms are excellent for perch. Shown here are typical colour choices for clear water (Arkansas shiner) and murky water (bright green). A good choice for drop shotting.

5. Z Man TRD HogZ (3-inch): A special buoyant plastic lure. It's always worth having a few floating arificials to try the deadly Ned Rig.

6. Keitech Swing Impact Fat (3.3-inch): Paddle tailed lures give out plenty of vibration, working in most types of water.

TROUT

Tackle: For smaller waters, where trout would average under half a pound, a featherlight outfit would suffice, with lines as low as 4lb. Short, finer rods help cast light lures on streams where tree cover is tight. On bigger waters, where there are stocked rainbows or larger wild brown trout to a good size, stronger tackle is sensible. A rod in the 1-5g stamp, or 2-10g for larger waters, coupled with reel lines of 4-10lb, is ideal.

Other kit: Waders are often essential to access trout streams properly, while polarising glasses are a huge help in spotting fish before they spot you.

Season: Dependent on location. For most wild fishing in the UK, for example, the season runs from 15 March to 30 September. Other countries, and indeed artificially stocked trout fisheries, sometimes allow fishing all year.

Still sometimes frowned upon by the game fishing elite, catching trout on lures has enjoyed a big resurgence in popularity lately. Nor is it difficult to see why. They are aggressive fish with good looks and athletic fighting qualities. Cast aside the clichés of delicate rises and hatching mayflies and these are finely-tuned predators in the truest sense of the word too, testing not only the angler's ability with a lure, but their cunning and stealth.

In spite of the upper crust reputation of trout fishing, these fish are also more common than most people imagine.

Above: Trout are scrappy, sporting fish that not only bend an ultralight rod well, but greatly improve the angler's ability to read running water.

Water quality is key, but you'll find them everywhere from deepest Cornwall to the most northerly parts of the British Isles, including a great variety of urban locations. Highly intolerant of pollution, they thrive best in clear waters from tiny streams to giant lochs.

All larger trout feed heavily on smaller fish, making them a prime target for the lure angler.

Superbly adaptable, trout eat a wide range of food. Flies are a mainstay, but all trout predate on smaller fish, including their own kind. And with the biggest trout in any water very much confirmed fish eaters, they are a great target for the lure angler.

Trout Fisheries

As for the trout in your area, it may take a little homework to track down affordable, accessible fishing. In many places, the old 'fly only' mentality dominates and lure anglers are prohibited from many river beats and reservoirs. However, there is an awful lot of fishing still out there that is cheap and readily available. From wild, bushy streams to urban fisheries, trout are not always as exclusive as you might think. In fact, some of the very best and most accessible fishing is found right in our towns and cities, sometimes free of charge. Wild

Left: Some of the best open-to-all trout fishing for the lure angler exists in our towns and cities, as Dan 'Esox' Fois demonstrates in urban Yorkshire.

lakes are not always exclusive or private either, however, especially when it comes to more remote locations, whether you travel to Cumbria's moorland tarns or the Scottish lochs.

Even in the depths of winter, there are various stocked lakes where you can fish for rainbow trout. And while most are fly-only, an increasing number are allowing lure fishing for trout through the winter – and with the average 'stockie' rainbow trout weighing a couple of pounds, these make for great sport.

Stealth and Aggression

Trout are not the cleverest of fish, but their aggression is equalled by an instinctive wariness. This is a creature that will lash out through hunger or territorial instinct, yet scurry away at the first sign of danger. They do not suffer clumsy anglers gladly, so before we even begin to discuss lures and techniques, it is important to have some understanding of trout behaviour and how best to approach them.

On many streams and rivers, wading is essential. You might find a few places you can cast from the bank, but wily trout have a habit of lying in the most inaccessible places. So the challenge is to wade carefully, so that you have a shot at the fish without spooking them.

Some of the trout's habits help us get round their shyness. For one thing, they

Aggressive they might be, but wild trout are also sharp-sighted and cautious. The shallower and clearer the water, the more wary they are, so the successful trout angler must often be stealthy.

are lovers of flow and oxygen, and will nearly always sit facing upstream (i.e. with their noses pointing directly into the flow). Hence you'll get closer to the fish by approaching from behind them, in effect, and casting upstream or across the current.

That's not to say that casting downstream isn't an option; in fact retrieving lures 'against' the flow can give them a really vigorous movement, while you can also use the flow to float a buoyant lure such as a plug into tricky bits of water you wouldn't dare cast. But fishing downstream tends to mean the fish are looking directly at you, so you must keep your distance and move with even more caution.

A typical trout river

Exploring a twisting, turning stream is a joy with a light rod and a minimum of tackle. Leopard-spotted and well concealed, the fish of clean, fast flowing rivers are perfectly adapted to their environment. They are alert to any presence around them, whether it is the plop of a spinner, or a clumsy step from the angler.

Anglers used to stillwater and coarse fish might find a fast-running river daunting, but it is an exercise that will teach you a lot about currents and rivercraft. Some of the best places to find trout are often where there is a steady or strong pull of water, with shelter nearby, whether that is a natural obstruction, for example, or a 'crease' where fast and slower water meet.

Larger trout, especially, thrive on ambush points. Look for rocks and boulders that provide shelter from the flow; look for bends, bushes and places where the riverbed drops a little deeper, or an obstacle provides a break in the current. Every trout river is unique, but they all have similarities, which can be broken down into a typical sequence of features, from shallow riffles to steady runs to deep pools. Generally, small, shallow running lures are great for steady glides and shallower water; but dropshot presentations can be excellent for deep, tricky pools. Our rivercraft diagram (*see p74-75*) has further useful notes on typical features.

Don't fear the fast water!

TIP

Anglers who don't regularly fish clear, rushing streams are often inclined to head straight for the slacks and slower areas on any river. This is not a preference shared by trout, however, which love pacy, well-oxygenated water. Not only does the current bring their food to them, but means they don't get the luxury of time to study your lure very carefully. Give them only a second or two to decide whether they want your lure, and you'll often get a proper car crash hit.

In fact, tumbling, swirling or creased water is always a likely spot for a trout. For one thing, little weirs, waterfalls, corners and steps create hustle and bustle. Trout not only love the oxygen, but are less aware of our presence. On hot days, for example, when the river is low and the fish easy to spook, these spots can be a life saver. Some snags and features are obvious, others less so, but from a single boulder to a submerged shopping trolley – they are all worth investigating.

Right: Take a tip from the game angler: rough water is not to be feared, but explored. Any suitable 'pocket' or break in the current could hold a large trout.

Lures for trout

There are many lures that work for trout, depending on the fish and where you find them. For wild fish a policy of 'match the hatch' is a good one. Small prey such as minnows, bullheads and bleak are found on many rivers. Fry and fingerlings are another staple, including young coarse fish such as perch. Trout will also eat their own kind.

Above: Like for like. Trout react aggressively to smaller rivals, so a spotty lure can work.

Small spinners are perhaps the classic way of fishing for trout that many of us used as kids. They can still be highly effective today, but can be terrible at twisting your line. Modern, inline spinners totally solve this problem and make excellent trout lures.

We must not forget classic metals either, though. Spoons have been long synonymous with trout and salmon fishing. Models such as the Abu Toby and Kuusamo Rasanen are still commonplace in trout fishing but the new breed of Japanese metals are also worth a cast, with beautifully subtle actions that work wonders in moving water. You can always remove excess hooks to leave just a debarbed single, too.

Trouble with Trebles?

With any species it is important to consider conservation, and this is never truer than with precious stocks of wild trout. Debarbing hooks is a great idea, but we would also strongly recommend using single rather than treble hooks, especially on any water where the trout are smallish. Plugs and spinners can easily be converted to single hooks via a split ring fitting. If you find your hook-up rates suffering, you might consider tiny, single hook rubber lures, which tend to collapse more on the strike and are harder for the fish to lever free than the traditional hard lures.

Small crankbaits are another classic, and there is something about the wiggle of a plug that can bring an aggressive response in trout. Smaller lures from 1.5-2.5in (4-6cm) are ideal. Plugs will wiggle especially aggressively when worked against the current, so these lures are also useful for working downstream and swinging across

Above: Poppers are excellent fun for trout, providing explosive takes. Many strikes are pure territorial aggression.

the flow – sometimes you'll need to do little other than hold the lure against the current to feel it weaving and throbbing. Even when they're not hungry, trout are hard-wired to react aggressively to rivals and intruders, and seem to hit vigorously wiggling plugs out of sheer spite!

For a more realistic route and the advantages of single-hook presentations though, modern soft plastic lures are the perfect size and shape to mimic natural food and catch trout. Pin and paddle tails are brilliant – go for silvers, greens and golds to mimic fish like minnows and roach fry. As we've already mentioned, drop shotting can be handy in deeper pools, weirs and other vantage points close in, but the most versatile presentation is with a small jig head and soft body. The upturned single hook of a jig is much less prone to snagging compared to traditional trebles hooks, so you can cast into deep or rocky spots with less fear, too.

Last but not least, we should also mention the world of artificial bugs, flies and other creatures. Lure makers craft some delightful surface poppers and imitations of everything from beetles to crickets. These can be excellent fun for summer fishing. Streamer-style trout flies are even deadlier, and are easily cast on ultra-light lure tackle with the addition of one or two split shot.

Stillwater Fishing

Away from the rivers you will also find trout in many clear wilderness lakes and stocked waters, from farm ponds to major reservoirs. And while some waters prohibit lure fishing, there are plenty of options out there if you look around. Some of the more enterprising small stillwater fly fisheries offer winter lure fishing too, where you can enjoy a great day's sport with an ultra light lure rod.

Smaller lakes are an ideal starting point, whether natural or stocked. Look for smaller fish around the edges at first, such as sticklebacks, and any fry. Find trout striking into prey and you won't have too much trouble getting follows and takes.

As on the rivers, you should look for cover and vantage points and depth changes. Don't always expect trout to be far from the bank though. Browns, especially, can be found feeding very close to the shore – and while waders are useful on many waters, you should always look and have a cast before plunging in.

Small plugs have their day, but various metals (enabling longer casts to cover lots of water) are also good. Trout can be frustrating followers though, shadowing the lure for a distance before turning away. Perhaps the best trick to turn these fish into takers is to tip your lure with a jelly or artificial worm. A single hook rig, much as you might use for mullet, is effective and kind to the fish too.

Lastly, if you want to combine lure fishing with a more traditional twist, most flies can be cast and retrieved using lure tackle. Small streamers such as the Woolly Bugger, Appetiser or Humungous are all worth a try, weighted with a couple of split shot, or even dropshot rigged. Is it strictly cricket? Not for the traditional fly fisher, perhaps, but it works!

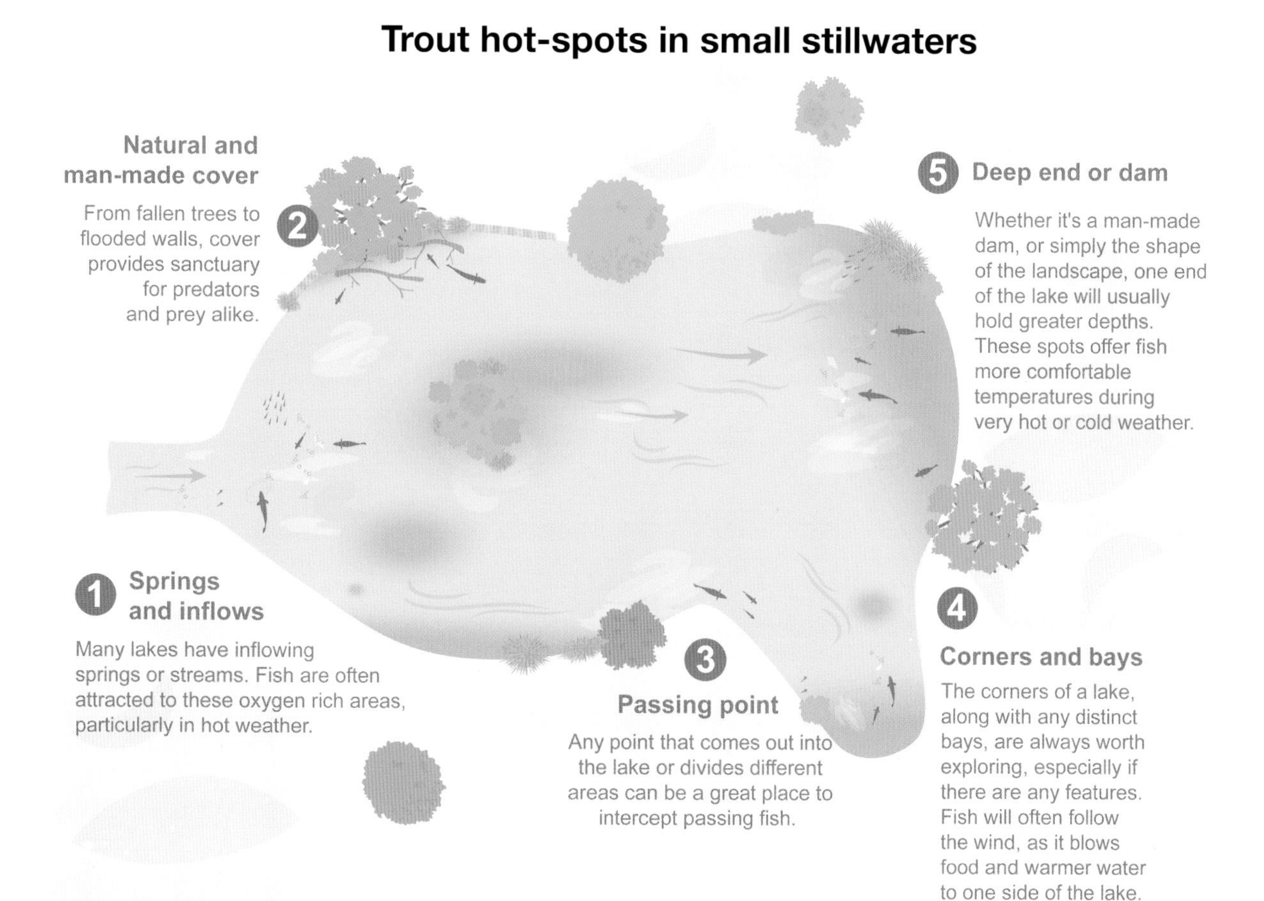

Territorial Browns v Roving Rainbows

The two main trout species have quite stark differences in basic behaviour. Our native trout, the brownie, tends to be a territorial creature on stillwaters. This isn't to say they won't move to feed, but each trout will often have a definite home patch. Once you have had a few casts in an area, it is much better to keep moving and explore as much water as possible – and they also love snags and holding spots really close to the bank so always take your time and keep back at first.
Rainbows, on the other hand, tend to keep on the move in the manner of more pelagic fish species. Catching these on many stillwaters is as much about getting the right depth as covering miles of water, so always pay close attention to the level your lures are fishing and make a mental note when you get bites.
On just about any stillwater, features such as inflows, bays and even aerators are also well worth singling out. Point swims, where the bank juts out, are a particular favourite as you can guarantee that trout will pass on their travels. Other areas, such as bays or corners where a protruding tree or rock formation offers sheltered, slightly warmer water are worth exploring. If the fish are cruising rainbows, as opposed to browns, you might continue to get bites in one suitable spot too, as new fish move into the area even after you've caught one or two.

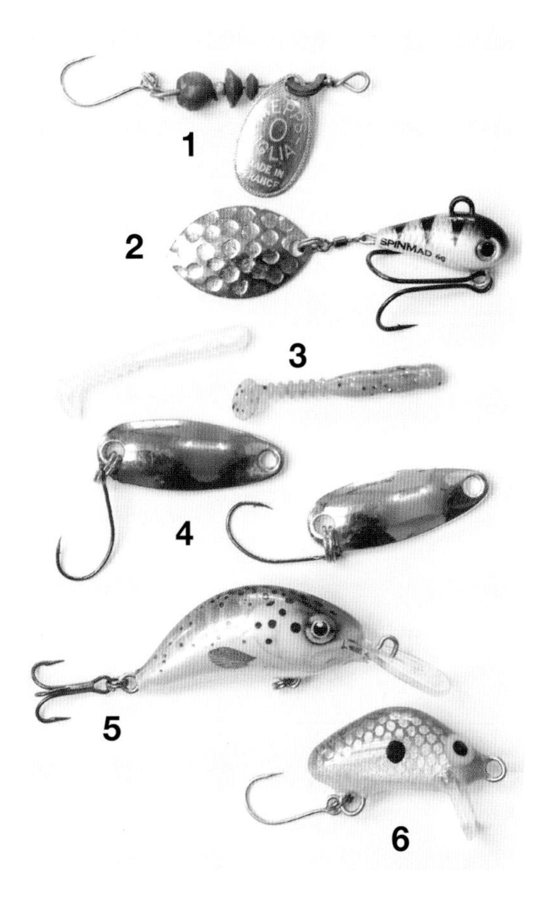

Six great lures for trout

1. Traditional spinner: A small, flashy spinner is still deadly for trout, especially for river fishing. Even deadlier when you tip the hook with a plastic grub or worm section!

2. Modern, inline spinner: Modern bladed lures are arguably better still than the originals, as they tend to have more casting weight and avoid line twist.

3. Miniature paddle tails: Tiny soft plastics of just 1-2in (2.5-5cm) are another safe bet for trout, usually fished on little jig heads.

4. Spoons: Classic smaller metals in gold, silver or copper still have their uses, working beautifully on a straight retrieve with the occasional pause to let them tumble and flash.

5. Shallow diving crankbait: Trout often react aggressively to hard lures. A model of just 1-3in (2.5-7.5cm) with a lively wiggle is ideal.

6. Snap bean: Hard lures that splutter on the surface, or run just beneath it like this little fellow, provide some of the most exciting sport of all for trout.

ZANDER

Recommended tackle: Light tackle is a joy to use, but with a bony mouth these creatures demand a rod with some backbone and a fast recovery to set hooks on the strike. Tubular tipped rods are best, perhaps with a test curve as light as 5-15g on small waters, but deeper stillwaters and rivers may require something heavier, say 10-30g. Braided mainlines are best, typically from 12-20lb. Zander don't have as severe teeth as the pike, but still require some abrasion resistance, so fluorocarbon traces are the norm, usually from 10-15lb. If pike are a risk, you may want to switch to fine wire.

Other kit: Quality long nose pliers are a must, as is an unhooking mat. Zander may look fierce, but are quite fragile and demand respect.

Season: Can be caught most of the year, although the fishing tails off in the spring, when many anglers leave the fish to spawn.

The fierce, enigmatic zander is highly prized with lure anglers wherever it is found. These fish provide savage bites and rod-bending sport, but are also a test of skill and patience. There are many reasons for this, but suffice to say the habits of zander are idiosyncratic and quite different to other predators.

You only have to look at the design of this fanged killer to get a few clues about its sinister lifestyle. One giveaway are its large eyes, covered in light-sensitive

Into the deep: Andy feels for takes on Rutland Water, during the golden last hour of light.

receptors: these render it something of a specialist at hunting in dark places, whether that means the extreme depths of a huge lake or the muddy waters of an urban canal.

As for the mouth of the zander, it is smaller than the pike but still viciously armed. The teeth are not as numerous as the pike, but pronounced and razor sharp. These fangs are not only used for grabbing hold of prey, it seems, but also wounding and disabling for a kill at the second or third attempt. Most smaller fish are fair game, whether they be roach, perch, bream, bleak, gudgeon or even young of their own kind.

Perhaps the greatest magic of zander fishing is the mystery of it. Owing to the deep, distant or murky places they operate, it is exceptionally rare to witness a take at first hand and so much of our knowledge is based on feel or educated guesswork. In one sense, therefore, our task is like that of a blind explorer, feeling their way into another world.

From what we do know, their modus operandi when it comes to hunting is subject to both definite habits and curious variations. One look at the abundant, large fins that line their lower bodies tells us that zander are fish that love to lurk around contours, snags and depressions on the bottom of any water. They are usually found in the deeps by day, although they will come into the shallows or maraud higher in the water during periods of low light, such as early or late in each daily cycle, to chase their prey. For the most part though, zander require precision and thought to catch, because they are not as trigger happy as pike and other predators.

Zander can be both active feeders and opportunists, with definite moods. There are times and areas where they will lurk and sit dormant. But they also have regular feeding times and patrol routes. If you can identify these hot spots and magical hours, the going can occasionally be easy. However, these spells can be brief and the real skill in catching zander is the art of enticing them to take when they are dormant or 'off'. More often than not, this is the state you will find them in, after all.

Smaller zander are definite pack hunters, moving in groups. These fish can vary quite a bit in size: on a narrowboat canal a typical zander may weigh just a pound, but on a giant lake or river, the average could be three to four times that.

As the fish get bigger, they shoal less and it is believed the real monsters are loners. Any zander of 5lb is a decent one, while even the most meager zander water is capable of producing the odd fish of 10lb or greater. Regardless of size, they are a delightfully sinister challenge, and immensely enjoyable to battle on light gear.

Indeed, as more lure anglers discover this, attitudes are changing towards this species which is a non-native arrival in many areas. Once seen as vermin or a threat to fisheries, populations appear to have stabilised in the English Midlands, for example, and they have become a popular and naturalised presence to a large extent.

Tackle and Lures

Although it's great fun to target zander on the lightest gear, they can be tricky to hook on the strike. A tubular tipped rod with a firm, through action is best – and if there was ever a species that merited a little more investment in quality tackle, it's the zander. We've seen anglers miss bite after bite, for instance, simply through having a rod that is too soft to penetrate that bony mouth efficiently.

Below: Lures for reservoir fishing: sizes of 4" (12cm) are typical for bigger stillwaters, along with jig heads weighing from 25-50g, depending on depths and conditions. The stingers in the box are optional, but can help convert tail strikes into fish.

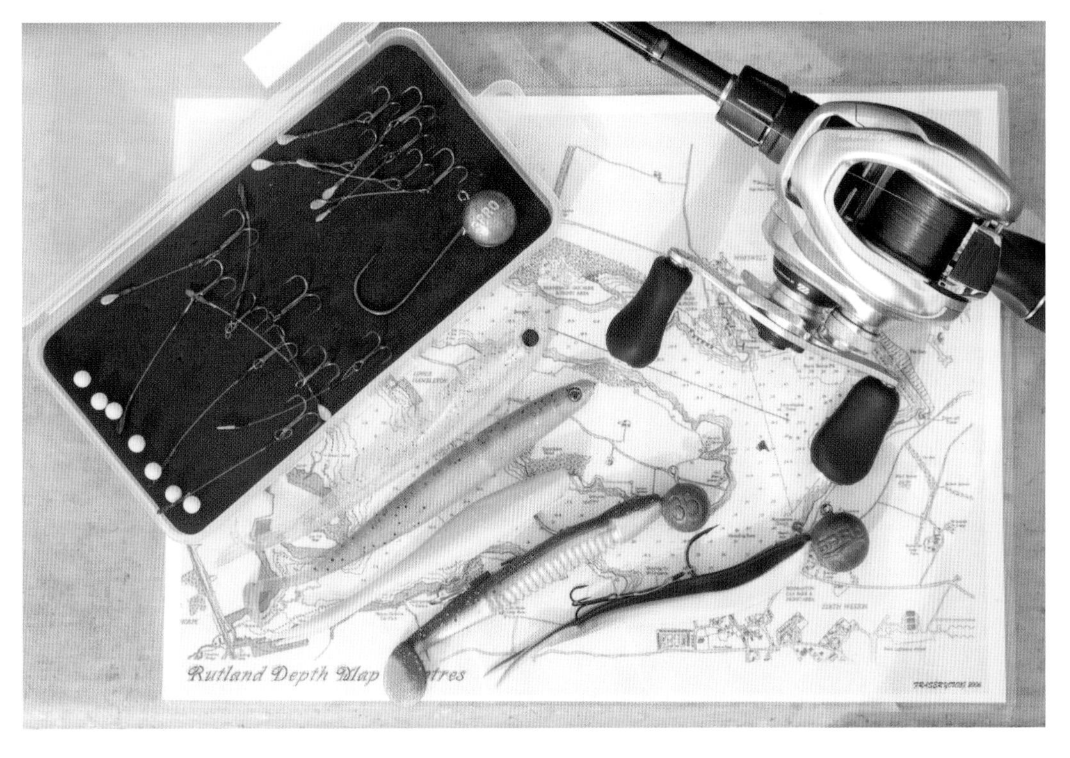

In terms of lures, there are many that will work, but we should start with soft plastics. Fish-style jigs are always effective, whether you choose a paddle or split tail. In river currents or waters with poor clarity, a paddle tail will give off more vibration, but slimmer tails have more subtlety for vertical fishing and greater depths.

Moving on to more traditional lures, crankbaits are very effective at times too, especially those which dive to greater depths. Spinners and spoons tend to be more limited, until you can find zander feeding in shallower water – although our friends in Scandinavia catch them fishing ice jigs and similar metals vertically! Evenings are perhaps the best time to try classic metals, when you might find zander coming closer to the bank to attack shoals of fry and prey fish.

Whichever lures you use, super sharp, fine-gauged hooks are a must. Large, thick-wired models simply don't penetrate as well – and zander can be tricky to hook at the best of times. Do also take a hook sharpener, because those points will lose their edge with the wear and tear of being bounced around the bottom and subjected to bony jaws.

Finally, there is a lot of debate as to the best colours for zander. They can be enigmatic fish, preferring a different colour one session to the next, for no obvious reason. It's worth experimenting to find the best fit for your local venues and the conditions; while for fishing deeper waters our colour chart (*see p69*) might provide some useful food for thought.

Canals, Drains and Small Waters

From the fens of England to the streets of Amsterdam and Ghent, zander seem to thrive particularly well on man-made channels. Even in the murkiest city waters, zander thrive, lurking around bridges, bends and locks, and raiding shoals of roach, bleak and other victims. These waters provide a lot of our bread and butter fishing, with easy access and a good head of fish at low cost. Indeed, they

Below: John Cheyne and Andy Mytton search the Grand Union Canal. Fishing with a friend is ideal for searching water quickly on canals and drains; once a bite occurs you can then slow down and cover an area more thoroughly, as smaller zander will often be found in groups.

are an excellent starting point for anyone who wants to start zander fishing, as you don't need any specialist gadgetry, boats or privileged access.

Travelling light and trying lots of spots is the way to locate the fish. It's a case of explore everything. You are looking for anything out of the norm: snags, bridges, boatyards, walls, wide sections and locks are all worth a go. Zander are seldom spread evenly, so it's always worth loitering and covering the water thoroughly any time you get a knock or suspect that fish are present. Don't always expect the best spots to cry out to you, either. In murky waters, for example, you will often find zander right by the bank.

Sometimes the best places to fish will be obvious, but other key features can be those you cannot see. The so-called 'shelf' on either side of a canal or drain, where the water slopes away, is a classic holding spot. Zander will patrol the bottom of these drop-offs, lashing out at small fish that are daft enough to pass too close. In fact, a slightly longer rod is often quite welcome on typical canals to allow the angler to tease a lure right along the shelf, with the lure working right under the rod tip. They will also sometimes be found right next to, or even underneath the bank, on murky canals. Anglers in the know catch lots of fish by effectively 'trolling' on foot, keeping their lure tight to the bank, or running it right along a drop off! If the fish are finicky, this can be done by walking very slowly along the towpath and tapping or shaking the lure (sometimes referred to as 'creeping' or 'dibbling' for zander!)

With the fish on these waters being modest-sized, typically, it can be fun to use light tackle and slightly smaller lures.

A typical canal zander; even at modest sizes they offer great sport on light tackle.

Soft plastics between 2-4in are ideal, whether you choose fish-shaped designs or 'creatures' such as worms or crayfish. Bigger lures might require a stinger hook in the tail, although much of the time it is simple and enjoyable to try lures of 2-3in with a single hook, which are easily grabbed whole by all but the smallest zander.

Jigs work superbly on the canals then, but there is also a case to be made for drop shotting. This is partly because the method keeps your hook and lure above the worst of bottom debris and lowers losses on some debris-strewn waters. It also allows a much slower, more thorough presentation around structures such as jetties and harbour walls. The key on smaller waters is to keep the lure in the bottom foot or so of water and to search the swim thoroughly. Keep that lure lively, but never rush it.

Bites can come at any time from within seconds of the lure hitting the water to right under your feet, so vigilance is key. Feel the lure down on a tight line, never let it fall on a slack as the majority of the bites will come on the pause, as the lure sinks. Generally, you will be fishing with the rod tip at shoulder height when casting across a canal, before gradually lowering it. Do note that while some fish will hook themselves against the rod or the weight of your dropshot sinker, many others will need a prompt, firm strike.

Lakes, reservoirs and large stillwaters

For any angler who wants to tangle with bigger zander, the answer is usually to find a larger stillwater. More expansive waters provide more food for fish to thrive and grow; but they can also be vast, daunting places, and even the experienced angler can feel like they are looking for a needle in a haystack.

This is where some homework, and perhaps some help from technology come in. You'll get some idea of features and depth changes just by counting down and working your lures, but if you can find a depth chart (many lakes have public maps and records available) this can be invaluable. Of course, if you regularly fish large waters, a fish finder is an even better way to map out your venue and identify the best spots.

The most obvious places to find zander are around significant depth changes. Whether it is a sudden deep hollow, an old riverbed or an excavation pit, they thrive at depths of over 60ft (20m). Indeed, the real chasms can hold good numbers of zander – although there are sometimes ethical questions about catch and release fishing in such spots.

Wherever you find zander, they can be quite tightly focused in small areas. Where you find a concentration, they are

Anglers fish around a submerged tower, textbook territory for reservoir zander.

Above: Larger reservoirs offer your best chance of a big zander.

not always hard to catch, but they do need some precision, because unlike pike they are not as willing to travel several metres to clobber a lure.

Further notes on lures and techniques

Deeper, larger lakes involve quite a different armory of lures to fishing smaller waters. A lot can depend on conditions, but you are often looking at 30g or more of weight, and as much as 50g on a really blustery, challenging day. With these weights you will inevitably overload the rod a little, as you still want to be using fairly sensitive tackle; you just have to be careful when you cast and not force the issue.

With lure and tackle choices, the greater the wind and waves and the greater the depths, the more weight you will need to keep control and retain that crucial 'feel' of the lure. Similarly, for drop shot tactics you might find yourself using much heavier weights than the standard few grams.

Whether you cast around or fish vertically from a boat is up to you, but working lures from a drifting boat is often successful, because it searches more water. Controlling the drift speed is also important, because if you travel impatiently, you may work the lures too fast or drag them to the wrong depth. A drogue, which essentially acts like a parachute in the water, is a useful tool to slow you down.

Special Zander Techniques

1. Fishing on the hang

As well as jigging under the rod tip, casting can be a great method for zander. A slow sink and draw style retrieve can be especially effective, whether done from a drifting or stationary boat. Here's how it's done:

1. The lure is cast out from the boat smoothly. If you are using a heavy jig of an ounce or more, don't force the cast. Resist applying too much tension, but just feel the lure down as you allow it to sink right to the bottom. You'll feel this as a gentle thud, or sense the line slackening off.
2. Engage the reel until you can feel the lure, before lifting the rod tip up three or four feet. You can incorporate a twitch or two here, or just use single sweep. Then keep the tip high for a second or two.
3. Feel the lure down, slowly reeling as you gradually lower the rod tip. Done correctly, it should take between ten and twenty seconds for the rod to drop back to the horizontal.

The idea here is that you are letting the lure drop and flutter down under gentle tension – and you'll find that under the weight of a fairly substantial jig head, that lure will continue working away as it kicks back towards you – there is no need to reel or jerk the rod! While it is the initial lift that gets the attention of a fish, the zander will tend to hit as the lure drops, so be ready to strike and keep the tension on immediately by keeping the rod up.

One last tip is to experiment with lure designs and jig weights. Often a slightly lighter jig (say a 30g rather than 40g) will sink a little slower, allowing you to achieve a slower drop with the lure, 'hanging' it off the bottom for longer and giving the zander more time to hit. Try the lightest model you can still 'feel' down, that the conditions will permit. If your jig is hitting bottom within 20 seconds, you are fishing too heavy.

'Hands-free' zander fishing: the subtler approach of the so-called 'dead rod' can be a winner.

2. Fishing the 'Dead Rod'

Zander don't like their prey doing cartwheels; quite often a subtle action works, and less is more. It doesn't get much more laid back than putting the rod in a rest and letting the rocking of the boat do the rest! With a dropshot or standard jig set-up, even the gentlest wave action will cause the lure to tremble, and you might be amazed at how many fish will hit a lure fished in this style. Dropshot is especially good here, because some fish will hook themselves against the weight of the sinker if your attention wanders.

Fishing guides in Holland like our friend Pim Pos in Amsterdam often use this 'dead' or sleeper rod method while guests are stopping for lunch – and it is surprising how many times the best fish of the day succumbs to the subtle appeal of a lure left to work itself with no more than the ripple of wind and waves on the boat. He also softens his lures in hot water, so that they have extra wiggle, even with the slightest of movements.

On big stillwaters anywhere, this method is worth a look and you can always set up a dead rod to fish securely in a rest, while you try casting around. Another good dodge if you are struggling is to try the dead rod 5-10ft off the bottom, to pick up any fish holding a little higher in the water. If you have space for it as a second rod, it is a 'free hit' in any case, and very often buys you a couple of extra bites on a difficult day. Just remember to keep one eye on it and to secure it in a decent rod rest in case the big one takes hold!

On the subject of presenting lures with minimal interference you can also try a similar trick when you are doing regular vertical jigging. Try lowering your jig to the bottom, fishing a slightly over-weighted lure to keep your artificial directly below the boat. Then simply wind it up about a foot (30cm) from the bottom and hold your rod as still as possible. The drift and natural movement of the boat will do the rest: hold on, as the bites can be explosive!

3. Pelagic sharpshooting

Not all zander are nailed to the deck. You will quite often find fish cruising or hunting higher in the water. With developments in fish finders, it's now possible to identify and target individual specimens that are on the move or just hanging in the mid to upper layers. So-called 'sharpshooting' is the specific approach of singling out these customers by quickly dropping a lure as close as possible and looking for a quick reaction.

For this technique, larger lures are preferable, often a 6in (15cm) or larger artificial that can be worked vertically, such as a forked tail soft plastic with a rattle. After spotting a predator on the fishfinder, the angler lowers the lure to the appropriate depth and literally shakes the rod to attract it, before pausing 10-15 seconds at a time, to let the fish home in. With more sophisticated fishfinders, it's possible to actually see the fish move for the lure! This can be a great way to single out large, lone specimens.

Depth issues & zander conservation

One of the most contentious issues of tackling deep water is the issue of fish welfare. Being reeled up from very deep water can cause fish problems as they deal with pressure changes, especially after a vigorous fight. Could some of these fish be damaged or die as a result?

The evidence is split – and it is difficult to ever know the truth without more extensive studies. Advanced fish finders have shown released fish shoot straight to the bottom and continue swimming. But divers have also reported casualties on lake beds after fishing tournaments.

So, with sport anglers looking to conserve stocks, what can we do to help? It looks a little rough, but one way the angler can encourage a returned fish to swim strongly back to the bottom is to throw it downward on release, which appears to help the fish to plunge back down into the depths. The evidence also suggests that time is of the essence and that the longer we take to release fish, the more risk there is. In other words, any fish caught at 20ft or more should be released post-haste to give it the best chance of recovery. Regulars on the deeper reservoirs have certainly had recaptures of fish caught at depth, so conscientious handling and speedy return are clearly the way forward – rather than taking ages to weigh and photograph the catch.

It is up to the individual then, but you may decide not to fish the deepest water if you are concerned about the fish. You could avoid the more bottomless areas, not to mention trying the mid-layers of the water, because zander are not always glued to the bottom. As a general rule, we would not typically fish deeper than

30-40ft. This can depend on the season as well as time of day, of course, with zander holding deeper the colder it gets.

Tails you lose

With smaller mouths than pike, zander will not always devour bigger prey whole. At times they will disable first by delivering a strike intended to stun or cripple before the kill. Nipping the tail off is a favourite trick – and all keen zander anglers are familiar with the experience of getting a sudden attack and then reeling in a lure with its tail missing! On other days, the fish also seem to hit the tail and miss the hooks.

Whatever the reasons for this behaviour, a few golden rules apply. First if you get a tap or miss the first hit, don't be too quick to reel in – just hang the lure for a moment and quite often a second strike will follow.

As for the problem of tail hits, the best solution is to incorporate a second 'stinger' hook just before the tail of a soft plastic lure. These can be bought or made up, and fastened to the back of your main, single hook. The commercially sold versions have a little loop with a little brightly coloured plastic to fit snugly, but you can achieve the same effect by coating your own in liquid plastic. *More on this subject on p55.*

River Fishing

Besides the many miles of canal and larger lakes, zander also do well on running water. Vilified as an invasive menace in some quarters, they have managed to populate several major English rivers, including sections of the Severn, Trent and Thames, growing to a good size.

River zander love drop-offs and snaggy depths every bit as much as their stillwater cousins. Obstructions and pits on the bottom of a river offer more than just a hiding place, but a break in the current, and zander will travel between and rest up in these areas, for safety and cover. Anything from bridge supports to a discarded shopping trolley can provide sanctuary, while they also love places like flood arms, weirs, deep slacks and any diversion from the norm.

As much as they like depth and cover, zander are not the lazy fish some think they are, and can actually handle quite strong currents when they are feeding. You only have to look at their long and sturdy lower fins, not unlike a barbel, which give them power and control even in quite meaty flows.

For some reason, zander like dirty rivers and difficult spots. They are aggressive and sometimes easy to catch, but not always easy to locate. They are certainly fitter than their stillwater counterparts, and the bigger ones can patrol quite long distances. So, if a spot looks excellent but you get no bites, it could be a case of the wrong time of day rather than the wrong location; and a revisit at another hour could well pay off.

Indeed, you will find river zander highly sensitive to time and conditions. For many years, a lot of anglers only night fished for them, but they can also be caught in the middle of a sunny day in murky water. That said, they tend to feed best on dull days and the first and last hours of light are the two times in each cycle you can be confident that fish will be on the move and feeding.

Searching the flow

Fishing a river for zander demands a different approach to still or slow moving water. With currents sweeping through,

River zander love sloping shelves, snags and any obvious vantage points.

you need to work with the elements to get your lures down to the fish. Medium to heavy jigs are one option, cast upstream (into the flow) or across the river and allowed to sink and work back to the bank with a lift and fall motion. The power of the current will dictate how heavy your lure must be.

Losses can be high with jigs, however, and many zander anglers who fish particularly snaggy rivers prefer to use plugs to get down to the fish. There is certainly some logic to this; working from the top down rather than the bottom up results in fewer lost lures. And if you can find a plug that dives well down without snagging, zander love a crankbait.

On the subject of hooking up, there are one or two other ideas that may also help. Zander certainly seem to thrive in really snaggy rivers such as the Severn. Given that zander don't fight incredibly hard, one method is to use strong tackle (say 30lb braided mainline) and use slightly finer hooks that will bend out under great pressure. Plugs with a pronounced front lip are another possible way to cut losses, as they tend to 'bounce' off cover, rather than getting hooked up; should you feel contact with a solid snag, you can pause and let the lure float up and out of danger. Even so, you might want to find an effective, cheap pattern to cover losses.

In short though, there will always be some degree of risk when getting down to river zander and in terms of floating plugs, you should pick a model that will dive well and try casting downstream and across (i.e. in the direction the current is travelling). When you retrieve it, the lure will then work against the flow, wriggling strongly under the pressure and diving steeply. This can be ideal for running a lure along the inside shelf of the river, where the bottom drops away. Try to find a particular make that dives to a certain depth (just off the bottom, or over the typical 'drop off' depth), and no further, and you might have a winning formula.

Another neat trick with a plug is to cast out and let it travel down with the current a little, before holding it steady against the flow. A long rod is useful here for better reach and control. If the push of water is sufficient you'll only need to hold it in place to feel the lure pushing and kicking on the rod tip. In this way you can search slowly and even simply hold position in likely areas, allowing the flow of the river to do the work. A sudden pluck or presence often means a fish; many will hook themselves when fishing downstream in this manner, but be ready to strike at any sudden tug or knock!

Typically, you will find zander where the river bed drops away from the shallows, to 6ft (2m) of depth or more. The only time you will commonly find them higher in the water or coming into the shallows is late in the day or at night, as they get bolder and hunt nearer the surface. It's then that a shallower diving plug or a even a spinner can work, this time not teased but worked quickly, like a fleeing prey fish.

Lastly, for those who have access to a boat, there are huge advantages to be gained. The process of searching a river becomes easier still, while a fishfinder can be used to map out miles of water and record successful spots via GPS. This, however, is another topic altogether – although the same lures and tactics will set you in good stead.

Six great lures for zander

1. Realistic Shad R7: Perfect for canal and river fishing, where the zander are feeding on smaller fodder (this pattern is actually designed to imitate a ruffe).

2. Reins S-Cape Shad: A good example of a versatile paddle-tailed shad. 3-4in (7-10cm) is a good all-round size for zander.

3. Lunker City Fin-S Fish: Forked tail lures are an excellent option, not only for casting but for vertical fishing or drop shot. In a 5-inch size, this is Andy's favourite lure for reservoir zander, usually with a heavy jig head and stinger hook.

4. Reins Bubbling Shaker: Versatile worm type lure, can be used for jigging or drop shot presentations.

5. Rapala XR-10: A great lure to try later in the day, when the zander are active. Running at 3-5ft, they are particularly good for canals, drains and shallow rivers.

6. Floating soft plastics (Z-Man Finesse TRD & Trick Shot): Ned rigging with soft plastics is an excellent tactic for zander lying dormant on the bottom. These will sometimes work when all else fails!

Kayaks represent a fun, affordable way to get afloat

THE WIDER WORLD OF LURE FISHING

One of the greatest joys of lure fishing for different species is the endless array of possibilities. Put simply, you will never exhaust the vast number of lures, species and techniques that are out there, waiting to be discovered. For those with wanderlust, there is scarcely a corner of the world where lures won't catch you fish. For those who love other outdoor pursuits, such as wilderness hiking and kayaking, these are also perfect to combine with fishing. Meanwhile, those with a competitive streak might enjoy the growing number of exciting matches and events to sharpen your skills and learn from other enthusiasts. So, while we couldn't hope to cover all of these diverse avenues in the pages of a single book, the aim of this section is to whet your appetite and broaden the ever wider horizons that modern lure fishing and species hunting bring.

Kayaks, boats and all that floats...

One obvious way to get a wider range of species on lures is to do so from a boat or small craft. Where possible, this can immediately remove the issue of long casts or lack of bank access. In fact, once you get directly above your quarry you can use much lighter tackle than you might from the shore. On many reservoirs and natural lakes it's possible to hire a craft;

Float-tubing in Finland.

meanwhile at sea, more sea skippers than ever are offering lure fishing trips and steering anglers towards a wider variety of species, too.

Perhaps the largest growth area of all, however, has been kayak fishing. Indeed, this is a fairly cheap and versatile way to boat fish. Just think: no mooring fees, no temperamental motors or other nonsense – and good exercise to boot! Whether you cast from one, or simply drift and use vertical presentations such as drop shot, they are excellent. While we don't have the space to go into huge detail here, those keen to find out more would be well advised to hire one first and take a guided trip. Safety is of course vital on any water, so it's important to practise falling in and getting back on board, not to mention investing in a buoyancy aid and a means of getting help should you encounter difficulties. A professional instructor can give you the perfect head start.

Finally, float tubes are another great option for those who want a portable means to tackle lakes. For many of Europe's great lakes, for example, a tube and flippers will fit neatly into a suitcase and can provide access to remote waters, even where no boat hire is available.

Have rod, will travel

With typical lure fishing tackle so portable, the world is your oyster when it comes to trying your hand at fishing overseas. While our natural bias in this book is perhaps towards the UK and Europe, all the methods and techniques will work for different species in less

With lure fishing so portable, the world is truly your oyster, whether you find yourself in Greece (top) or San Francisco (bottom).

familiar territory, too! Nor do you need to book an exotic destination or dedicate weeks of time; in fact, short sessions and light, quick-to-assemble tackle are ideal for species hunting with any free hour or two on family holidays, wherever you happen to be.

Above: With giant drills, strange jigs and garden gnome-size rods, ice fishing has to be one of the weirdest forms of lure fishing.

Travel tips for the lure angler

- Laws vary between countries, so do check things like licences and rules before you travel if possible.
- The majority of tackle is best stowed in your main luggage, away from the prying eyes of officials who love to go through your hand luggage! They will often object to things like lures, hooks and tools.
- Some rods come in special protective cases; but the rest can easily be stored in a cut down section of the plastic tubing most rods are delivered in these days. A four piece lure rod should fit in most large suitcases!
- Make a checklist before you leave, taking into account that there may be no means of buying tackle where you travel to. Always include spare line, unhooking tools and other basics.
- A good local guide is worth every penny, so treat yourself! They can cut out days of trial and error, not to mention giving you advice on which tackle and lures to bring before you've even packed your bags.
- Not everything that swims in the sea is safe to handle. Most fish are harmless but there are others that can bite or sting you or that have venomous spines, so it is always worth researching the ones to avoid. As a rule, if you are unsure, shake it off or use your forceps and try not to handle it!
- You don't need to pack the kitchen sink to catch fish. A pot of scented worm will catch pretty much anything that swims but it is worth taking a small selection of soft plastics, pin and paddle tails, jigheads, metals, drop shot weights, split shot, hooks and some small crankbaits and you will have every eventuality covered.

Above: Will Pender and Andy Mytton enjoy fish spotting from a harbour wall. More than a competition, the Cornish Lure Festival is a celebration of camaraderie, where all fish species great and small count!

Colourful competitions and events

With a growing community of lure fanatics better connected than ever, the digital age has spawned an amazing scene for meet ups and competitions. These can vary from sociable species hunt events, to national and international matches. You don't need to be a world class angler to have a heck of a lot of fun, however. In fact, they're great places to hone your skills and learn from like-minded anglers. Here are just some of the events that are now a classic annual fixture.

The Cornish Lure Festival

Split into two main categories, this fabulous event sees visitors travel hundreds of miles to hit the Cornish coast for an action packed, and often sleepless, weekend of action. In the course of 48 hours, there are two categories to go for: the best bass and the finest tally of different species. It is not uncommon for the latter to be won with sixteen or more types of fish- and is a grand opportunity to learn from the locals and LRF fanatics from all over the country.

Gunki Iron Tournament

Building to a street fishing phenomenon in recent years, this tournament began as much as an endurance competition as a fishing competition. Even for the dedicated, 24 hours of solid fishing is an epic test for the anglers, who come from all over Europe. The best 15 of various species count for each team, from pike and perch to flounder and even catfish on tidal waters! A collective effort, it involves groups of three, with two

Above: Cool, colourful and sociable, Open Street Gent has been a game changer for friendly yet competitive lure fishing. Similar events are also held in Amsterdam and other European cities, as the lure fishing bug spreads!

anglers to be fishing at all times unless walking to the next spot! Andy's gang claimed a respectable 6th place finish recently.

Open Street Gent

Cool, colourful and sociable, Open Street Gent has been a game changer for friendly yet competitive lure fishing. Similar events are also held in Amsterdam and other European cities, as the lure fishing bug spreads!

The coolest street fishing competition in the world! Famed for its social scene as much as the fishing, OSG is an event anglers flock to. With an average attendance of over 200, it puts fishing at the heart of the beautiful city of Gent, Belgium, with prizes to be won for the biggest zander, pike and perch. It's also awash with street art inspired trophies, barbeques, beer and music, making it a superbly laid back event whether you have your eyes on a prize, or just a brilliant experience.

All fish great and small – to be found on your travels

It's well worth investing in a compact go-anywhere lure fishing kit. Who knows what you'll find on your next trip?

Giant Goby *Gobius cobitis*

Mainly found close to rocks or hiding in snags on the sea bed, this mother of all gobies is an opportunistic predator that will hit and run you to ground at lightning speed. Protected in the UK, they are far more common in the Mediterranean.

Reticulated leatherjacket
Stephanolepis diaspros

Having migrated to the Mediterranean via the Suez Canal, this fish mainly feeds on algae but can be caught using small scented lures such as Isome on a tiny hook, often found grazing on harbour walls in the shadows.

Snappers

The Cubera Snapper is one of a mean extended family of toothy, reef-loving fish. With relatives in warm seas all around the world, the snapper family are voracious predators that react aggressively to lively lures, especially poppers and large plugs fished close to cover – and will provide a stern test on medium to heavy lure rods.

Jacks

The 'Jack' family includes many bony fishes: horse eye Jack, giant trevally and amberjack. Found from African seas to the Caribbean, they are voracious predators that tend to hunt in packs. The author had this Jack Crevalle on a large surface-popping plug, after which it stretched a 20lb class outfit to the limit!

Sea Bream

This family of fish might be rather limited in British waters, but foreign seas have a wide variety of fish in this family including two-banded, saddled and white bream (above). Don't be fooled by the harmless appearance; these are tenacious predators and while those you'll find close to shore tend to be small, even a modest one will really bend a light rod.

Wide-eyed Flounder

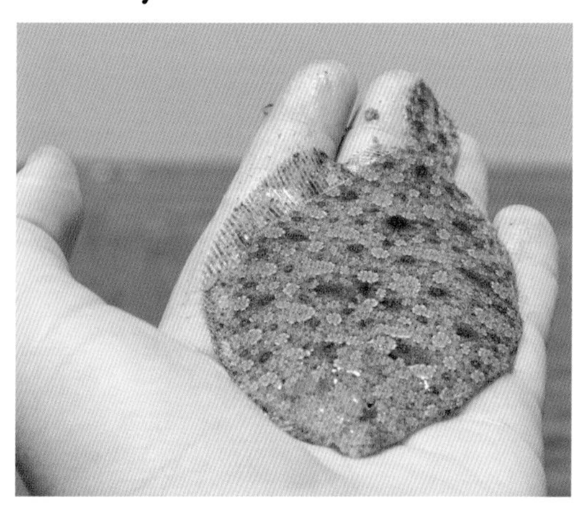

Flatfish are another family of predators represented all over the world. The wide-eyed flounder has a completely different eye position to our own flatties; but is just as susceptible to small lures.

Cardinal Fish

This brightly coloured little predator is found in the Mediterranean and feeds best at night. Note the enlarged eye and strange double dorsal fin.

Barracuda

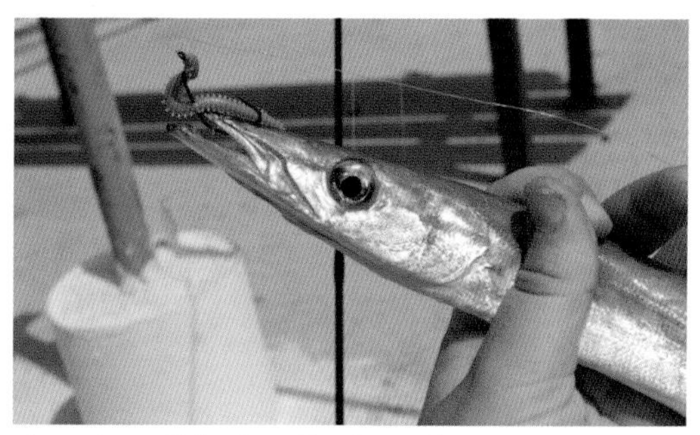

Another fish family found in most warm seas, barracuda are famed for their arrow-like design and crazily sharp teeth. Those found in European waters are often smaller than other species at 30-60cm in length typically, although they have been known to grow in excess of 6kg and 165cm. They're a favourite with Andy, who has caught barracuda on all types of lure, from ragworm look-a-likes to small minnow plugs. Be warned though; the larger ones may require wire leaders, while even their younger siblings can easily cut a soft plastic lure in two! The fish shown here was caught from a grass bed during the middle of the day, an area they hang around whilst the sun is at its highest.

A LURE ANGLER'S GUIDE TO FISH SPECIES

Welcome to our lure angling species guide. This section of the book is intended to be a handy means of identifying and targeting the various types of fish that can be caught on lures in salt and freshwater around the UK, although many are also found in Europe and beyond. Whether you use it as a fun way to track down and ID different catches, or it becomes something of an obsession to catch as many as possible, we hope you find it enjoyable and useful.

With the evolution of ultra fine lures and methods, a vast range of species are now possible on lures. We know anglers who travel the entire UK and beyond to develop their list, much as twitchers do with birds! That part is up to you; but all we will say is: keep that sense of wonder and embrace your inner kid. Fascination is healthy; one-upmanship is not.

We've mined all manner of sources to provide clues, notes and tips on each fish here. Some are more rare than others and in this respect, size really isn't everything. Some are so unlikely on lure tackle that we omit them or only mention them in passing (rays and dogfish are typical examples).

Just for fun, we've also included an 'I-Spy' style tickbox for each fish, so you can challenge yourself to see how many you can catch.

Distribution/ location and further tips

You will notice that rather than an exhaustive A-Z guide, we've tried to list fish in order of importance/frequency, broadly bunched into 'types' from blennies to the cod family. Where possible, we've also indicated typical locations and geographical trends. For much greater local detail, we'd also recommend the excellent database at **nbnatlas.org**

Discovering which species are present in any given venue is half the fun in itself! By challenging yourself to try new locations, or fish familiar places at night, for example, you may well be surprised! There are some great sources on species hunting online, too, not least of all our friend, ace species hunter Scott Hutchison's blog: **something-fishy-going-on.blogspot.com**

Protected Species and Conservation

With our waters and fish stocks under more pressure than ever these days, we'd always remind readers to check local rules and fish responsibly. For all the different species listed, sustainable catch and release fishing is preferred; we are all ambassadors for angling!

Careful handling (with wet hands and a soft unhooking mat) along with de-barbed hooks are highly recommended. With sustainable sea species, you may of course want to keep the occasional fish for tea where rules permit; but be sure to stick to legal size limits. You'll find current measurements in the food and farming section of **www.gov.uk**

Finally, we've also indicated some species that are protected and should be swiftly and carefully released if caught by accident.

SALTWATER SPECIES

☐ BASS *Dicentrarchus labrax*

The silver, spiky-finned bass is perhaps our most iconic sport fish in saltwater. A fearsome all-round predator that eats everything from tiny shrimps and sandeels to baby pollack, it's no wonder the bass is a favourite for the imitative lure angler. School bass are common around estuaries and harbours, but the bigger fish can be a genuine challenge.

Size: Typically 30-50cm (Specimen: 70cm+)

Season: These days small bass are an all year round catch. They spawn in the spring, so you're more likely to catch a better bass from late summer into autumn.

BLENNY & GOBY FAMILIES

For all but the most cynical of anglers, these little fish are an absolute delight, taking us back to childhood days searching rockpools. Surprisingly aggressive and strong for their tiny size, they are great candidates for the very lightest of tackle. Don't assume that all blennies are tiny or common, however. Some varieties will reach the vast length of 20cm or more, while others are so rare and beautifully marked they tend to get keen species hunters all of a lather.

These species have a wide diet, including marine worms, shrimps, tiny crustaceans and fish. We suspect a lot of the time they lash out from sheer aggression, too. Split shot rigs armed with tiny hooks (right down to sizes 20 and below) and fragments of soft scented plastic worm are as good a place as any to start.

Because of their highly localised, non-migratory nature, blennies and gobies tend to be found in the same places every season, making them a godsend for year-round anglers. We should also note that some of these little critters have quite sharp teeth, so watch your fingers! A coarse angler's disgorger is a handy tool to have, and barbless hooks are kinder and easier to remove.

☐ COMMON BLENNY
Lipophrys pholis

Also known as a 'shanny', this is first blenny most of us encounter, whether with rod, net or bare hands. They are kings of the rockpool and the harbour wall where they are most likely to remain hidden in cracks in the wall or under rocks until coaxed out. They mainly feed on marine invertebrates or worms. They can live in intertidal zones and can survive out of water as long as the environment is cool and damp and can at times be found moving from rockpool to rockpool as the tide retreats.
Size: 5-12cm

TIP

To see if there's a blenny or three at home in a rockpool, drop a lively worm-style soft plastic lure right into the middle and watch them dart out.

DID YOU KNOW?

In case you're wondering, the tompot blenny's name comes from their curiosity, which often gets leads them into crab pots. Hence local fishermen nicknamed them 'Tom of the Pot'.

☐ TOMPOT BLENNY
Parablennius gattorugine

Often found competing with the common blenny for top spot in harbour walls and rockpools, this inquisitive customer is a firm favourite with many lure anglers. Among the biggest of the common blennies, they are quick to react to small lures and are strong fish for their size, identified by their cheeky looks, reddish coloration and the strange little appendages above the eyes. Anything but fussy, they provide bites on the hardest days. Bristling with attitude, they will quickly snag a lure if you let them, or indeed take a nip at your fingers.
Size: Average 10-15cm, occasionally as big as 20cm+

☐ VIVIPAROUS BLENNY
Zoarces viviparus

Also known as the viviparous eelpout, this rock-hugger is easily identified by its elongated body, wide head with protruding lips. A thin dorsal runs right along the back and joins with the caudal (tail) fin. They give birth to live young. Quite localised in distribution, they can be found around most of the UK but sightings are far more common in the northern British Isles. **Size:** Typically 15-25cm, occasionally 30cm+

□ BUTTERFISH
Pholis gunnellus

A well-camouflaged, eel-shaped creature, this blenny family member is a favourite for kids delving into rockpools – and this setting represents your best chance of finding one. Presumably the 'butter' tag is due to their slipperiness rather than the colour! Marbled brownish in colour, with an elongated mouth and tiny 'lips', they also have black dots down the dorsal. They can be shy at times, but with extra fine tackle and a tiddly little hook (as small as a 24) along with the merest sliver of plastic worm or perhaps a bright, tiny fly, you might just get one! **Size:** 10-20cm

□ YARRELL'S BLENNY
Chirolophis ascanii

Named after British zoologist William Yarrell, this creature has an elongated form and an extended, ribbed dorsal, along with tiny appendages above the eye. More common in northern England, Northern Ireland and Scotland. Sightings suggest that they may prefer slightly deeper water than other blennies and are often found further back in the intertidal zone. **Size:** 15-25cm

□ MONTAGU'S BLENNY
Coryphoblennius galerita

A tiny, but beautiful-looking blenny, this one is highly prized with species hunters. Preferring warmer water, most sightings

DID YOU KNOW?

Like several other blennies and smaller rockfish, the butterfish can actually breathe air. This comes in useful when they get stranded in tiny crevices and rockpools at low tide.

tend to be across south-west England and west Wales. At first glance it looks like it has two dorsal fins, which join on the back. Colour varies, but usually a light brown ochre, with barred sides and quite often light blue or creamy spots. A real collector's item if you can catch one!
Size: 6-10cm

☐ BLACK FACED BLENNY

Tripterygion delaisi

Not so much a rarity as a collector's item for many species hunters, this blenny is highly coveted to put it mildly. For large parts of the year, they are a speckled brown colour and can be hard to recognise; this all changes around the breeding season, however, when the male turns yellow with a black face (Jack Perks captured the above on camera. It bit Andy's lure twice but evaded capture!) If you are lucky enough to catch one of these fish, treat it with the utmost care and return it quickly because they are quite rare and in many areas of Europe are protected by law.
Size: 5-10cm

☐ BLACK GOBY

Gobius niger

A typically dark coloration gives this goby its name, although some can be much lighter depending on the locality. Perhaps the easiest way to identify it is via the triangular first dorsal fin. Sandy and muddy areas are prime, close to shore and in estuaries. They are great scavengers, capable of taking live or dead food.
Size: 8-10cm, occasionally to 15cm

DID YOU KNOW?

As tiny as they might seem, gobies tend to make better parents than much larger fish. Many species not only pair up using strange dancing rituals, but will create or dig a little refuge for their eggs, which are laid amongst stones and even inside empty shells. Males will vigorously eject unwanted visitors, rather like disgruntled nightclub bouncers.

□ ROCK GOBY

Gobius paganellus

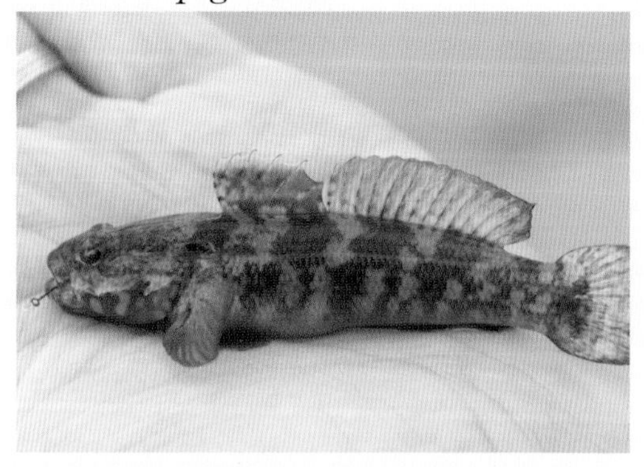

A supremely successful little devil, this chap has colonised waters from Scotland to Africa. The rock goby has a yellow, gold or creamy white band that runs horizontally along the tip of the first dorsal fin. These fish are common and often one of the first to be caught when species hunting.

Size: Average 5-8cm, sometimes 10cm or more.

□ SAND GOBY

Pomatoschistus minutus

A small but greedy customer, the sand goby is a miniature ambush predator found all around the UK coast. A mottled, light brown almost translucent coloration lends it perfectly to concealing itself on sandy bottoms, ready to strike at small prey, with sand shrimps a firm favourite. Not that he'll turn his nose up at a worm, provided your hook is small enough. Similar in appearance to the common goby, the sand goby's eyes are higher on its head & closer together and its body is more elongated.

Size: 4-6cm typically.

□ TWO SPOT GOBY

Gobiusculus flavescens

This goby can often be found higher up in the water along weed lines. Prolific around Plymouth's Barbican area, these fish can often be mistaken for other species as they act differently to other gobies. They have a wide distribution too, from Portugal to Scandinavia. Identifiable by two black spots, one directly below the first dorsal fin with a second spot at the root of the tail, they are also sometimes called 'Crabeye Goby'. The spots simulate eyes and are for defence. These fish also have light blue spots along the lateral line.

Size: 2-6cm typically. 10cm would be a big one.

☐ LEOPARD SPOTTED GOBY

Thorogobius ephippiatus

Like many mini species, locating these fish is the main challenge. Once you have done that, simple finesse tactics should work. This distinctive goby is an LRF angler's favourite; fresh out of the water it has a bluish sheen and distinctive spots like leopard skin, although the markings are often a pinkish colour. The LSG has a slightly larger mouth in proportion to its head than other gobies, and if you can catch one, others often follow as they tend to move in groups and return to the same spots year on year.

Size: Average 8-10cm.

☐ COMMON GOBY

Pomatoschistus microps

Contrary to the name, this titchy little goby can be quite hard to find. Perhaps the miniscule size explains why they are so seldom captured? They are often found in rockpools and broken ground, but also like sand and estuarine areas. A truly minute predator, they feed on sand shrimps and crustacea. A size 24 hook wouldn't be too small, rigged with the tiniest fragment of scented lure.

Size: Typically 3-5cm.

☐ PAINTED GOBY

Pomatoschistus pictus

Even by goby standards, this is a tiny critter, rarely exceeding 6cm in length. It's a pretty little customer, however, brownish overall, with a line of four twin dark brown spots along the side. Its dorsal fin is perhaps the real giveaway which has rows of dark brown to black spots, with bright blue and pink woven in. They are not incredibly common on our shoreline – and are more associated with the Mediterranean. Global warming could change all that, however!

Size: Typically 3-6cm.

☐ GIANT GOBY *Gobius cobitis*

PROTECTED SPECIES

A goby to dwarf the others and you might accidentally come across this fish when looking for other gobies and blennies. Speckled brown to yellow-green, it lacks the wide distribution of other species, although Devon and Cornwall have turned up several over the years. For holidaymakers, they are also found throughout the Mediterranean. They can thrive equally well in shallow, rocky water or in greater depths. Like other blennies, it is an aggressive opportunist. Do note that they are a protected species and if you're lucky enough to catch one, it should be released with great care – as current UK law states!

Size: 15-30cm

DID YOU KNOW?

This fish is the true heavyweight champion of gobies, with a record weight of 262g!

SEA BREAM FAMILY

If we count every type of sea bream found occasionally in UK waters including the remotely possible, there are at least eight sea bream. However, only two or three of these are a regular enough target to be worthy of serious pursuit. You are unlikely to get bigger bream from the shore, so the key to netting a good one might be to explore inshore boat fishing. Perhaps bream are one of the more under-appreciated species though, because during our tank tests they were among the most inquisitive and aggressive customers, attacking lures with real gusto. Whichever bream you find, they respond best to lures kept moving, so work yours with vigour.

☐ BLACK BREAM *Spondyliosoma cantharus*

While fish of greater than a few ounces are seldom found within casting range of light lure tackle, juveniles of this pretty, hard fighting species can quite often be caught around piers, breakwaters and other structures. Boat fishing in calm conditions is the way to get a specimen on light tackle. They tend to eat shellfish and crustacea, but worm style lures seem

to bring out their curious, aggressive tendencies. Females are silvery, while males darker and often with vivid blue 'warpaint' on the face.
Size: Typically 5-20cm from shore. Better fish in deep water can be 2lbs or more.
Recommended lures: Any small, lively worm or fish shaped soft plastic.
Season: Summer to autumn.

☐ GILTHEAD BREAM
Spondyliosoma cantharus

Once a rarer find in British waters, warming seas have extended the range of this strong, deep-flanked fish, which is a staple for Mediterranean anglers. Juveniles are now becoming increasingly common all along the south coast of England. Like the black bream, they are naturally at home eating food items like hermit crabs and starfish, but are naturally aggressive and will take most smaller worm or fish style lures. Like black bream, 'gilts' are aggresive and can be caught on small paddle or straw tail lures fished on a jighead.
Size: Typically 5-20cm from shore. Better fish in deep water can be 2lbs or more.
Recommended lures: Small scented worm or crab.
Season: Summer to autumn.

OTHER BREAM
The following can all be filed under 'possible but unlikely'. Among various occasional catches, red bream are one possibility. Couch's bream *(above)* are perhaps the more likely; they are a common catch in the Channel Islands, although smaller fish occasionally congregate in southern UK estuaries. As for the rest, the likes of Axillary, Pandora's, White and Saddled Bream are very occasionally caught in UK waters; but could be a more realistic target if you sneak a lure rod onto your next Mediterranean holiday.

THE COD FAMILY

Featuring a good spread of species from the small and common-as-muck pouting to rarer catches and deeper water species, the cod family is a staple of sea fishing. Realistically, the typical size of fish caught from the shore will be modest, although occasionally, calmer conditions will permit light tackle fishing for their bigger relatives from a boat. Many of this family are more comfortable in cold waters, hence their distribution tends to be more northerly.

DID YOU KNOW?

One of the more curious characteristics of the cod and some related fish is its ability to communicate using a range of grunting sounds (a bit like the Cornish).

☐ COALFISH

Polachius Virens

Often confused with the pollack, this cold water species is especially prominent in the north of Britain, where in Scotland it is sometimes called the 'saithe'. Similar in habits and appearance to the pollack, but the coal fish has a smaller eye, the lower jaw protrudes only slightly longer than the upper and its lateral line is often pale.
Size: Typically 25-40cm (Specimen: 60cm+)
Season: As with pollack, small fish show up year round, but summer to autumn is best for the specimens.

☐ COD

Gadus morhua

Weighty metal lures have long been a favourite way to catch these formidable, bucket-mouthed predators over wrecks. For the lighter lure fanatic, smaller 'codling' are also found from the shore and

provide some great sport- and they love lures such as metals and soft plastic shads. Whether you use the vantage point of a pier, rocks or even a kayak, these beautiful fish make a fine catch. A cold-water species, distribution is northerly across the UK and Europe.
Size: Typically 20-45cm from the shore (Much larger possible by boat)
Season: Year round for smaller fish. Summer through to autumn for the better samples.

☐ POLLACK

Pollachius pollachius

An active predator with a mean burst of speed, pollack can be found all over the British Isles. Unlike the coalfish, the bottom jaw extends well beyond the top, while the lateral line tends to be darker.

Small ones of two pounds and under often haunt piers, rocks and harbours. Deeper reefs and waters slightly further offshore tend to hold the better samples, with kayak or boat fishing a good bet. They take sandeel and shad-style lures well, besides classic metals. Weedless lures can be very useful too, over kelp-strewn marks and rocky reefs. Pollack fight tremendously hard, but are quite fragile when landed; keep handling to a minimum if you want to release them.

Size: Typically 25-40cm (Specimen: 60cm+)

Season: Year round for smaller fish. Summer through to autumn for the better samples.

☐ POOR COD

Trisopterus minutus

A miniature cousin of the mighty cod, this fish might exist on a much smaller scale, but is just as predatory. A lover of rock marks and broken ground, he'll eagerly devour shrimps, small crabs and other life. Also fairly common from many harbours and piers.

Size: Typically 10-20cm typically

Season: Year round

☐ POUTING

Trisopterus luscus

Not so much common as hard to avoid and a real sucker for just about any lure, 'pout' are a regular catch from Cornwall to Caithness. They tend to be school fish that like any structure, from broken ground to harbour walls. Most tend to be small, and yet they are still a welcome species on light tackle, taking lures aggressively when other species refuse to bite.

Size: Typically 10-20cm (Specimen: 25cm+)

Recommended lures: Anything that moves, but especially bite-sized metals and soft plastics.

Season: Year round.

☐ WHITING

Merlangius merlangus

A pale, smaller member of the cod family, these ravenous fish hunt in packs across broken ground. Commonly found from beaches, but they can also be caught from man-made structures. Look out for the large eye, single chin barbule and tiny, sharp teeth! Like other cod family members, they are quite brittle so handle as little as possible when releasing. That said, the larger ones make excellent eating.

Size: Typically 10-20cm (Specimen: 30cm+)

Recommended lures: Small metals and soft plastics

Season: Hordes of these fish come close inshore through late autumn and winter and stick around into spring.

□ HADDOCK

Melanogrammus aeglefinus

Another cod family member of more northerly waters, haddock are highly prized right across the cooler seas of Europe. It has the cod-like design of its cousins, but with silvery sides and a distinct 'thumb' mark by the pectoral fin. Juvenile haddock are fairly common on the west coast of Scotland, in particular and, much like cod and coalfish, respond well to metals and soft jigs wherever encountered.

Size: Typically 20-40cm from the shore

Season: The most likely time to catch this fish is in winter when they come closer inshore.

□ TADPOLE FISH

Raniceps raninus

The black sheep of the cod family, this odd fish is dark and tadpole-shaped with long fins running most of the way along the back, and virtually from the tail to the middle of the belly. Its blunt head, which is wide and full of backwards-facing teeth, earns it the nickname 'Jelly Nose'. Although not a common catch at all, rocky areas might just produce. It seems more common in northern England and Scotland; and they appear to be quite seasonal. They are not fussy eaters, whether that means nabbing a shrimp, small worms or even the odd small fish.

Size: Usually 10-20cm, but can be surprisingly big (the UK shore caught record is 1lb 5oz!)

Season: Summer.

ROCKLING FAMILY

These are another set of species in the cod family you might come across on the shore. With a long, eel-like body and a whiskery head, they are a somewhat rarer yet highly desirable catch from rocky and broken ground.

Scaleless and slimy to the touch, they are voracious little predators and scavengers, which eat worms, inverts and crustaceans, so we'd recommend a small scented lure in shallow rocky water.

Night fishing is a must, as all are nocturnal, not to mention more common in the north of England and Scotland than the south. Here, we include the three most likely varieties:

☐ THREE-BEARDED ROCKLING

Gaidropsarus vulgaris

This nocturnal bottom feeder is a rare prize for the keen species hunter. Not extremely common, but catchable by night, usually fishing right under the rod tip in a rocky location, although they also frequent sea walls and some harbours. Distinguishable from the shore rockling by a larger typical size and its bold pinkish and brown spotted colours.

Size: Typically 15-25cm, occasionally much bigger (up to 60cm!)
Recommended lures: Scented worm, small scented soft plastics.
Season: Spring to autumn.

☐ SHORE ROCKLING

Gaidropsarus mediterraneus

Found in broken, rocky ground and also occasionally large rockpools, this is a highly prized fish for the LRF species hunter. They eat quite a range of worms and crustaceans, so should in theory be catchable, if challenging. They are smaller than three-bearded rockling and more dull brown, although they also have three barbules on the head.
Size: Typically 5-20cm.
Season: Year round.

☐ FIVE BEARDED ROCKLING
Ciliata mustela

Another possible night catch, although rare on any lure, this rockling is easy to identify by its five barbules. As with other rocklings, it likes broken ground and is occasionally disturbed when turning stones or going through rockpools at low tide.
Size: 10-20cm.
Season: Year round.

OTHER COD FAMILY FISH:

Unlikely, although plausible from a boat in the right conditions, we might also include the Ling *Molva molva,* at a pinch. You're most likely to catch these fish from a boat, as they rarely venture into shallow waters – unless you can find a deepwater rock mark to try.

FLATFISH FAMILY

While they might not seem like the most obvious target, several flatfish are catchable on light lures. Flounder are the most common you'll find, closely followed by dabs and plaice, fish that you might need to find a special mark or get out in the kayak for. Brill, sole and others in the family may also be possible with luck and effort.

☐ FLOUNDER
Platichthys flesus

Common throughout Britain, this fish can be found around sandy harbours, beaches and estuaries. Prey is more wide ranging than you might think, from shrimp and prawn to crabs, sandeels and even baby mullet and bass. We deal with flounder tactics in detail on pp87-95.
Size: Typically 10-30cm (Specimen: 40cm+ or over 2lbs/1kg).
Season: All year, apart from early spring (March/April) when the larger fish go out to sea to spawn. The best fishing for the bigger samples tends to be winter.

☐ DAB
Limanda limanda

These delightful little fish venture close inshore through the autumn and winter. Well concealed, they are smaller than plaice and flounder typically; they look similar to both but lack the orange spots of the plaice, while the body is translucent unlike the flounder and there is a bigger curve in the lateral line. Their diet is made up mostly of smaller organisms found on the sea floor, such as marine worms, crabs and small molluscs.
Size: Typically 10-20cm, occasionally 30cm+
Season: Year round for smaller fish. Summer through to autumn for the better samples.

☐ PLAICE
Pleuronectes platessa

Another common flatfish, these are a scarcer catch for most lure anglers because they tend to be found further from the shore than flounder. That said, you will find smaller plaice in some estuaries, or by casting into sandy water from a suitable vantage point such as a pier or breakwater. The other answer could be to get out in a kayak or small craft, where they can be susceptible to jigged worm-style soft lures, or even small metals dragged across the sand.
Size: Typically 15-30cm (Specimen: 45cm+)
Season: Spring and summer are best, with post-spawn fish especially hungry in spring.

☐ TURBOT
Scophthalmus maximus

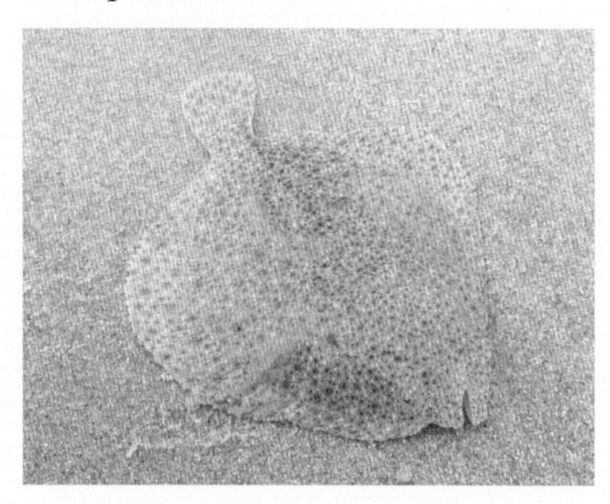

These fish were once common and reached sizes of over thirty pounds. Today, stocks are a shadow of former levels, but you may often find smaller samples on sandy beaches, where they will sometimes feed in mere inches of water. Boat fishing might be the best chance of a larger one, but you might also find them where you get deeper, sandy habitat near the shore (such as beneath cliffs and promontories). They are voracious predators, with comparatively bigger heads than the other flatties.
Size: Inshore samples typically 15-30cm
Season: Spring to autumn

> **TIP**
> Turbot love to pounce on sandeels, and it's perhaps no coincidence that slender metals tripped across the bottom have been highly successful for our friend Ben Bassett and others.

□ TOPKNOT
Zeugopterus punctatus

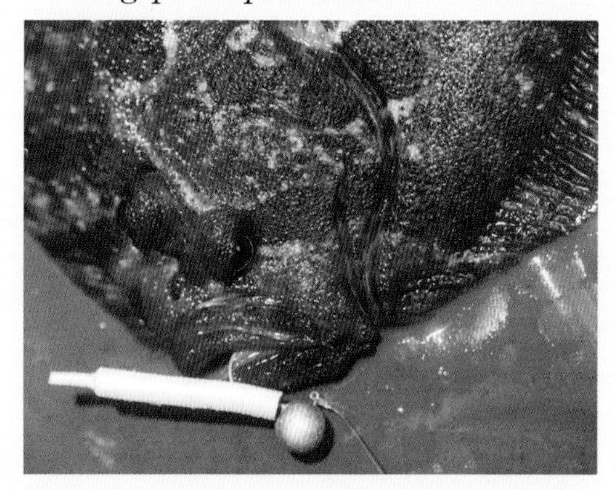

A truly unique little flatfish, the topknot is not a fish of sand flats, but one that prefers rocks and the stony sides of walls, pilings and pillars. He is an ambush expert, a sinister, oversized mouth seizing any prawn or small fish that comes too close. With soft yet rough skin and mean looks, it is a favourite of LRF anglers. Not always common, or easy to catch, populations are often highly localised – but these always include harbours, marinas and other structures offering cover at every stage of the tide.

Size: 10-40cm.
Recommended lures: Smaller soft plastic fish and worms.
Season: Year round, but summer best.

GURNARD FAMILY

The gurnard family are quirky-looking fishes with exaggeratedly big eyes and huge mouths for their size that can bulge open in an instant to devour prawns, small fish and other unfortunates. They prefer more open ground, such as sandbanks, and often frequent the same spots as plaice and other flatfish. It must be said that they are a challenge to catch on lures, not because they lack aggression but because getting at them from the shore can be tricky, although you will occasionally find the smaller ones within casting range with a little local knowledge; they'll even occasionally be found from harbours and piers. Alternatively, inshore boat fishing could be worth a try in calm conditions. Their quirky looks and habits make them quite highly coveted by lure anglers and species hunters.

□ TUB GURNARD
Trigla lucerna

Instantly recognisable by its large, ornately decorated pectoral fins, they are often reddish in colour with blue pectoral fins. They can be found inshore around the coast from midsummer into late autumn and are sometimes caught by bass anglers bouncing fish or sandeel imitation lures or metals across the bottom. They are increasingly being targeted by LRF anglers on species hunts too though.

Size: Typically 25-30 cm, occasionally as big as 50cm.

Recommended lures: Try worm-style soft plastics, jigged or drop-shotted
Season: Mid-summer to late autumn.

☐ GREY GURNARD
Eutrigla gurnardus

This smallest of the commonly found gurnards is light-bellied with grey sides and back. They run to over two pounds, but more commonly a fish of half that size would be a reasonable one. They are especially prolific in western Scotland- and our friend Scott Hutchison has enjoyed prolific success with scented drop shotted lures, landing as many as 50 greys in a session!
Size: Typically 20-30 cm.
Season: Summer to late autumn.

DID YOU KNOW?

The weird and wonderful gurnard is a fish family quite unlike any other. The finger-like fin appendages allow them to 'walk' across the bottom. Another unusual characteristic is their ability to communicate using odd guttural sounds. The name is thought to be a mangling of the French verb 'to grunt' (grogner).

OTHER GURNARDS

Although less likely, the Red Gurnard *Chelidonichthys cuculus* (right), is possible. Confusingly, these can be less red than the tub gurnard – and never have blue on their fins. The rare Streaked Gurnard *Trigloporus lastoviza* are also possible on light lure tackle, although it would probably require calm conditions, a boat and a slice of good fortune.

MULLET FAMILY

As far as lure fishing goes, mullet tend to be a fish of utter frustration. Not entirely surprising, because other than the occasional tiny shrimp or worm, they filter through the mud for things we cannot even see, let alone imitate with a lure. However, there is always an exception to the rule and every so often they will unexpectedly nab a small lure. Oddly, the thin lipped grey mullet also find spinners attractive and using special tactics as detailed on page 56 of our Rigs and Presentations chapter, could be an excellent way to get even.

TIP

Before these fish drive you to despair, a spinner tipped with a section of real or fake worm is the most reliable way of ticking off this frustrating species!

☐ THIN LIPPED GREY MULLET *Liza ramada*

Your best shot at a lure-caught mullet, period! As the name suggests, these slightly smaller grey mullet have thinner lips than their bigger cousins, the thick lips. They are not always easy to tell apart either, but the thin lips tend to have a dark spot at the bottom of the pectoral fin. The gill covers almost meet at the throat of the fish too, while there is a bigger gap on the thick lipped mullet. These fish can occasionally be tempted on tiny lures or specialised tactics, including the use of flies.

Size: 20-40cm (Specimen: 55cm+).

Season: Spring to autumn.

Spot the difference:

Thin and thick-lipped mullet can be tough to tell apart. Perhaps the easiest way is to look at the throat slits: viewed from below, this is much narrower on the thick-lipped mullet (left) than the thin-lip (right).

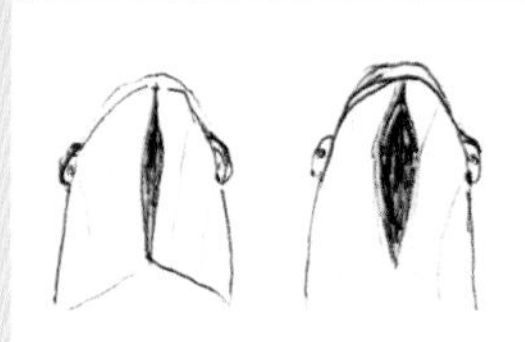

☐ THICK LIPPED GREY MULLET *Helon labrosus*

The biggest of the mullet family is also the hardest to catch. Most of the time they filter feed on things we can hardly see – although tiny shrimps or maggots are occasionally taken. This is an extremely rare catch on any lure and you can count yourself blessed if you tempt one! You would not only require luck, but great skill to land it, because they are one of the hardest-fighting fish you will find anywhere.

Size: 30-45cm (Specimen 60cm+).

Recommended lures: A section of scented plastic worm or perhaps an artificial shrimp?

Season: Spring to autumn.

☐ GOLDEN GREY MULLET *Chelon aurata*

A subspecies in the family, these are less common than the others. The golden grey only reaches smaller sizes and is easily identified by a gold mark on the side of the head. To touch they are often more slippery than other mullet with thick slime a usual tell-tale sign that you have just handled a golden grey. In behaviour, they are much like the thin lips. Very difficult with most lures but are known to

like harbour ragworm and small shrimp, so you might just trick one with a piece of Isome or perhaps even a split shotted fly.
Size: 20-30cm (Specimen: 45cm+)
Season: Spring to autumn.

☐ RED MULLET
Mullus barbatus

Quite different from the grey mullets, these fish have a lovely red coloration and an overslung mouth with two whiskers. Strictly speaking, they are not a 'true' mullet. Tricky to locate, they are more a fish of the Mediterranean, but with warming seas they are increasingly showing up in British estuaries in the summer months; and we've seen them around Plymouth and some Dorset beaches. They eat various small prey, such as marine worms and shrimps and it is sometimes possible to catch them with tiny lures. Andy has had some success sight fishing for them on the foreshore during calm seas using a simple split shot rig and Isome.
Size: Typically 15-30cm.
Season: Summer.

PELAGIC FISH

This family of fish are predators that live a life of constant movement and migration. There are many types worldwide, including giants such as tuna, but for the light lure angler, our most likely catches include the likes of mackerel, scad and herring. What these fish lack in size they more than make up for in speed and beauty. The mackerel, for example, can easily reach speeds in excess of 20 kmh (6+ metres per second), making them supremely athletic and aggressive quarry. Summer to late autumn tends to be 'rush hour' but with warming seas they can often be caught into winter- and our Cornish pals catch mackerel and sardines into January!

☐ MACKEREL
Scomber scombrus

Quick, aggressive and commonplace in summer seas, mackerel provide welcome sport on light lure tackle. There are actually two different types of mackerel caught in the UK. The most common is the Atlantic mackerel (*Scomber scombrus*), but in southern UK and Mediterranean, Atlantic Chub Mackerel (Scomber colias) are also caught. Fast growing and voracious, mackerel are pack hunters that gather in large groups to harry shoals of fry and sandeels.

Size: 20-30cm typically (Specimen 45cm or 1.5lbs/0.75 kg+)

Recommended lures: Just about anything, but spoons such as the classic Dexter Wedge still take some beating.

Season: From the shore, June to September is best, with the bigger fish arriving late.

□ SCAD

Trachurus trachurus

Sometimes referred to as 'horse mackerel' these seasonal visitors offer great sport, but can be quite hit and miss. At certain locations, such as several southern harbours, they can be found in vast numbers for just a few days at a time. They look like silvery, armour-plated mackerel and are just as aggressive and hot tempered when hooked. Curiously, the scad is also a key species in the roots of light rock fishing. In Japan they are known as 'Aji' and are a hugely important fish. In the UK, they are not a daily occurrence but can provide some of the most hectic fishing of the summer months. Evening and night fishing can be especially effective.

Size: 15-25cm typically.

Recommended lures: Small metals and darting jigs.

Season: Typically June to September, appearing in large numbers but short bursts.

□ GARFISH

Belone belone

One of the weirdest of summer predators, these fish hunt high in the water. Their weird, sharp bills stun the tiny fish they feed on – and you will often find them with the mackerel shoals, although they are more inclined to operate from the surface to 8ft (3m) deep. They are a fast-paced fish that love to chase any smaller lure – but the unusual beak or bill, can make them fiendishly hard to hook. The best tactic is to trim down to a tiny lure, or attach an additional stinger hook to the tail of a your lures; see our chapter on pelagic species (p125-132) for more details.

Size: 30-45cm (Specimen 60cm or 2lbs/1kg+).

Season: May to October.

☐ GREATER SAND EEL (or LAUNCE)
Hyperoplus lanceolatus

A sand eel that will take lures? Surely some mistake? Well, anglers might tend to think of these creatures as bait rather than quarry, but the largest of sand eels can be surprisingly predatory and we've even seen them caught on tiny sand eel imitations! Small metals work well.
Size: 10-25cm, occasionally as large as 35cm!
Season: Summer to autumn.

☐ HERRING
Clupea harengus

Another pelagic fish that was once common from Cornwall to the coast of Germany, this fish is not so widespread any longer. Of course, you will still find seasonal migrations in the UK, and there are still plentiful stocks in the Baltic Sea and other areas. They are not only tasty, but good fun on the lightest tackle.
Size: 15-30cm.
Recommended Lures: Small metals & darting jigs.
Season: Migrations vary on location
Tip: Try similar tactics for mackerel, but scale down lures and hook sizes.

OTHER PELAGIC SPECIES

With populations of pelagic fish roving big distances, you never quite know what will show up next. The humble pilchard or Sardine *Sardina pilchardus* is scarcer than it once was, but possible. Reports of sprats are not rare, but tend to be juvenile herring in reality.

There are also some other pelagic species which are protected under current law and so fishing by any method is not permitted at the time of writing. These include the European smelt *Osmerus eperlanus* (above) and two species of migratory shad, the Twaite shad *Alosa fallax* (below) and Allis shad *Alosa alosa*. As such, they score no points from UK waters. Handle as little as possible and release with great care if you catch one!

ROCKFISH, MINI BEASTS & ROCKPOOL DWELLERS

Size truly isn't everything in the world of light lure fishing. In fact the new mania for many anglers is not to see if they can catch the biggest fish, but to see how many different species can be caught. Aside from the blennies and gobies we've already covered, here are some of the stranger critters you are likely to encounter.

☐ LONG-SPINED SEA SCORPION

Taurulus bubalis

Another rock and wall hugger found across the UK and beyond. Small fish with bags of attitude, their exaggerated mouths allow them to swallow prey almost as big as they are! They are well camouflaged ambush predators, whose main diet is other fish, prawns and small crustaceans. That said, they are especially partial to worm-shaped lures. Despite the deadly looks, they are safe to handle. They look angry when caught, bristling defensively, and will sometimes give off a curious vibration similar to a mobile phone in the palm of your hand (try it, we dare you!). Any sea wall is likely to have a few 'scorps' as LRF anglers affectionately call them, while they can also sometimes be found in rockpools.

Size: 8-15cm typically, occasionally 20cm+

Recommended Lures: Worms, grubs & creatures in small sizes.

Season: All year round.

☐ SHORT-SPINED SEA SCORPION

Myoxocephalus scorpius

Sometimes confused with its close relative the long-spined sea scorpion, this aggressive little fish is a much rarer catch from the shore. A survival expert, it can survive brutally cold waters and likes greater depths than its long-spined cousin. Boat fishing or especially deep shore marks can produce them though, and they can grow to an astonishing two pounds plus! The obvious way to tell them apart is that, as the name suggests, this fish has shorter spines and also lacks the white barbules found on the long-spined scorp.

Size: 8-15 cm (Record: 2lbs 7oz!).

Recommended Lures: Worms, grubs & creatures in small sizes

Season: All year round.

☐ COMMON DRAGONET
Callionymus lyra

A creature of small size and strange beauty, the dragonet is found among rockpools and shallow, sandy water. Easily identified by an elongated body, sail-like, short-spined dorsal fin, and an impressively big triangular head. The males have the bigger fins and can have brilliant blue markings. In behaviour, this miniature big-mouthed monster, will ambush and devour most shrimps, invertebrates and small fish given half a chance. Sand patches between rocky outcrops are one of the best places to tick this fantastic little freak off your list.

Size: Typically 10-18cm, males up to 25cm

Recommended Lures: Tiny plastic worms, grubs or soft fish.

Season: Spring and summer.

☐ RETICULATED DRAGONET
Callionymus reticulatus

A rare and special catch, you'll find this dragonet in many of the same locations as the common. They can be tricky to catch though – and sometimes hard to separate

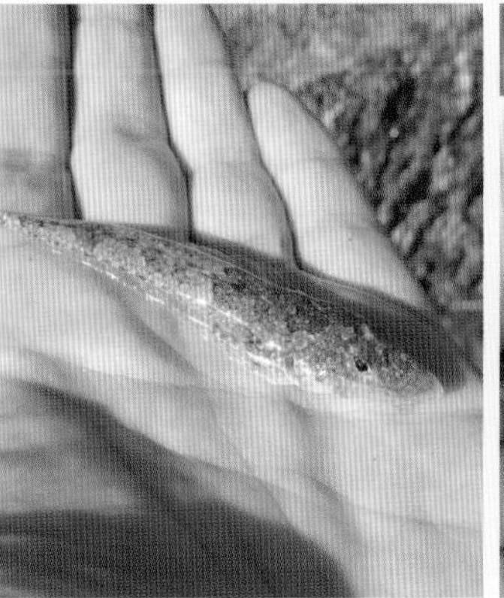

from the common dragonet. Size is one denominator, with the reticulated only usually growing to around 11cm maximum. Snout length is perhaps the other most reliable way to split them apart: the common dragonet has a much longer snout of two to three times the diameter of their eyes, whereas the reticulated dragonet's tends to be only slightly longer than the diameter of the eyes. The real expert in the matter is our friend Scott Hutchison – and you'll find more notes on his *Something Fishy Going On* blog.

Size: Typically no larger than 11cm.

Season: Spring and summer.

☐ SAND SMELT
Atherina presbyter

Small and silvery olive in colour, perhaps a majority of anglers are unaware that these fish even exist. They are prolific all around the British coast however, shoaling in groups that can number hundreds of individuals for safety. Particularly common in the shelter of harbours and breakwaters. For the species hunter, they are an easy fish to tick off the list, provided you tackle up fine enough. They are also a key food source for predators such as bass; so if you find them by the score you might want to find a decent match for one in your lure collection.

Size: 4-8cm typically.

Recommended Lures: Tiny soft lures and scented worm tails, hooks down to 18 or less.

Season: All year round.

☐ FIFTEEN-SPINED STICKLEBACK

Spinachia spinachia

For some childish yet serious fun, these beautiful fish are both striking in appearance and much coveted by LRF nuts. There are scattered groups right around the UK, although they are not especially common. Marinas and harbours are probably your best bet, where they can be observed in calm conditions. Resembling a cross between a pipefish and a pike, their good looks belie fierce instincts – and this is never truer than in the spring when they pair up and nest. To stand any serious chance at catching one you'll need tiny hooks and the merest fragment of a soft lure or Isome type worm.

Size: 8-12cm.

Season: Most visible in late spring, April to June.

☐ CLINGFISH

Lepadogaster lepadogaster

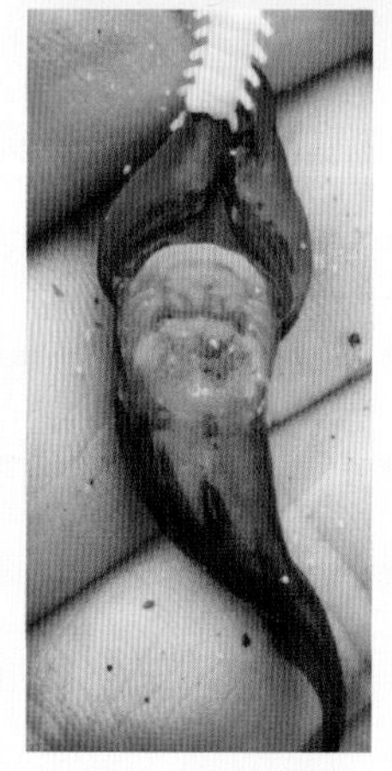

Also known as the Cornish Lumpsucker (how charming!) this little critter is about as strange as species get. As the name suggests, they are designed to stick to rocks via specially adapted pelvic fins. A real collector's item for the species hunter, you might just tempt one with careful local detective work and the tiniest fragment of Isome on a miniscule hook!

Size: Up to 5cm.

Season: Summer.

For some of the rarer rock-loving fish, your best chance might be the deeper and more inaccessible rockpools. On the largest tides of each cycle, you might be able to find deeper and usually hidden areas at the low water mark; but you'll need to be quick and map an escape route on hairy marks!

TIP

WEEVER FAMILY

Beauty is truly in the eye of the beholder when it comes to fish; but in spite of the bad reputation, weevers are beautifully designed and can have eerily lovely colours. They also come with a health warning however! With poisonous spines, weavers demand great care. Avoid handling and, if you target them by design, do take barbless hooks and long forceps!

WARNING!

Getting spiked by a weever can be a painful experience. The best immediate remedy is to soak your hand or foot in hot water (as hot as you can bear it), FAST, before seeking medical attention.

☐ LESSER WEEVER FISH

Echiichthys vipera

If you have ever bathed on a sandy beach in the UK you will have heard stories of people being stung by a weever fish. This is normally a lesser weever. The fish is most commonly found inshore on sandy beaches and if disturbed it will sting using its venom glands on the first dorsal fin or on the gill cover. Species hunters are perhaps the only fans of these fish, and they can be caught on small metals, split shot rigs or drop shot presentations across sandy beaches. It often happens, once one is hooked, that multiple fish follow their struggling neighbour.

Size: 5-10cm, specimen 15cm

Season: All year round

☐ GREATER WEEVER FISH

Trachinus draco

The greater weever is the lesser seen of the two weever fish in the UK. They are generally caught offshore by lure anglers fishing metal or soft plastic lures close to the bottom. Fish of 1-2lb have been known but due to them preferring deeper water they are a difficult species to tick off.

Size: 15-30cm (Specimen, 45cm)

WRASSE FAMILY

Bold in both appearance and nature, wrasse are powerfully-built lovers of rocky and weedy spots. The UK has several different types of wrasse to target and they are relatively common, although populations are now under threat where salmon farms harvest them for parasite control.

All the wrasse family are eager biters and offer great sport on light tackle (see our dedicated chapter on pp117-124 for tips and methods). For the species hunter, catching the full set is a must – and the diversity of colours and locations makes these fish endlessly interesting. Worms, fish and creature-style lures catch the big ones, although tiny grubs and Isome worms can be best for winkling out those beautiful smaller specimens.

☐ BALLAN WRASSE

Labrus bergylta

Bold biting and fantastically varied in colour, ballan wrasse are one of the hardest fighting fish you'll encounter on lure tackle. Found around rough ground, weed and structures all over the UK with the exception of the far north, they are widespread and respond well to a variety of lures. The main

challenge tends to be presenting artificials in the craggy places they live.

Size: 15-40cm typical (2kg or 50cm+ is a specimen).

Recommended Lures: Bigger plastic worms, fish and creatures. Weedless rigging can be vital.

Season: Year round, but June to September best.

DID YOU KNOW?

Ballan wrasse will occasionally change sex to ensure successful breeding! Most of the time this happens when a dominant male dies and a female switches sides to fill his niche.

☐ CORKWING WRASSE

Symphodus melops

A smaller member of the wrasse family, the corkwing has similar habitats and diet. Often confused with small ballans, but this species is easily identified by the dark spot just before the tail fin. Like other wrasse, these fish nest during spawning and have an interesting sex life to put it mildly: fish will pair up, but small males known as 'sneakers' disguise themselves as females to fertilise eggs.

Size: 8-12 cm typically (specimen: 20cm)

Recommended Lures: Small soft plastic worms, grubs and creatures.

Season: Year round, but summer best

Don't strike too soon for smaller wrasse when using worm lures, but let bites develop.

☐ CUCKOO WRASSE

Labrus mixtus

A vibrantly coloured fish that can be any hue from pink to red, although the males are often distinguished by their bright blue heads. Also a species that can change sex if there is an imbalance of males and females in an area. Like all wrasse, they are armed with formidable teeth, and while their main diet is shellfish and crustaceans, they also eat shrimps and other invertebrates and can be fiercely territorial in their reaction to lures. A rare catch from the shore, boat fishing is often your best option.

Size: 10-15 cm typically, occasionally up to 25cm.

Recommended Lures: Small soft plastic worms.

Season: Year round, but summer best.

☐ GOLDSINNY WRASSE

Ctenolabrus rupestris

The baby of the British wrasse family, these fish are not often caught by the shore angler unless fishing small hooks and lures. The body is usually reddish brown, but can vary on location. Identifiable by a black spot on the top of the tail and one at the front of the dorsal, along with a slightly more slender profile and forward-reaching teeth. A less well known fact is that smaller wrasse love eating sea lice, and have been targeted by commercial salmon farming for this reason.

Size: 5-10cm typically.

Recommended Lures: Small soft plastic worm sections or grubs.

Season: Year round, but summer best.

Scale down your hooks if you want to catch these fish. A size 14 isn't too small.

☐ ROCK COOK WRASSE

Centrolabrus exolentus

One of the most striking of the wrasse species, these fish are often caught but can be easily mistaken for a corkwing wrasse. The rock cook has a smaller head and mouth in proportion to the body. This fish has a rounder profile than the others mentioned, they will mostly have a brownish back with a pale underside and gold flanks. They also have a distinct line before the caudal fin (tail), and males have blue tips to all of their fins.

Size: 5-15cm typically.

Recommended Lures: Small soft plastic worms and grubs.

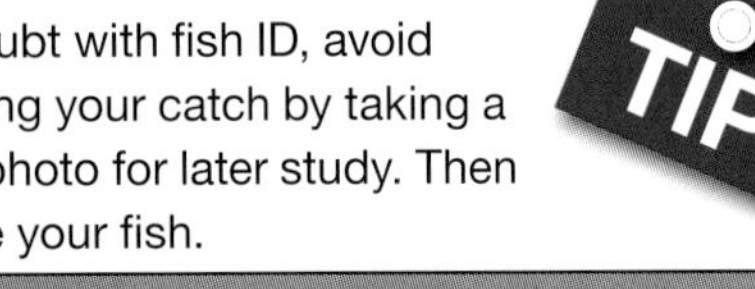

If in doubt with fish ID, avoid stressing your catch by taking a quick photo for later study. Then release your fish.

☐ BAILLON'S WRASSE

Symphodus bailloni

Similar in shape and appearance to the corkwing the baillon's wrasse can often be mistaken for their more common cousins. Perhaps the most distinguishing feature of this rare wrasse is a dark spot around two thirds of the way along the dorsal fin.

Female baillons have a grey or blackish urogenital pappilla (the egg laying part of the fish), while males are distinguished by reddish or pink fins and gorgeous sea green facial markings, sometimes also with pink spots.

Size: 10-20cm typically, but occasionally up to 40cm.

Recommended Lures: Small soft plastic worms or grubs.

☐ JOHN DORY

Zeus faber

One of the most exotic looking of the summer species, the John Dory is a fish of myth and legend for most. They tend to show up across wrecks and inshore during summer months and can be caught freely on lures. The John Dory is slim in profile and round in the body. It has a black spot resembling an eye as a distraction on the side of its body. It can often be found whilst drifting alongside kelp beds from a boat and is sometimes caught by anglers targeting bass or wrasse on lure from the shore. Most are caught more by luck than judgement, but when they do arrive they are highly prized.
Size: 12-24inch (30-60cm)
Season: Warmer months.

RARE, ACCIDENTAL & BORDERING-ON-IMPOSSIBLE SPECIES

As enthusiastic as we are for unusual captures and difficult challenges, there are some sea fish that won't often play ball with lures. Granted, you might very occasionally catch these species by using something like a scented lure, fished static. But this is neither a common occurrence or drastically different to bait fishing, so we have chosen to leave a certain number of species out of our list. These include the rays and dogfish, both of which feed in their own unique way and are not a viable target with lures (until someone develops a battery-powered artificial that releases electrical distress signals; now there's an idea!).

However, there are species that will take lures but are just too random or scarce to merit your attention on any consistent basis. Just about anything can show up in British waters, and we know of at least one lure-caught amberjack, amongst scores of weird catches, rumours and wild tales!

Some other species, such as the shads, are protected by law these days too and while the capture of one would make a pleasant surprise it would be wrong of us to promote their deliberate capture, which is not permitted at the time of publication. However, there are a few types of fish that might be viable.

☐ TRIGGER FISH
Balistes capriscus

Trigger fish often show up on the beaches of the south–west during the late summer months. They are caught frequently on bait and savvy anglers in Plymouth have tracked their movements around the harbour and are now consistently catching them on lure. Famed for their strength, the trigger can be caught using worm type lures with drop shot presentations.
Size: Typically 20-40cm
Season: Summer through to autumn.

TIP

Use strong hooks, the trigger fish has been known to bite through fine gauge models!

☐ RED BAND FISH

Cepola macrophthalma

Few anglers can claim to have caught this strange beast, which tends to like water at least 20m deep. It is found across the Mediterranean and African coast, but has become more common in British waters in the past two decades – and some charter boats on the south coast quite regularly find them. Elongated, ribbon like and with fierce teeth they are quite unmistakable. They sometimes make burrows and feed on small crustaceans, although they will also swim up off the bottom to hunt for other food sources.

Size: Typically 25-50cm.

Season: Summer and autumn.

EELS

Snake-like and slender in appearance, eels are fish that that will never generate a huge fan club. In fact, there are probably more anglers who try to perfect how not to catch them than would ever seek them out with any method, least of all a lure. However, for the keen species hunter they are quite fascinating creatures and a genuine challenge. As we increase our admittedly limited knowledge we are also learning that these fish are more than merely bottom-dwelling scavengers; in fact they are also active hunters. Only two types of eel are available to UK anglers.

☐ SILVER EEL

Anguila anguila

Above: Yes, it really can happen. The shock is visible on Andy's face with this 1m plus eel, caught during the International Lure Fishing match at Rutland in 2018. As thrilling as it was for all of 20 minutes, the fish didn't count!

Usually just referred to as eels, these fish undergo huge migrations to make their way into fresh water. They are mysterious, ravenous creatures that will eat a wide variety of prey, from tiny crustacea and leeches, to small live or dead fish. They are silver-hued when at sea, before darkening to a brownish colour as they come inland. The largest are those that end up staying in one location for years. They grow slowly and a five-pounder could be over 50 years old.

It is indeed possible to catch eels on the lure. Andy has had two momentous captures, one at the World Predator Classic. In some locations, drop shotting for eels works, usually under torch light. The hardest part can be spotting them. Once

you've located your eel, it can be teased into biting by presenting a drop shotted lure into their sight line. They tend to adopt a 'strike' position rather like a snake! As adept at swimming backwards as forwards, they put up quite a fight on light lure tackle.
Size: 20-60cm typically. Exceptionally 1m+
Recommended Lures: Try a slowly fished jig or dropshot a scented lure.
Season: Usually migrate from late spring to summer.

□ CONGER
Conger conger

Have we well and truly lost the plot with conger eels? Not necessarily, because on occasion they will attack lures presented close to structures such as harbour walls and boulders. This usually happens at night, when these creatures will strike out at small fish, besides scavenging and grabbing at all manner of other prey. To stand a chance of landing even a small 'strap' conger would take a slice of luck, a steady nerve and fairly strong tackle.
Size: 50cm to 2m+
Recommended Lures: Scented soft plastics.
Season: Year round – if you're crazy enough.

FRESHWATER FISH

There might be less variety when it comes to catching coarse fish on lures, but for the adventurous angler, there are some great challenges. Almost all of our coarse fish are predatory at times, such as when fry are abundant, while even the most docile-looking fish can be curious or aggressive enough to bite a lure occasionally. Please note that in many areas the 'close season' (15 March-15 June) will apply to protect spawning fish.

PREDATORY COARSE FISH

Even the most friendly-looking freshwater fish will occasionally snack on their smaller kin. It's not unheard of for big fat carp to go hoovering up fry, or for barbel to snack on minnows. However, we'll start with those species with a genuine taste for meat.

□ CHUB
Squalius cephalus

Sometimes shy, but greedy and highly predatory, chub are an occasional capture for many anglers – but with a dedicated approach these fish yield rewarding fishing. Trying surface-waking lures or spinners is a delightful way to tackle them

in the summer, while jigs can work when searching snags and pools through the winter. For more in depth tips and advice, see our chub chapter (pp133-138).
Size: 20-45cm (Specimen 60cm or 5lbs/ 2.5kg plus).
Season: All year except closed season/ spawning period of 16 March-25 June.

☐ PERCH
Perca fluviatilis

The most common and catchable of freshwater predators, this is likely to be the first you ever tick off the coarse species list. This is a fish to catch on a huge variety of smaller lures (start with 1-3" (2-7cm) and enjoy the very lightest of tackle. The larger fish take some catching, however, and are a real test of skill. More about perch on pp139-152.

Size: Typically 10-25cm (Specimen 35cm or over 2lbs/1kg).
Season: All year.

> **TIP**
> Be prepared to take a risk for a big chub! This species loves tricky, inaccessible spots.

☐ PIKE
Esox lucius

Our largest native predator is a fearsome fish that grows rather large to be landed on the lightest tackle. For this reason (and the fact that so much has already been said and written about pike!), they didn't get their own chapter in this book. You might encounter them by accident on many freshwaters though, so if they are expected it might pay to use stronger tackle and a fine wire trace, as 'pike insurance'. Even a modest pike will give you an epic battle on light tackle.

Size: 30-90cm (Specimen 100cm /10kg)
Season: All year, although anglers may avoid spawning time, as well as warmer waters in high summer temperatures to protect these deceptively fragile fish. Close season often also applies on natural waters.

□ RUFFE

Gymnocephalus cernuus

This small, pepper-spotted perch lookalike is a nostalgic favourite for many anglers. They have become less common in recent years, but you will still find them on some rivers and canals. Spiky, and sneaky, you will often catch one close to the bank, where they hunt prey such as sticklebacks, fry and invertebrates. Scale your tackle right down and they are a delightful challenge.

Size: 5-10cm typically (specimen: 15cm).

Recommended lures: Tiny soft jigs, worms and grubs.

Season: Year round, but summer best.

Use lures down to just an inch, with small, sharp hooks.

□ WELS CATFISH

Silurus glanis

You would not want to tangle with one of these on ultralight gear at all, but for the sake of completeness, this is an exciting target for the dedicated species hunter. Our understanding of this fish is changing, and they are active predators as much as scavengers. Anglers across

Western Europe are now catching some serious catfish on lures.

Size: Anything up to 3 metres plus!

Recommended lures: Large soft plastics, rattling plugs or any lure with strong vibration.

See a psychiatrist.

□ ZANDER

Sander lucioperca

Remarkable eyesight, fearsome teeth and a fickle nature make zander one of the most popular and challenging species for lure anglers. Easily identified by the elongated shape, oversized eye and spiky fin. Masters of hunting in low light,

they can pick out a lure in amazingly poor visibility. See our dedicated chapter (pp171-172) for further notes and tips.
Size: 20cm-50cm (specimen: 75cm or 10lbs plus)
Recommended lures: Soft plastics of 3-6' (10-15cm) typically , jigged or dropshotted.
Season: Year round.

□ IDE
Leuciscus idus

Found throughout much of Europe, this fish is an obliging all-round predator that loves small lures. They have a colour rather like rudd, but behave more chub-like, actively corralling and hunting in groups for smaller fish. In Britain, there are one or two isolated populations of breeding fish; most of the time we only have its more colourful cousin, the ornamental varieties of orfe.
Size: 20-45cm.
Recommended lures: Small spinners, crankbaits and jigs.
Season: Summer to autumn.

Try lures of 2-4" (5-10cm) with a slow, steady retrieve.

OTHER COARSE FISH & OCCASIONAL PREDATORS

□ GRAYLING
Thymallus thymallus

A fish of almost ethereal beauty, the 'lady of the stream' is not usually thought of as a predator. However, a lesser known fact is how readily larger grayling predate on minnows and smaller fish. On the continent, anglers certainly target them with tiny spinners and other lures – which tend to attract quite big fish. Flies could also be cast on ultralight lure tackle – or presented using a float. They like the same waters as trout, with clean, steady paced water ideal.
Size: 15-40cm typically (specimen 2lbs or 45cm).
Recommended Lures: Spinners, soft plastic worms and grubs.
Season: Summer to winter (UK coarse fishing season).

TIP

Try a tiny spinner, perhaps tipped with a section of soft plastic worm.

☐ RUDD

Scardinius erythrophthalmus

Beautiful gold-sided fish, you'll find rudd on natural waters such as lakes, slow-flowing rivers, drains and canals throughout most of the UK and across Europe and even far flung New Zealand. They are playful, inquisitive fish and occasionally aggressive too, especially when there are fry in the water or close to spawning time. Although few anglers would try it, they can sometimes be caught on tiny spinners or the smallest of soft jigs.

Size: 10-40cm

Recommended Lures: Small soft plastic grubs or tiny spinners.

Season: Late spring to summer best.

☐ ROACH

Rutilus rutilus

Silver-sided and red finned fish, common on all kinds of freshwater. Not traditionally thought of as predators, but they will feed on quite a wide variety of prey, from worms washed into the river, to small fry and insect larvae. Experiments with very small lures will definitely produce roach. Try a split shot rig with a tiny soft lure and be prepared to go right down with hook sizes (14-18).

Size: 10-40cm

Recommended Lures: Tiny grubs and soft plastics.

Season: Year round (closed season applies on many waters)

> **TIP**
> Scale right down with lures and try hooks in the 14-18 range.

☐ COMMON BREAM

Abramis brama

These big dozy fish look like they should have little interest in a lure, but many anglers will have a tale or two about an accidental capture. These fish do occasionally prey on fry, besides bottom dwelling prey, and they display an unusual aggression in the late spring and early

summer, when they get ready to spawn and become decidedly less gentle-natured.
Size: 30-60cm.
Recommended lures: Small soft jigs and worms.
Season: Late spring and summer.

A worm-style lure, fished nice and slowly on the bottom, is a lure as likely as any to tempt a bream.

☐ CARP

Cyprinius carpio

Not a fish you would often target deliberately on a lure, but there are nevertheless scenarios where they will take a lure (and we've lumped commons, mirrors and the rest in together here). An explosion of fry will sometimes see them switch on, while large carp will also eat prey such as gudgeon and other small bottom dwellers from dragonfly larvae to crayfish. A formidable challenge – but all it takes is your lure in the right place, and a little luck! If it's a big fish it could be some time before you even see it...
Size: 30-80cm
Recommended Lures: Small soft plastics.

☐ BARBEL

Barbus barbus

These large, bottom-feeding fish of running waters are tricky on a lure but possible, especially in the early season (16 June onwards), when they not only eat invertebrates of all types (including caddis larvae and leeches), but will sometimes gorge on smaller coarse fish such as minnows and gudgeon. On certain rivers, crayfish are also fair game.
Size: Typically 45-75cm
Recommended lures: Smallish, heavy 'creatures' such as crayfish and minnow lookalikes.
Season: Summer best (from 16 June)

OTHERS & OCCASIONALS

Among the other coarse fish there are various other species you might just catch if you are lucky or dedicated enough. Hybrids of various species are possible – and we've had roach/bream hybrids on lures. The Tench *Tinca tinca* is a bottom-feeder with poor eyesight, occasionally caught as a random accidental catch on a lure. A soft plastic worm or grub could just be your ticket. Crucian carp *Carassius carassius* is perhaps even harder as it feeds on tiny invertebrates most of the time.

Ironically some of the tiniest coarse fish you might catch on tiny LRF style lures, are more aggressive and obliging. These include the Bullhead *Cottus gobio* (right) sometimes known as a sculpin, and Gudgeon *Gobio gobio* (below). Meanwhile, the Three-spined

stickleback *Gasterosteus aculeatus*, is wildly aggressive and will sometimes hold onto a tiny lure even when not hooked. If you are really crazy and use the tiniest of hooks, you might even catch species including the Minnow *Phoxinus phoxinus*, Bleak *Alburnus alburnus* and perhaps even the rare Bitterling *Rhodeus sericeus,* which lives in a tiny handful of venues including the Llangollen Canal, or perhaps even the Pumpkinseed Sunfish *Lepomis gibbosus*, a feisty little non-native species that occurs in a small handful of southern fisheries, including Silver Springs, Somerset. You'd need a miniscule hook and tiny fragment of soft plastic worm – or even perhaps a tiny fly or piece of red wool. This is the danger of species hunting – it can get seriously addictive, especially when you restrict yourself to lures.

GAME FISH

From the outset, sadly, lure anglers in much of the UK face a lot of exclusion when pursuing so-called game fish. Some waters will allow spinning for salmon and trout, for example, but there are countless rivers and lakes where it's strictly 'fly only', so you might have to do some homework to find sport. Do also be aware that game fish have different seasons to coarse species, and migratory fish have a different licence! While we don't have acres of space to explore salmon and sea trout fishing here, you'll find plenty of advice on trout fishing in our dedicated chapter (pp153-160)

☐ BROWN TROUT

Salmo trutta

A fish of clean, swift flowing rivers and wild lakes, the 'brownie' is a sharp-sighted and versatile predator. Easily recognised by its golden or bronze sides and vivid spots. Wild fish can often be told from those stocked by a milky white edge to their lower fins. Smaller trout subsist mainly on invertebrates and fly life; but all trout will also eat other

fish, including minnows, sticklebacks and the fry of other species and their own kind.
Size: Typically 15-40cm (specimen 45cm+ or over 2lbs/1kg).
Season: Depends on location. 15 March - 30 Sept for most of the UK.

☐ SEA TROUT
Salmo trutta

No, the above isn't a misprint; the Latin name is the same because genetically this is the same fish as the brown trout. It's just that for some reason, a proportion of these creatures venture out to sea for the richer pickings on offer. Hence as well as being more silvery, they are often a lot bigger than their stay-at-home pals. Like salmon, they will migrate and grown much bigger in size, before returning to spawn. Night fishing is most successful, in conditions when a river has plenty of water to help travellers get up weirs and other obstacles.
Size: 30-70cm.
Season: Depends on location. 15 March - 30 Sept or most of the UK.
Recommended lures: Small spinners, crankbaits or soft plastics in the 1-3" (2-8cm) bracket.

☐ ATLANTIC SALMON
Salmo Salar

The silver king of fishes is one for any keen angler's bucket list. It is a beautiful, torpedo-shaped fish that travels thousands of miles to sea and back, returning to spawn. It's a bit of a myth that they do this once and die though, because larger fish might make two or even three such journeys. You will of course also find small parr in rivers, which look a little like juvenile trout, but these should be left well alone.

When 'fresh run' and newly arrived in a river, they tend to be bright silver; whereas the longer they stay in freshwater, the darker and more brown they become, looking a bit like very large brown trout. They have a different head shape though, with a more

pointed head, typically, and the eye further forward. Males will develop an elongated, mean-looking jaw or 'kype'. With wild populations under more threat than ever, we would urge you to return ALL salmon you catch, even if rules state you can keep one.
Size: Typically 45-100cm.
Season: Depends on locality and rules. Usually March till September.
Recommended lures: Spinners, plugs, Devon Minnows, worm-shaped soft plastics.

DID YOU KNOW?

When returning to their home rivers, salmon and sea trout feed on... absolutely nothing! They've already done all their fattening up at sea. They don't fancy so much as a wafer thin mint and it's a minor miracle they will take any kind of lure. Is it pure instinct then, or perhaps curiosity or even annoyance that makes them want to smash a fly or lure? Who knows.

☐ RAINBOW TROUT

Oncorhynchus mykiss

Brasher and more aggressive than our native brown, the rainbow trout is actually an American import. With fine sporting – and eating – qualities, however, they are long established as a favourite with anglers. The vast majority of these are infertile triploids, although a couple of rivers including the Derbyshire Wye now have accidental breeding populations.
Size: Typically 40-60cm.
Season: Year round on stillwaters; same as brown trout on rivers (typically 15 March-30 September).
Recommended lures: Smallish spinners, crankbaits & soft plastic fish or grubs.

☐ ARCTIC CHAR

Salvelinus alpinus

Natural populations of this fish are limited in the UK, mostly to northern England and Scotland. As such, they tend to be a protected species. However, in other waters, they are now stocked and sometimes possible to catch. In their native habitats, the char is a fish that thrives at great depth. On the continent trolling, deep jigging and even ice fishing are popular. On smaller, stocked waters, tactics are similar to rainbow trout, although they will tend to feed lower in the water.
Size: Typically 20-35cm.
Season: Depends on location.
Recommended lures: Tiny jigs, baited spoons or deep diving plugs.

OTHER TROUT & GAME FISH

In addition to the above, there are also several strains of stocked trout that may occasionally be angled for with lures. All of the following are sterile triploids, like rainbow trout. Some of these are technically char or hybrids too, such as the red-bellied Brook trout *Salvelinus fontinalis* (a type of char) and the Tiger trout which is a mix of brook trout and brown trout.

Blue trout and Gold trout are both specially bred colour variants of the rainbow trout. There is also the recently bred Sparctic trout (a cross between Arctic Char and brook trout, first stocked in 2018), which is already spreading across several fisheries. Getting access to waters that allow lure fishing is often the challenge.

The new 'Sparctic trout' is actually a cross between brook trout and char.

Jack Perks films a large pike in the clear water of Stoney Cove, a flooded old quarry near Leicester

The Contributors

Dominic Garnett is an acclaimed angling writer and guide based in Devon. Well known for his lively, wide-ranging articles and weekly *Angling Times* column, The Far Bank, he is also the author of six books, including Amazon bestseller *Flyfishing for Coarse Fish* and the collection of angling tales *Crooked Lines*.
Read more from him at: **dgfishing.co.uk**

Andy Mytton is a fanatical lure angler based in Dorset. His innovative, light tackle approach has accounted for a spectacular array of fresh and saltwater species in the UK and abroad, as well as TV appearances on BT and Sky Sports. On the competition and events scene he has also fished far and wide, representing England at international level and winning the Zandermasters title.

Jack Perks is a pioneering photographer and filmmaker with a special passion for aquatic wildlife. Underwater shots are a particular speciality, with his breathtaking work featuring in the likes BBC SpringWatch, The One Show and various articles and books, including the fascinating hardback *Freshwater Fishes of Britain*. You can see more of his stunning footage and images at:
www.jackperksphotography.com

Neil Sutherland is an award-winning graphic designer and keen lure angler. Having worked on a huge range of projects from website design to children's TV, his other great love of lure fishing reveals itself in the beautifully conceived diagrams of this book. You'll also spot some of his stylish fish and lure designs in these pages, which can be found on a range of t-shirts, prints and stickers at:
aquazidstudio.com

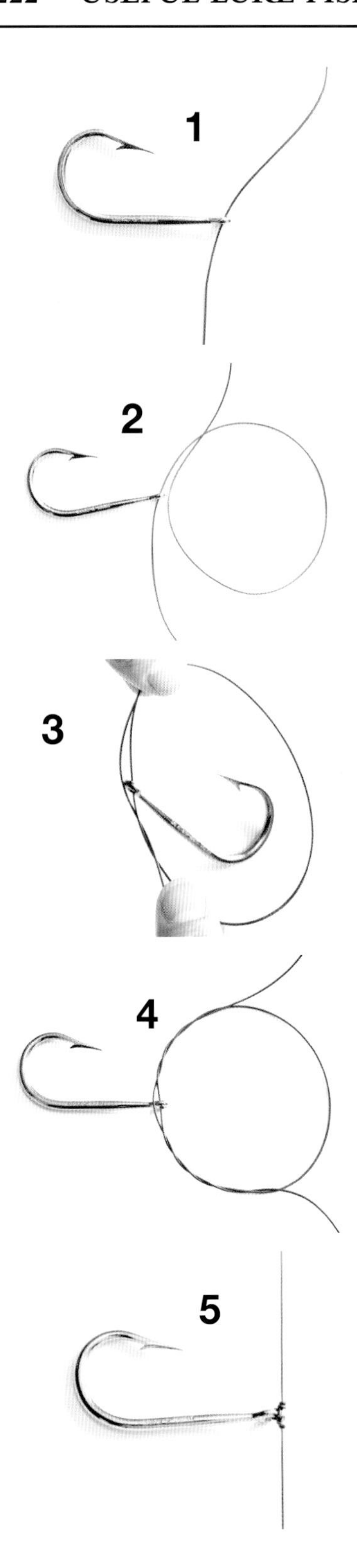

There are many handy knots for the lure angler to learn. It's easy to be lazy and stick to only two or three that you already know. However, with special rigs and presentations, not to mention fine, slippery braids, it really does pay to expand your repertoire! When you hook that special fish, you are only as secure as your weakest link, after all. Here are four especially useful knots for the lure angler.

DROP SHOT KNOT

This is a great knot for drop shot rigging, or any scenario where you want to tie a hook or fitting parallel to the line. In our opinion, it is a better and tidier knot for drop shotting than the often recommended palomar knot.

1. With your hook positioned 'point up', thread the line through the eye.

2. Form a loop, doubling the line in front of the eye of the hook as shown.

3. Now pass the hook through the loop. You might need to pinch the doubled up line in place to keep everything tidy.

4. Pass the hook through the loop 3-4 times to form neat turns of line above and below.

5. Wet the turns with a little spit and carefully tighten up. You might need to use your thumb nail to keep everything tidy and ensure that line doesn't wrap round the back or sides of the eye.

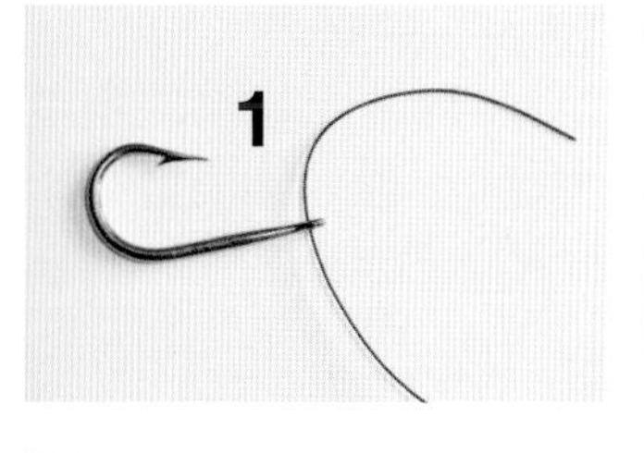

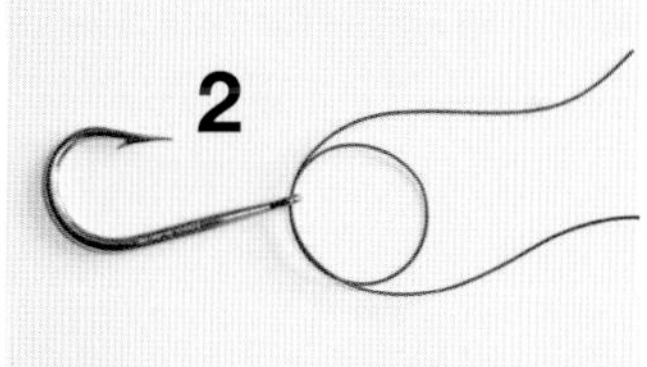

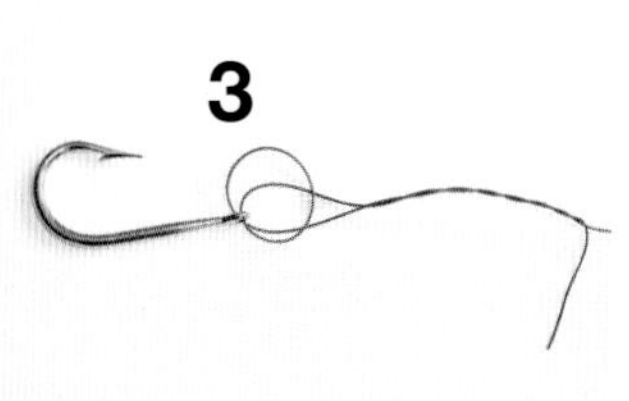

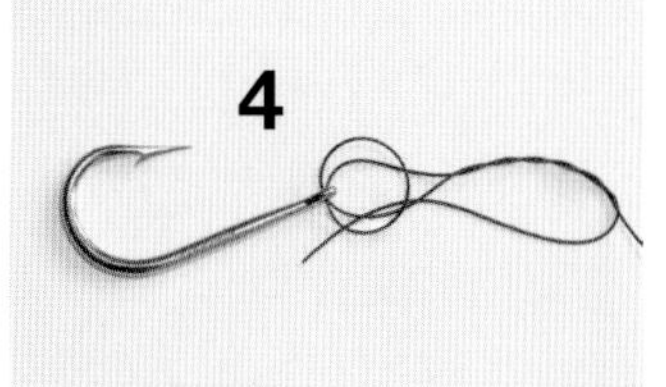

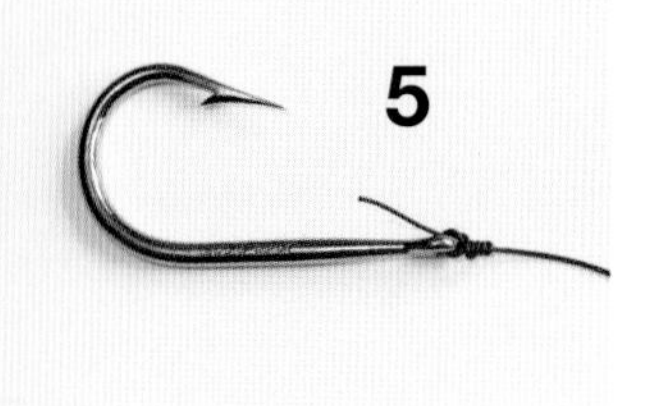

DOUBLE CLINCH KNOT

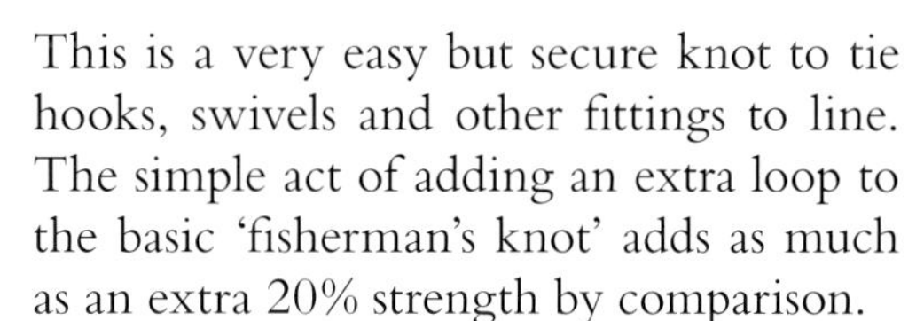

This is a very easy but secure knot to tie hooks, swivels and other fittings to line. The simple act of adding an extra loop to the basic 'fisherman's knot' adds as much as an extra 20% strength by comparison.

1. Run your line through the eye of the hook.

2. Now run your line through the eye a second time, to form an additional loop.

3. Keeping the two loops together, form a series of turns of line above them, as shown. For thicker lines, five or six turns may suffice. For lines of 6lbs (3kg) or less, make at least eight.

4. Pass the free end through BOTH loops you formed earlier.

5. Wet the knot with some saliva and carefully draw tight before trimming the free end.

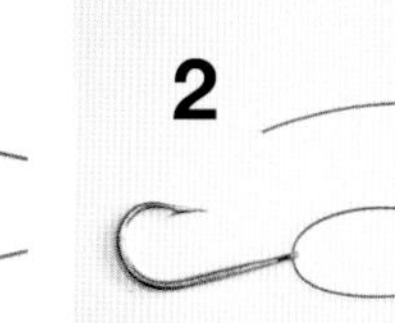

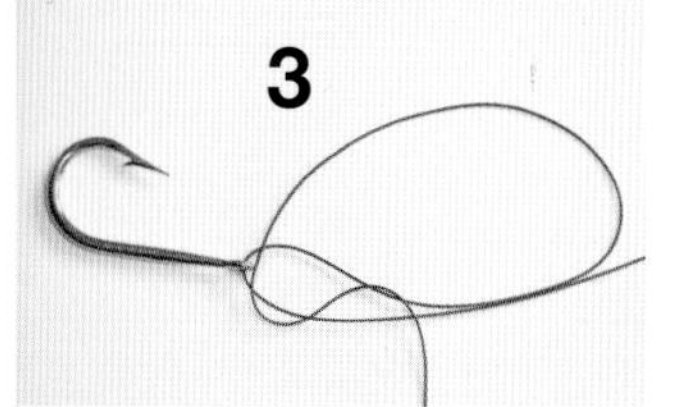

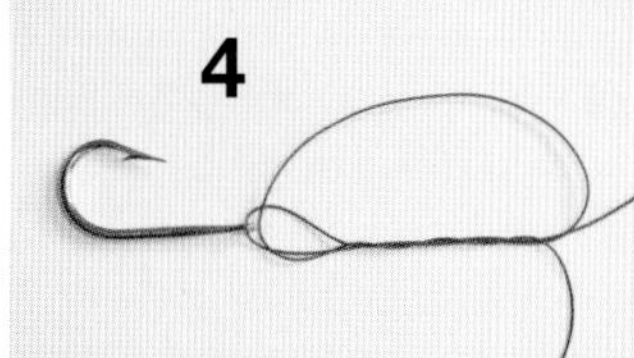

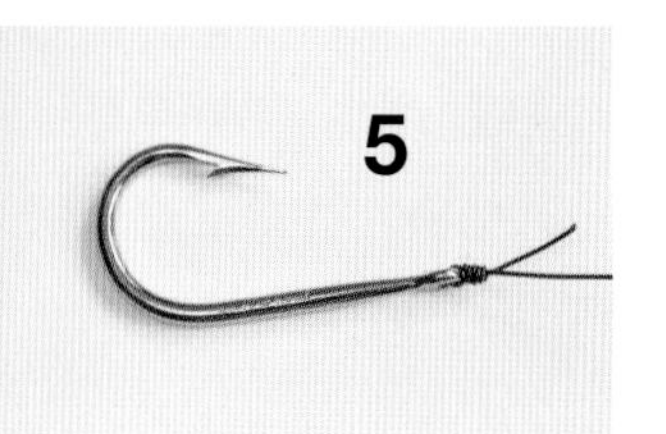

GRINNER KNOT

Modern braids are excellent for lure fishing, but their fine diameter and slippery coatings render some of the basic fishing knots rather insecure. The grinner knot is a very strong and secure answer to this problem, whether you are tying on a swivel, snap link or a hook. A trustworthy knot for any demanding situation!

1. Pass the line through the eye of your hook, swivel or fitting.

2. Run a generous amount of line through and then double it back on itself, as shown, like a flattened, backwards 'S' shape.

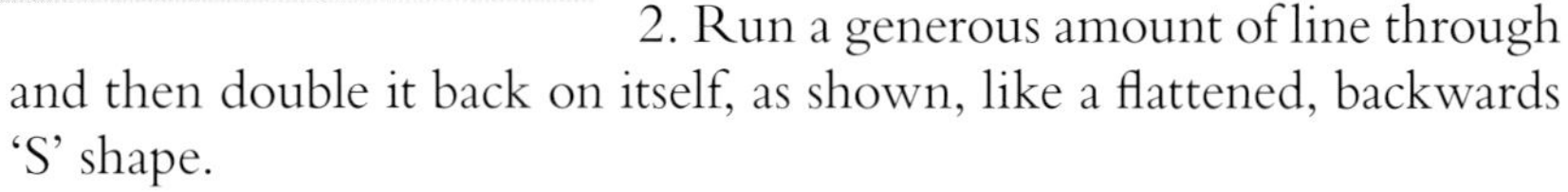

3. Now take the free end, taking it underneath the lower two strands of line and passing it through the gap in the middle.

4. Build up 6-10 turns of line, like this, passing the loose end through the central loop to form each turn. The finer the line, the more turns are required.

5. Wet the knot before carefully drawing tight and trimming off.

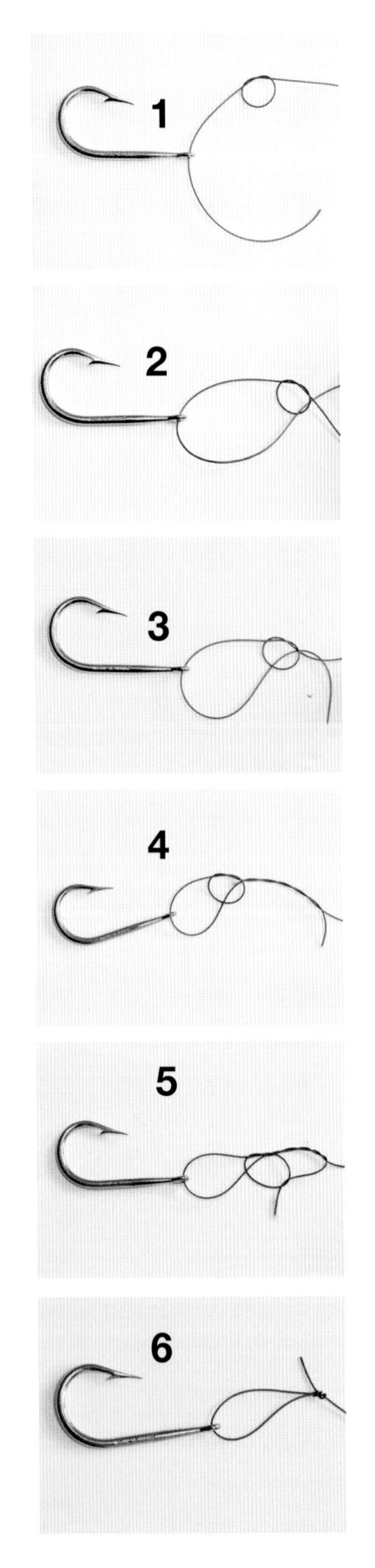

NON-SLIP LOOP KNOT

This is another underused but very handy knot for the lure angler. It really comes into its own when you are tying any lure direct to a mono or fluorocarbon trace. Having your lure trapped in a loop will give it far more natural movement compared to having it pinched in place by a tight standard knot. This is especially the case with thicker fluorocarbons, which can behave rather stiffly.

1. Start by tying a simple overhand thumb knot in the line, like this, before passing the free end through the hook eye.

2. Now pass the end of the line through the thumb knot, passing from the front to behind.

3. Take the free end and bring it from over the front of the main line, like this.

4. Now form five or more turns of line above.

5. Pass the free end of the line through the bottom of the central loop, from back to front like this.

6. Wet and tighten carefully. You now have a strong, fixed loop that will keep your hook or lure secure, but provide free movement.